Traffic Flow Theory

A Monograph

DANIEL L. GERLOUGH and MATTHEW J. HUBER

TRB

SPECIAL REPORT 165

Transportation Research Board
National Research Council
Washington, D.C., 1975

Transportation Research Board Special Report 165
Price $20.00
Edited for TRB by H. P. Orland

Subject area
54 traffic flow

Transportation Research Board publications are available by ordering directly from the board. They may also be obtained on a regular basis through organizational or individual supporting membership in the board; members or library subscribers are eligible for substantial discounts. For further information, write to the Transportation Research Board, National Academy of Sciences, 2101 Constitution Avenue, N.W., Washington, D.C. 20418.

The project that is the subject of this report was approved by the Governing Board of the National Research Council, whose members are drawn from the councils of the National Academy of Sciences, the National Academy of Engineering, and the Institute of Medicine. The members of the committee responsible for the report were chosen for their special competence and with regard for appropriate balance.

This report has been reviewed by a group other than the authors according to procedures approved by a Report Review Committee consisting of members of the National Academy of Sciences, the National Academy of Engineering, and the Institute of Medicine.

The views expressed in this report are those of the authors and do not necessarily reflect the view of the committee, the Transportation Research Board, the National Academy of Sciences, or the sponsors of the project.

LIBRARY OF CONGRESS CATALOGING IN PUBLICATION DATA

National Research Council. Transportation Research Board.
Traffic flow theory.

(Special report—Transportation Research Board, National Research Council; 165)

An updating and expansion of An introduction to traffic flow theory, by the Special Committee on Publication of Selected Information on Theory of Traffic Flow, National Research Council.

1. Traffic flow—Mathematical models. I. National Research Council. Special Committee on Publication of Selected Information on Theory of Traffic Flow. An introduction to traffic flow theory. II. Title. III. Series: National Research Council. Transportation Research Board. Special report—Transportation Research Board, National Research Council; 165.

HE336.T7N38 1975 388.3'1 75-42259
ISBN 0-309-02459-5

FOREWORD

This publication is an updating and expansion of *Highway Research Board Special Report 79,* "An Introduction to Traffic Flow Theory," published in 1964. This updating was undertaken on recommendation of the Highway Research Board* Committee on Theory of Traffic Flow in 1969 after the original printing of Special Report 79 had been exhausted. The Federal Highway Administration (FHWA) funded the project and in June 1970 awarded to the University of Minnesota a contract to write this publication. The project was carried out under supervision of an Advisory Committee that included the following prominent individuals:

Robert S. Foote, Port Authority of New York and New Jersey
Dr. Adolf D. May, Jr., University of California
Richard W. Rothery, General Motors Research Laboratories
K. B. Johns, TRB Liaison Representative

In addition, Sidney Weiner, of FHWA, served as committee secretary until the final stages of the project and Barry Benioff, of FHWA, assumed responsibility as committee secretary in mid-1973.

Our appreciation is extended to the authors of this report, Drs. Daniel L. Gerlough and Matthew J. Huber, of the University of Minnesota, for their commendable efforts in writing this monograph and for their cooperative attitude throughout the project. We would also like to acknowledge the time spent by the members of the Advisory Committee in providing guidance and direction on the form of the final monograph and for their extensive reviews of the early versions of the report. Additional acknowledgment is made for the guidance and counsel received from K. B. Johns, Engineer of Traffic and Operations, Transportation Research Board.

It is sincerely hoped that the present publication meets its objective of synthesizing and reporting, in a single document, the present state of knowledge in traffic flow theory.

Advisory Committee Co-Chairmen:

William W. Wolman, Chief
Traffic Systems Division
Office of Research
Federal Highway Administration

Donald G. Capelle
Vice President for Research
Alan M. Vorhees & Associates

*On March 9, 1974, the Highway Research Board became the Transportation Research Board to reflect the actual scope of its activities.

CONTENTS

DEFINITIONS AND NOTATIONS

Symbol	*Term*	*Definition*
a	Acceleration	The time rate of change of speed, d^2x/dt^2.
a_n		The acceleration of the nth vehicle.
c	Wave speed	The speed at which a wave of differing concentrations is propagated in the traffic stream. See also u_w.
D	Density	See "concentration" (k).
h	Time headway	The time interval between passages of consecutive vehicles moving in the same lane (measured between corresponding points on the vehicles).
h_n		Headway between the (n–1)st vehicle and the nth vehicle.
H_n	Total headway time	The time interval between passages of the first and the nth vehicle moving in the same lane.
k	Concentration	The number of vehicles occupying a unit length of a lane at a given instant; often referred to as "density" when expressed in vehicles per mile.
k_j	Jam concentration	The maximum concentration of vehicles when jammed at a stop.
k_m	Optimum concentration	The concentration when flow is at a maximum rate.
L	Car length	The length of a vehicle.
n		The vehicle number.
N		Total number of vehicles.
$p(x)$, $P(x)$	Probability	The likelihood of occurrence of an event.
q	Flow	The number of vehicles passing a point during a specified period of time; often referred to as "volume" when expressed in vehicles per hour measured over an hour. The rate of change of count.
q_m	Maximum flow	The maximum attainable flow.
r	Correlation coefficient	A statistical measure of the association between data and a regression line.
s	Spacing	The distance between consecutive vehicles moving in the same lane (measured between corresponding points on the vehicles).
s_n	Spacing	The spacing between the (n–1)st vehicle and the nth vehicle.
t	Time	An interval or index of time.
T	Time	Total time.
u	Speed	The time rate of change of distance, dx/dt.
u_f	Free-flow speed	The speed when traffic is flowing freely on the facility.
u_m	Optimum speed	The speed when flow is at a maximum rate.
$\bar{u}_s$	Space mean speed	The arithmetic mean of the speeds of the vehicles occupying a given length of lane at a given instant.

Symbol	*Term*	*Definition*
$\bar{u}_t$	Time mean speed	The arithmetic mean of the speeds of vehicles passing a point during a given interval of time.
u_w	Wave speed	The speed at which a wave of differing concentrations is propagated in the traffic stream. See also c.
V	Volume	See "flow" (q).
x, y, z, X, Y, Z	Position	An index of position; coordinates.
Δ	Increment	
η	Normalized concentration	The ratio k/k_j.
σ	Standard deviation	A statistical measure of the dispersion of data from the mean.
$\dot{x}$	First derivative (speed)	The differentiation of x with respect to some independent variable; i.e., dx/dt.
$\ddot{x}$	Second derivative (acceleration)	The second differentiation of x with respect to some independent variable; i.e., d^2x/dt^2.
Var (•)		Statistical variance.
E(•)		Expected or mean value.
$\exp(x-y)$		e^{x-y}.
$p((x\|a, b)$		Probability of x given conditions a and b.
	Dynamics	Action of force on bodies in motion or at rest (includes kinetics, kinematics and statics).
	Kinematics	Motion in abstract without reference to force or mass.
	Kinetics	Motion of masses in relation to forces acting on them.
	Phenomenological	Describing a phenomenon without explaining it.
	Statics	Treatment of bodies, masses or forces at rest (or in equilibrium).

Chapter 1
INTRODUCTION

1.1 BACKGROUND

With about 110 million registered vehicles in the United States and this number expected to increase, street and highway traffic will continue to be an important social and economic problem. Accident rates continue at high levels. Although there is a growing interest in public transit systems, the automobile may be expected to continue as the dominant form of urban transportation for many years to come. One reason is that "intersuburban travel is becoming a steadily larger proportion of all metropolitan travel because greater numbers of people live in one suburb and work in another. . ."[1]

It is thus important to be able to design and operate streets and highways with the greatest possible efficiency. Traffic flow theory is one tool to aid in attaining this end. It serves the dual purposes of aiding the understanding of traffic behavior and predicting future performance.

1.2 HISTORY

One may take as a working definition of traffic flow theory, "The description of traffic behavior by application of the laws of physics and mathematics." At present, as indicated by the definition, there is no unified theory of traffic flow; rather there are several theoretical approaches to describe the phenomena. It is hoped that as our knowledge of traffic theory increases, such a unified theory will be developed.

Traffic flow theory had its beginnings in the 1930s with applications of probability theory and the first models relating volume and speed. During the 1940s there were few theoretical developments, presumably because of the interruption caused by World War II. The early 1950s saw theoretical developments based on a variety of approaches, principally car-following, traffic wave theory (hydrodynamic analogy), and queueing theory. Investigators came from a variety of disciplines, including theoretical physics, applied mathematics, control theory, psychology, economics, and, of course, engineering. Although many of these investigators were on the staffs of universities, others came from a variety of research institutes, as well as various levels of government.

By 1959 traffic flow theory had developed to the point where it appeared desirable to hold an international conference. Accordingly, at the invitation of General Motors Research Laboratories a symposium was held in Detroit, Michigan, December 7–8, 1959. More than 100 persons gathered for the presentation of 15 technical papers and many lively discussions. It was generally agreed that the benefits of this symposium were substantial and that additional symposia should be convened in the future. As of this writing there have been six symposia, as summarized in Table 1.1.

During the Detroit meeting some of the discussions disclosed the fact that the papers on traffic flow theory were being published in various journals, many of which were not readily available to the traffic engineer. As a result, it was recommended that a publication be prepared to bring together papers summarizing the (then) state of the theory of traffic flow. It was suggested that the Highway Research Board might undertake such a publication. The Highway Research Board (of the Division of Engineering and Industrial Research of the National Academy of Sciences) together with the NAS Division of Mathematics appointed a joint committee to assemble such a publication. Highway Research Board Special Report 79 *An Introduction to Traffic Flow Theory* was issued in 1964.[2]

In the intervening years several books have been published on various aspects of traffic flow theory (Table 1.2), and several journals have been started to carry papers on traffic flow theory and related topics (Table 1.3). The U.S. Bureau of Public Roads (now the Federal Highway Administration) has in the past years commissioned several publications reviewing work in traffic flow theory. The principal items are listed in Table 1.4.

1.3 BACKGROUND OF CURRENT MONOGRAPH

By fall 1968 the original printing of Special Report 79 was exhausted. At its meeting in January 1969, the Highway Research Board Committee on Theory of Traffic Flow discussed what action should be taken. Although taking into account the books that had been published, the Committee deemed it worthwhile to have *An Introduction to Traffic Flow Theory* up-

TABLE 1.1 International Symposia on Theory of Traffic Flow

Symposium	Proceedings
Symposium on the Theory of Traffic Flow, Detroit, Michigan, Dec. 7–8, 1959	*Theory of Traffic Flow* (ed. by R. Herman), Elsevier Publishing Co., 1961
Second International Symposium on the Theory of Road Traffic Flow, London, England, June 25–27, 1963	*Proceedings of the Second International Symposium on the Theory of Traffic Flow* (ed. by J. Almond), OECD, Paris, 1965. Available from OECD Publications Office, 1750 Pennsylvania Avenue, N.W., Washington, D.C. 20006
Third International Symposium on the Theory of Road Traffic Flow, New York, New York, June 1965	*Vehicular Traffic Science* (ed. by L. C. Edie, R. Herman, R. Rothery), American Elsevier Publishing Co., 1967
Fourth International Symposium on the Theory of Road Traffic Flow, Karlsruhe, Germany, June 18–20, 1968	*Beitrage zur Theorie des Verkehrsflusses* (ed. by W. Leutzbach and P. Baron), Strassenbau und Strassenverkehrstechnik, n. 86, Bundesanstalt für Strassenwesen, 5 Köln Raderthal, Bruhler Strasse 324, West Germany
Fifth International Symposium on the Theory of Traffic Flow and Transportation, Berkeley, California, June 16–18, 1971	*Traffic and Transportation* (ed. by G. F. Newell), American Elsevier Publishing Co., 1972
Sixth International Symposium on Transportation and Traffic Theory, Sydney, Australia, August 26–28, 1974	*Transportation and Traffic Theory* (ed. by D. J. Buckley), American Elsevier Publishing Co., 1974

TABLE 1.2 Books on Traffic Flow Theory

Frank A. Haight	*Mathematical Theories of Traffic Flow* Academic Press (New York, London)	1963
Winifred D. Ashton	*The Theory of Traffic Flow* Methuen & Co. (London) John Wiley & Sons (New York)	1966
Donald R. Drew	*Traffic Flow Theory and Control* McGraw-Hill Book Co. (New York)	1968
Ilya Prigogine and Robert Herman	*Kinetic Theory of Vehicular Traffic* American Elsevier (New York)	1971
Denos C. Gazis (ed)	*Traffic Science* Wiley-Interscience (New York)	1974
Wilhelm Leutzbach	*Einführung in die Theorie des Verkehrsflusses* Springer-Verlag (Berlin)	1972

TABLE 1.3 Journals on Traffic Flow Theory [a]

Journal	Publisher	Year First Published
Traffic Engineering and Control	Printerhall Ltd., 29 Newman Street, London W1P3PE, England	1958
Transportation Science	Transportation Science Section, Operations Research Society of America, 428 East Preston Street, Baltimore, Maryland 21202	1967
Transportation Research	Pergamon Press Inc., Maxwell House, Fairview Park, Elmsford, New York 10523	1967
Transportation Planning and Technology	Gordon and Breach Science Publishers, Inc., 440 Park Avenue South, New York, New York 10016	1972
Transportation	Elsevier Publishing Co., Journal Division, P.O. Box 211, Amsterdam, The Netherlands	1972

[a] Abstracts of articles in a variety of journals may be found in *International Abstracts in Operations Research,* published by the International Society of Operational Research Societies. Available from Operations Research Society of America, 428 East Preston Street, Baltimore, Maryland 21202.

dated and rewritten. The Highway Research Board Department of Traffic and Operations authorized the Committee to seek funding and an author to carry out the rewriting. As a result of the Committee's action the Federal Highway Administration (FHWA) agreed to provide funding, and in June 1970 FHWA entered into a contract with the University of Minnesota under which D. L. Gerlough and M. J. Huber would write a monograph on traffic flow theory. It was agreed that partial funding would be provided by the University of Minnesota.

TABLE 1.4 Principal Review Studies of Traffic Flow Theory Sponsored by Federal Highway Administration

Analytical Models of Unidirectional Multi-Lane Traffic Flow: A Survey of the Literature, System Development Corp. Technical Memorandum TM(L)-379/001/02, 31 January 1969. NTIS No. PB 183075
Traffic Systems Reviews and Abstracts, issued monthly by Federal Highway Administration, from September 1967 to 1971. Available from NTIS

1.4 CONTENT OF MONOGRAPH

Although it follows the general content of HRB Special Report 79, this monograph is somewhat expanded and updated. It was agreed at the outset that models pertaining strictly to transportation planning would be excluded inasmuch as they could by themselves fill a comprehensive treatise. Although it is primarily addressed to graduate students in traffic engineering, it is hoped that this monograph will serve as a reference for anyone wishing an introduction to traffic flow theory. It is assumed that the reader has a background in the fundamentals of statistics.

Models of traffic flow may be classified either according to the theoretical approach used or according to the characteristics or phenomenon they describe. This monograph mainly follows the former practice.

Notation and definitions used are found on special pages in the front of the monograph.

The following paragraphs describe the contents of the individual chapters.

Chapter 2: Measurement of Flow, Speed, and Concentration

Before one attempts to model the fundamental traffic characteristics of flow, speed, and concentration, it is important to have unambiguous definitions of these characteristics in relationship to the methods of measurement, as well as the appropriate methods of computing averages, etc. Historically, the definitions of traffic characteristics were related to the methods of measurement. Unfortunately, methods of averaging often used were influenced by a lack of clear understanding of the processes involved.

The earliest students of traffic behavior had as the only instruments available to them stopwatches and manual counters. Thus flows past a point and headways were the important measurements. These were supplemented by measurements of speed by means of timing each car's transit across a "trap." As other methods of measurement have been developed, it has become evident that the numerical results depend on the method of measurement.

In Chapter 2 various methods of measurement currently available to the traffic engineer are considered, and the definitions of characteristics are related to these methods of measurement; methods of averaging are also considered.

Chapter 3: Statistical Distributions of Traffic Characteristics

In designing new traffic facilities or new control plans, it is necessary to predict the performance of traffic with respect to some particular characteristic (e.g., the frequency of headways of a particular size, the number of cars likely to arrive in an interval, and speeds exceeding a certain value). It is often desirable to be able to make a prediction with a minimum amount of data available or assumed. For instance, it may be necessary in designing a pedestrian control system to predict the frequency of headways of greater than 10 sec; in designing a left-turn pocket it may be necessary to predict how many times per hour the number of cars arriving during one signal cycle will exceed four. Statistical distribution models enable the traffic engineer to make these predictions with a minimal amount of information.

Statistical distributions are useful in describing a wide variety of phenomena where there is a high element of randomness. In traffic the most important distributions are counting distributions—those useful in describing the occurrence of things that can be counted—and interval distributions—those useful in describing the occurrence of the (time) intervals between events. Distributions are also used, however, in describing such phenomena as speeds and gap acceptance. The uses of distributions in traffic are treated in Chapter 3.

Chapter 4: Traffic Stream Models

It requires little more than casual observation to detect that as traffic flow or concentration increases, there is a decrease in speed. In fact it is often possible to estimate one unknown flow characteristic from another that is known or easily measured. For example, the ratio of flow to capacity may be estimated using observed values of speed or concentration. In Chapter 4 models that relate pairs of the basic traffic flow characteristics (e.g., speed/flow, speed/concentration, and flow/concentration) are examined, as are models that consider travel time as one of the variables.

Chapter 5: Driver Information Processing Characteristics

In attempting to devise models to represent traffic behavior we are, of course, indirectly dealing with human behavior. Several traffic flow models to be discussed in later chapters contain parameters that are used to account for various characteristics of the driver in the driver–vehicle system. Some models deal with traffic as a deterministic phenomenon, even though the driver portion of the system, at least, is highly stochastic.

The whole field of human factors in traffic could, of course, be the subject of a very lengthy treatise. Nevertheless, it is hoped that Chapter 5 will provide some insight into the way drivers use the information they receive and that this knowledge can then be of use in various traffic flow models. Although a driver is continuously making decisions, his control actions are limited to control of heading (steering) and control of acceleration.

Chapter 6: Car Following and Acceleration Noise

From the drivers' action interpretations of information received, given that actions are limited to control of acceleration (braking and acceleration) and heading (steering), we now consider the dynamics of a stream of traffic. These dynamics result when a series of drivers attempt to regulate their acceleration to accomplish a smooth, safe trip.

In car-following analysis traffic is recognized as being made up of discrete particles, whose interaction is examined. The techniques of analysis are those of automatic control systems. A principal effort of such studies has been to try to understand the behavior of a single-lane traffic stream by examining the manner in which individual vehicles follow one another. Studies of this nature have been used to improve flow through tunnels, explain the behavior of traffic at bottlenecks, and examine control and communication techniques that will minimize the occurrence of rear-end chain collisions in dense traffic.

Acceleration noise is developed as a measure of the quality of traffic flow.

Chapter 7: Hydrodynamic and Kinematic Models of Traffic

Because traffic involves flows, concentrations, speeds, etc., there is a natural tendency to attempt to describe traffic in terms of fluid behavior. Although traffic was examined as the interaction between particles in Chapter 6, in Chapter 7 we apply to traffic those models that have been developed for fluids (i.e., continuum models); by this we are implicitly saying that we are more concerned with the over-all statistical behavior of the traffic stream than with the interactions between particles. Because the sample size for traffic includes only a few particles relative to a true fluid, fluid models of traffic are applicable to the behavior of a stream rather than individual cars.

Despite some writers' postulated analogy between traffic and a real fluid, it is preferable to begin with fundamental observations and postulates concerning traffic and, then, to identify analogies with fluids as these analogies appear. Thus, we start by developing the continuity equation for traffic, illustrating its analogy to the continuity equation for fluids. Next to be developed is the concept of waves in traffic, and examples of this application to practical problems are given. Such studies have been used to improve flow through tunnels and to explain behavior of traffic at bottlenecks. Thereafter the model of choice points toward a unification of the continuum and car-following theories. Finally, a Boltzmann-like theory of traffic is discussed in brief.

Chapter 8: Queueing Models

A desirable goal for transportation engineers is to design and operate facilities that minimize delay to the users. Delay resulting from congestion is a common phenomenon: Vehicles wait in line for an opportunity to enter a freeway with controlled access ramps; pedestrians queue up on a crosswalk in anticipation of a gap in road traffic or at a turnstile in a transit station; left-turn slots must be sufficiently long to store the maximum number of vehicles that can be expected to wait for a left-turn signal.

Interest may rest in the length of time a user must wait, or the number of units waiting in line, or the proportion of time that a facility might be inactive (an empty parking stall, for instance). Queueing models, which employ methods of probability and statistics, provide a means for predicting some of these delay characteristics.

The purpose of this chapter is to present some of the results of studies of probability models of traffic delay. After an introduction to some elements of queueing or waiting-line theory, examples concerned with delay problems that occur when all users pass through a single-channel control point, such as a left-turn slot or a single exit lane for a garage, are given. Next, the analysis is extended to consider several channels of service; for example, several parallel toll booths or the different stalls of a parking facility. Also considered is the case of a user who doesn't get served; for example, the person seeking a parking space who continues to another destination when none is found.

An analysis of delays at intersections is considered next, beginning with an analysis of unsignalized intersections. Queueing models for more complex intersection control, such as pedestrian control or traffic-signal control, are also considered in this section. Finally, queueing theory is applied to delay on two-lane roads.

Except for the detailed development of the formulas given in section 8.2, this chapter avoids detailed mathematical development, but does present the theorists' assumptions and some results of interest.

Chapter 9: Simulation of Traffic Flow

When traffic situations are so complex that they cannot be predicted by a model of one of the types discussed in earlier chapters, computer simulation often serves as a useful predictive tool. The Federal Highway Administration has supported various generations of programs to simulate a network of signalized intersections, the latest of which is known as UTCS-1. Several freeway simulation models have also been developed. Although some traffic simulations have made use of analog or hybrid computers, the emphasis in Chapter 9 is on simulation by digital computers.

1.5 REFERENCES

1. Rae, J. G., The mythology of urban transportation. Traffic Q., 26(1): 85–98, (1972).
2. Gerlough, D. L., and Capelle, D. G. (ed.), *An Introduction to Traffic Flow Theory*. Special Report 79. Highway Research Board, Washington, D.C. (1964).

Chapter 2

MEASUREMENT OF FLOW, SPEED, AND CONCENTRATION

2.1 INTRODUCTION

The three most important characteristics of traffic are flow, speed, and concentration. Before attempting to model these characteristics, it is essential to define them unambiguously. As will be seen later, the definitions are related to the methods of measurement, as well as to the methods of averaging the measurements.

The only instruments available to earliest students of traffic behavior were stopwatches and manual counters. Thus flows past a point and headways were the important measurements. These were supplemented by measurements of speed by means of timing each car's transit across a "trap." As other methods of measurement have been developed, it has become evident that the numerical results depend on the method of measurement.

Here, various methods of measurement currently available to the traffic engineer are considered, and the definitions of characteristics are related to these methods of measurement. Specifically, the methods of measurement discussed include measurements at a point, along a length (by photography), and by a moving observer. Methods of averaging are also considered. The definitions are mainly those resulting from the works of Wardrop,[1] Lighthill and Whitham,[2] and Edie.[3] Definitions in terms of stochastic processes have been given by Mori et al.,[4] Breiman,[5,6] and Foster.[7]

2.2 MEASUREMENTS AT A POINT

Because of the instruments used by early investigators, it was natural for quantities measured at a point to be adopted as important measures of traffic performance. Even today many operational decisions are based on "point" measurements. To obtain useful data at a point, it is necessary to cover extensive time periods. Thus all characteristics are expressed as averages. Figure 2.1 shows a series of vehicle trajectories; i.e., space–time plots of the paths of vehicles while in a specific space–time domain. Consider line AA′; an observer standing at point A along the roadway (or sampling traffic with a pneumatic-tube counter at point A) will obtain a count of the trajectories of vehicles that pass point A during time T. When only counts are being made, it is customary to record the counts for relatively short intervals even though the total counting period may be relatively long. It is possible, however, to make additional measurements such as the (time) headways between vehicles, or vehicle speeds (by a short trap or by radar).

2.2.1 Flow

Traditionally, the traffic engineer has used volume or flow * as one of the primary measures of traffic condition or state. This has been because flow is the easiest of all characteristics to obtain. Consider the situation portrayed in Figure 2.1. Then if N cars cross line AA′ in a time T, the flow is computed as

$$q = N/T \qquad (2.1)$$

In usual traffic engineering practice it is customary to start the timing of counting intervals at random with respect to traffic; e.g., at the start of an hour or the start of a 15-min period. The timer is then allowed to run continuously, recording counts as the hand sweeps past each timing point (e.g., each 20 sec).†

For a more general understanding of this phenomenon, consider that each vehicle has associated with it a headway, h_i, that is measured between the times of arrival of corresponding parts (e.g., front bumpers) of successive vehicles. From this the following relationship can be derived:

* For the distinction between volume and flow see "Definitions and Notations."

† An alternative method is discussed in Chapter 3.

$$q=\frac{N}{T}=\frac{N}{\sum_{i=1}^{N} h_i}=\frac{1}{\frac{1}{N}\sum h_i}=\frac{1}{\bar{h}} \tag{2.2}$$

where N = the number of headways measured and $\bar{h}$ = the mean headway. For some studies, it may be useful to think of an instantaneous flow associated with each vehicle. Thus,

$$q_i=1/h_i \tag{2.3}$$

When using Eq. 2.3, care must be exercised in computing an average flow. Eq. 2.2 shows that

$$q=\frac{1}{\bar{h}}=\frac{1}{\frac{1}{N}\sum h_i}=\frac{1}{\frac{1}{N}\sum \frac{1}{q_i}} \tag{2.4}$$

Thus the average flow, when computed from individual flows associated with each vehicle, is the *harmonic mean* of the individual flows.

2.2.2 Speed

The average speed is an important measure of the traffic performance at a particular point or along a particular route; in addition, it is one of the fundamental characteristics of traffic flow. There are two principal average speeds, the time mean speed and the space mean speed.[1] Unfortunately, in early traffic literature there was confusion as to the use of these two averages. (As an example of the confusion, the reader is invited to compare the discussions in various editions of the *Traffic Engineering Handbook*).[8–10]

2.2.2.1 Time Mean Speed (Spot Speed). In the past, it has been common practice among traffic engineers to report the "spot speed" for a given location. This is computed as the *arithmetic mean* of the observed speeds:

$$\bar{u}_t=\bar{u}_{\text{spot}}=\frac{1}{N}\sum_{i=1}^{N} u_i \tag{2.5}$$

In theoretical discussions of traffic flow this value is referred to as the "time mean speed." [1]

2.2.2.2 Space Mean Speed (Harmonic Mean Speed). Consider the situation where three cars, one at 20 mph, one at 40 mph, and one at 60 mph, are traversing a length D. At a point along D the spot speed would be reported as $(20+40+60)/3$ or 40 mph. Suppose, however, that we are interested in average speed as calculated from average travel time. Average travel time would be

$$\bar{t}=\frac{1}{3}\left[\frac{D}{20}+\frac{D}{40}+\frac{D}{60}\right]=\frac{1}{N}\sum_{i=1}^{N}\frac{D}{u_i}$$

The average speed representing this travel time would be

$$\bar{u}_s=\frac{D}{\bar{t}}=\frac{D}{\frac{1}{N}\sum_{i=1}^{N}\frac{D}{u_i}}=\frac{1}{\frac{1}{N}\sum_{i=1}^{N}\frac{1}{u_i}} \tag{2.6}$$

For the three cars this would be $\bar{u}=\frac{1}{1/3(1/20+1/40+1/60)}=32.7$ mph

Thus, whereas the spot speed (i.e., the time mean speed) is the arithmetic mean of the speeds observed at a point, the speed that represents average travel time is the harmonic mean of speeds observed at a point. Furthermore, it will be noted that the harmonic mean speed is lower than the time mean speed. For reasons that are discussed in section 2.3.2, the harmonic mean of speeds observed at a point is known as the space mean speed.

2.2.2.3 Relationships of Mean Speeds. The relationship for computing time mean speed from space mean speed was first recognized by Wardrop.[1] His relationship, derived in Appendix A-1, is

$$\bar{u}_t=\bar{u}_s+\sigma_s^2/\bar{u}_s \tag{2.7}$$

where σ_s^2 is the variance about the space mean speed.

In traffic engineering practice it is often desirable to convert spot (time mean) speeds to space mean speeds. For this purpose the following approximate relationship, also discussed in Appendix A-1, has been developed: *

$$\bar{u}_s=\bar{u}_t-\sigma_t^2/\bar{u}_t \tag{2.8}$$

where σ_t^2 is the variance about the time mean speed.

For the speeds crossing line AA′ in Figure 2.1 the arithmetic mean speed is 21.71 mph, while the harmonic mean speed is 16.18 mph. When the harmonic mean speed is estimated from the arithmetic mean speed and the variance about it, the result is 15.58 mph. Tables 2.1 and 2.2 illustrate the estimation of the harmonic mean (i.e., space mean) speed from

* Haight and Mosher[11] have presented a method for converting $\bar{u}_t$ and σ_t^2 to $\bar{u}_s$ and σ_s^2 by means of tables that assume that speeds follow the Pearson type III distribution (see Appendix B-2).

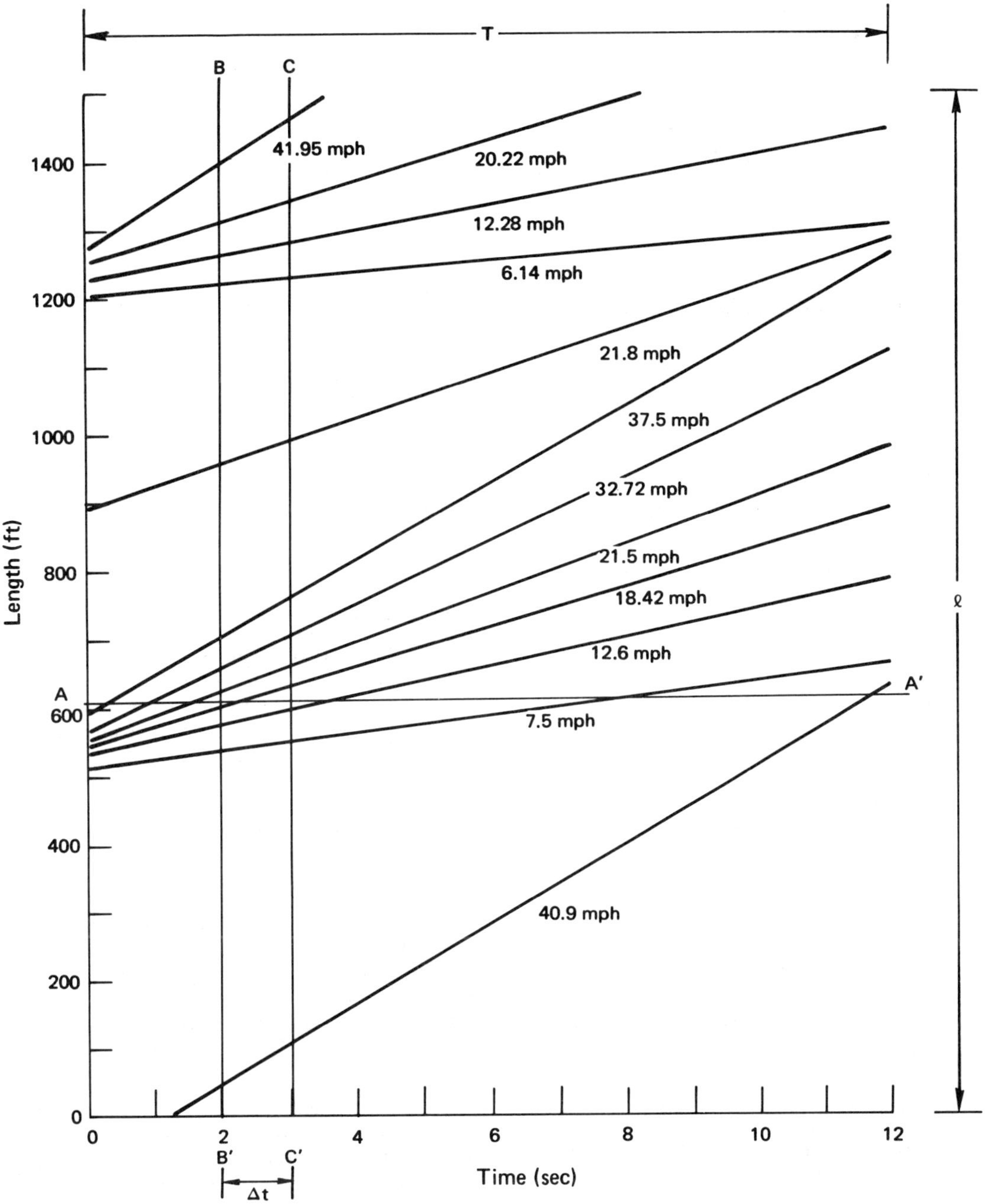

Figure 2.1 Vehicle trajectories: line AA′ represents a fixed point in space; line BB′ represents a fixed point in time.

point measurements and the estimation of the arithmetic mean (i.e., time mean) speed from the harmonic mean. (Of course, the variance about each mean is necessary in estimating the other mean.) Because space mean speed is applied to the models of Chapter 4, the term "speed" with no adjective implies space mean speed in all later sections of this monograph.

2.2.3 Concentration from Point Measurements

Although concentration (the number of vehicles per unit length) implies measurement along a distance, traffic engineers have traditionally *estimated* concentration from point measurements, using the relationship

$$k = q/\bar{u}_s \tag{2.9}$$

TABLE 2.1 Speed Measurements and Averages [a]

Miles per Hour	Frequency
57	2
58	7
59	1
60	14
61	5
62	25
63	24
64	32
65	13
66	22
67	9
68	13
69	4
70	10
71	0
72	2
73	1

[a] From observations on Interstate 94 at Prior Street Bridge, August 1970.

TABLE 2.2 Harmonic Mean Derived from Arithmetic Mean of Table 2.1 [a]

	Arithmetic Mean	Harmonic Mean
Direct computation	64.1902	64.0368
Estimate of arithmetic mean from harmonic mean	64.1925	
Estimate of harmonic mean from arithmetic mean		64.0353
Variance [b] about arithmetic mean = 9.94723		
Variance about harmonic mean = 9.97089		

[a] See also Appendix B-8.

[b] The variance computed here is the unbiased estimate of the variance based on sample measurements; i.e., Bessel's correction has been included.

which may be derived as follows: Assume the total stream is made up of substreams, with each substream having its own (constant) speed. Segregate flow into subflows according to speed [1]

$$\bar{u}_s = \frac{\sum k_i u_i}{k}$$

where k_i/k is the fraction of the total density in the stream having speed u_i, and $k_i u_i = q_i$ (from analysis of units). Therefore,

$$\bar{u}_s = \frac{\sum q_i}{k} = \frac{q}{k}.$$

Eq. 2.9 is of fundamental importance and is discussed further in Chapter 4.

2.2.4 Lane Occupancy

Although concentration or density is considered the fundamental characteristic to be measured in freeway surveillance, density cannot be measured directly by electronic means. Thus, during the early days of freeway surveillance various estimates of density were investigated. This led to development of a measure that is now called *lane occupancy*.

If one could at any instant measure the lengths of all vehicles on a given roadway section and then compute the ratio

$$R_1 = \frac{\text{sum of lengths of vehicles}}{\text{length of roadway section}} \tag{2.10}$$

This ratio could then be divided by L_m, the average length of a vehicle expressed in miles, to yield an estimate of the concentration or density in vehicles per mile.

For example, suppose vehicles having lengths (in feet) 17, 13, 20, 40, 17 and 20 are distributed over a length of highway one lane wide and 1,000 ft long. The ratio R is then 0.127. If the average length of a vehicle is taken as 21 ft (or 0.00398 mi), a computed value of 31.91 vehicles/mile is obtained as the concentration on this highway section.

It is not feasible to use on-line methods to measure the sum of the lengths of vehicles in a given roadway section. It is possible, however, to estimate this value by time measurements.

Several types of presence detector are available to the traffic engineer, including induction loops, magnetometers, ultrasonic reflectors, and photo cells. A presence detector has the characteristic that it remains in the "on" or "closed" condition as long as the vehicle remains within the zone of effectiveness of the detector. This characteristic is the basis for point measurements of lane occupancy.

It is not difficult to build an electronic instrument to measure the ratio

$$R_2 = \frac{\text{sum of times vehicle detector is occupied}}{\text{total time of observation period}} = \frac{\sum t_0}{T} \quad (2.11)$$

This ratio can then be used to estimate density and speed.

Two general types of error are present—instrumentation and estimation errors. Instrument errors depend on the design of the instrument and are not discussed here, except to remark that time is often measured by counting pulses from a generator and that the principal errors are related to events whose beginning or ending falls between pulses.

2.2.4.1 Estimation of Occupancy with Presence Detectors. The ratio of occupied time to total observation time, expressed as a percent, is called "lane occupancy," denoted by ϕ.

$$\phi = \frac{\text{total occupied time}}{\text{total observation time}} \cdot 100 = \frac{100 \sum t_0}{T} \quad (2.12)$$

To use lane occupancy for estimation of density and speed, it is necessary to know the *effective* length of a vehicle as measured by the detector in use. Because both the vehicle and the detector have finite lengths, the length of roadway covered by the vehicle during the "on" period of the detector will be different from the length of the car and may often be longer. Figure 2.2 illustrates this situation.

Density or concentration (for a given lane) may then be estimated by

$$\hat{k} = \frac{\phi}{100} \frac{5{,}280}{L_e} \quad (2.13)$$

where L_e is the effective length of the vehicle in feet and $\hat{k}$ is the estimated concentration.

For example, consider the following times a detector is occupied during a 60-sec period:

0.39 sec	0.46 sec	0.43 sec	0.47 sec
*0.50	*0.51	0.48	0.46
0.32	0.44	*0.50	0.45
			0.44

* Significance of asterisks is discussed following Eq. 2.17.

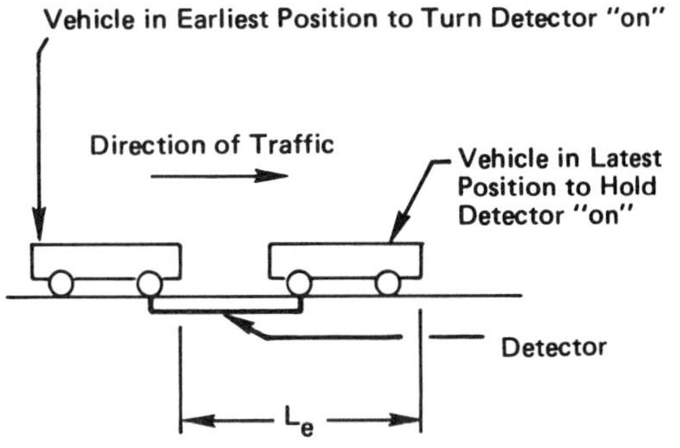

Figure 2.2 Effective length of vehicle at presence detector: $t_0 = L_e/\bar{u}_d$, where t_0 is the occupied time, L_e is the effective length of the vehicle, and $\bar{u}_d$ is the speed over the detector (time mean speed).

Thus, $\sum t_0 = 5.85$ and $N = 13$.

Then, $\phi = \frac{5.85}{60}(100) = 9.75\%$.

If the average *effective* length of a vehicle is taken as 26 ft, $\hat{k} = \frac{9.75}{100} \cdot \frac{5{,}280}{26} = 19.8$ vehicles/mile.

To estimate speed, it is necessary to count the vehicles crossing the detector during the observation period. Average speed can then be estimated by substituting into $u = q/k$ to give

$$\hat{u} = \frac{N L_e}{\sum t_0} \quad (2.14)$$

where $\hat{u}$ is the estimated speed (ft/sec).

Using the previous data, $\hat{u} = \frac{13\ (26)}{5.85} = 57.8$ ft/sec (39.4 mph).

Note, however, that Eqs. 2.13 and 2.14 make use of the average effective length of a vehicle. This can lead to serious errors when the mixture of cars and trucks in the traffic stream is varying. Weinberg et al.[12] have proposed a method for compensating for this situation by making use of the ability of some presence detectors (e.g., ultrasonic reflectors) to distinguish between high and low vehicles and to count them separately. If high vehicles are taken as trucks and buses and assigned an effective length, L_t and low vehicles are taken as cars and assigned an effective length L_c, a better estimate of speed is expressed:

$$\hat{u} = \frac{N_t L_t + N_c L_c}{\sum t_0} \quad (2.15)$$

where N_t is the number of trucks and buses during the observation period and N_c is the number of cars during the observation period. Extending this method to the estimation of concentration gives

$$\hat{k}=\frac{N_t+N_c}{N_t L_t+N_c L_c}\frac{\phi}{100}=\frac{1}{L_e}\frac{\phi}{100} \quad (2.16)$$

For $\hat{k}$ in vehicles/mile, Eq. 2.16 may be restated

$$\hat{k}=\frac{N_t+N_c}{N_t L_t+N_c L_c}\frac{\phi}{100}5{,}280 \quad (2.17)$$

Using the previous data and considering those occupancy times marked by an asterisk to be trucks with length 40 ft and the other occupancy times to represent cars with average length 20 ft, $N_t=3$; $N_c=10$; $\hat{k}=\frac{3+10}{3\cdot 40+10\cdot 20}\frac{9.75}{100}5{,}280=20.9$ vehicles/mile.

It should be noted that these estimates of concentration and speed from presence detectors are biased. Mikhalkin et al.[17] give methods for obtaining an unbiased estimate.

2.3 MEASUREMENTS ALONG A LENGTH

Either line BB′ or line CC′ in Figure 2.1 indicates the information one might obtain from an aerial photograph (or photograph from a tall building). From a photograph it is possible to scale a distance and count the cars in this distance. With two photos spaced a short time apart (as the lines BB′ and CC′ in Fig. 2.1), it is possible to get speeds and flows. (The aerial photo computations given here are true for an infinitely long road; Breiman[18] gives more accurate methods for finite lengths.)

2.3.1 Concentration Along a Length

From the scaled distance and vehicle count, concentration is

$$k=N/l \quad (2.18)$$

where N is the number of vehicles counted and l is the length of roadway section (in miles). It is possible, however, to be more specific: given that s_i is the distance of the ith car behind the car ahead (measured front bumper to front bumper),

$$k_i=\frac{1}{s_i}=\frac{1}{h_i u_i} \quad (2.19)$$

where N is the number of spacings counted, h_i is the time headway (as before) of the ith car, and u_i is the speed of the ith car (as before).

$$k=\frac{N}{\sum_{i=1}^{N} s_i}=\frac{1}{\frac{1}{N}\sum s_i} \quad (2.20)$$

or

$$k=\frac{1}{\frac{1}{N}\sum_{i=1}^{N} 1/k_i} \quad (2.21)$$

Here, the average concentration is the harmonic mean of the individual concentrations.

2.3.2 Speed from Measurements Along a Length

When two (or more) aerial photos are taken in sequence with a short time interval between them, the situation represented in Figure 2.1 by lines BB′ and CC′, separated in time by Δt, obtains. Although each vehicle traverses a different distance, all are observed for the same time, or

$$u_i=\frac{s_i}{\Delta t}$$

$$\bar{u}_s=\frac{1}{N}\sum_{i=1}^{N}\frac{s_i}{\Delta t}=\frac{1}{N\,\Delta t}\sum_{i=1}^{N} s_i \quad (2.22)$$

Because this average speed $\bar{u}_s$ is the mean taken along a distance, Wardrop[1] gave it the name "space mean speed." The relationship between this mean and its variance and the mean of point measurements is given by Eq. 2.7.

2.3.3 Flow from Measurements Along a Length

Having obtained values for (average) concentration and average speed, flow is computed from

$$q=k\,\bar{u}_s \quad (2.23)$$

2.4 TRAFFIC MEASUREMENTS BY MOVING OBSERVER METHOD

It has been found that an effective method of assessing traffic along an arterial (or in an area) is the measurement of traffic by one or more moving observers.[14,15] With this method, an observation car travels first with traffic being measured and then returns in the direction opposite to that of the traffic being measured. An observer in the car records the travel time for each direction. For the trip with the traffic stream he also records the number of vehicles that pass the observation car and the number of vehicles passed by it. For the trip opposite to the traffic stream the number of cars passed in the stream is counted. Further discussion of the procedure and the derivation of the computational equations is given in Appendix A-2.

The relationships are

$$q=\frac{x+y}{t_a+t_e} \tag{2.24}$$

$$t=\text{mean travel time}=t_e-\frac{y}{q} \tag{2.25}$$

$$\bar{u}_s=l/t$$

$$k=\frac{q}{\bar{u}_s} \tag{2.26}$$

where t_e=time observer is moving with stream of traffic; t_a=time observer is moving against stream of traffic; x=number of cars met while moving against stream of traffic; y=net number of vehicles that pass the observer while moving with the traffic stream (i.e., the number that pass the observer minus the number he passes); and l=length of roadway section.

For example, given $t_e=144.4$ sec, $x=102$ vehicles, $t_a=68.2$ sec, $y=4$ vehicles, and $l=6{,}000$ ft, then,

$$q=\frac{102+4}{68.2+144.4}$$
$$=0.499 \text{ veh/sec}=1{,}795 \text{ veh/ hr}$$

$$t=144.4-\frac{4}{0.499}=136.4 \text{ sec}$$

$$\bar{u}_s=\frac{6000}{136.4}=44 \text{ ft/sec}=30 \text{ mph}$$

$$k=\frac{0.499}{44}=0.0113 \text{ veh/ft}=59.9 \text{ veh/mile}$$

2.5 SUMMARY OF TRAFFIC MEASUREMENTS

Characteristic	Observed over Short Length during Long Time	Observed over Long Length during Short Time	Observed by Moving Observer
Speed (miles/hr), $\bar{u}_s$	$\bar{u}_s=\frac{N}{\sum\frac{1}{u_i}}$	$\bar{u}_s=\frac{\sum u_i}{N}$	$\bar{u}_s=l/t$, where $t=t_e-y/q$ and l=length of link
Flow (vehicles/hr), q	$q=\frac{N}{T}=\frac{N}{\sum t_i}$	$q=k\bar{u}_s$	$q=(x+y)/(t_a+t_e)$, where x=cars met against stream, y=net cars that pass observer with stream, t_a=travel time against stream, and t_e=travel time with stream
Concentration (vehicles/mile), k	$k=\frac{q}{\bar{u}_s}$	$k=\frac{N}{l}=\frac{N}{\sum s_i}$	$k=\frac{q}{\bar{u}_s}$

2.6 REFERENCES

1. Wardrop, J. G., Some theoretical aspects of road traffic research. Proc. Inst. Civil Eng., Part II, 1(2): 325–362 (1952).

2. Lighthill, M. J., and Whitham, G. B., On kinematic waves: II. A theory of traffic flow on long crowded roads. Proc. R. Soc. A229, No. 1178: 317–345

(May 10, 1955). Reproduced in *An Introduction to Traffic Flow Theory,* pp. 8–35. Special Report 79, Highway Research Board, Washington, D.C. (1964).

3. Edie, L. C., Discussion of traffic stream measurements and definitions. Proc. Second Int. Symp. Theory Traffic Flow (held in London, 1963), (1965), pp. 139–154.
4. Mori, M., Takata, H., and Kisi, T., Fundamental considerations on the speed distributions of road traffic. Transp. Res., 2(1): 31–39 (1968).
5. Breiman, L., Point and trajectory processes in one-way traffic flow. Transp. Res., 3(2): 251–264 (1969).
6. Breiman, L., Space-time relationships in one-way traffic flow. Transp. Res. 3 (3): 365–376 (1969).
7. Foster, J., An investigation of the hydrodynamic model of traffic flow with particular reference to speed-density relationships. Proc. Austr. Road Res. Board, 1: 229–257 (1962).
8. *Traffic Engineering Handbook,* First edition. Institute of Traffic Engineers (1941), p. 70.
9. *Traffic Engineering Handbook,* Second edition. Institute of Traffic Engineers (1950), p. 155.
10. *Traffic Engineering Handbook,* Third edition. Institute of Traffic Engineers (1965), pp. 159, 161, 167, 168.
11. Haight, F. A., and Mosher, W. W., Jr., A practical method for improving the accuracy of vehicular speed distribution measurements. Highw. Res. Board Bull. 341 (1962), pp. 92–116.
12. Weinberg, M. I., Deleys, N. J. and Schneeberger, R. F., *Surveillance Methods and Ways and Means of Communicating with Drivers.* NCHRP Report 28, Highway Research Board, Washington, D.C. (1966), pp. 25–28.
13. Breiman, L., *Space-Time Relationships in One-Way Traffic Flow.* Systems Development Corporation Report TM-3858/001/00 (July 15, 1968), plus further details provided by private communication.
14. Wardrop, J. G., and Charlesworth, G., A method for estimating speed and flow of traffic from a moving vehicle. Proc. Inst. Civil Eng. (London), Part II, 3 (2): 158–171 (1954).
15. Williams, T. E. H., and Emmerson, J., Traffic volumes, journey times, and speeds by moving observer method. Traffic Eng. Control, 3(3): 159–162, 168 (1961).
16. Berry, D. S., and Green, F. H., Techniques for measuring over-all speeds in urban areas. Proc. Highw. Res. Board, 29: 311–318 (1949).
17. Mikhalkin, B., Payne, H. J., and Isaksen, L., Estimation of speed from presence detectors. Highw. Res. Rec. No. 388 (1972), pp. 73–83.
18. Breiman, L., *On Estimating Traffic Distributions from Aerial Data.* Systems Development Corp., Report TM-4638/005/01 to Federal Highway Administration under Contract FH-1107623 (July 1972), PB 226135.
19. Yule, G. V., and Kendall, M. G., *An Introduction to the Theory of Statistics,* 14th ed. (1950) Hafner, p. 150.

2.7 RELATED LITERATURE

Breiman, L., *Using Time Data To Get Space Headway-Speed Distributions in Traffic Flows.* Systems Development Corp., Tech. Memo. TM-3858/018/00 to U.S. Bureau of Public Roads under Contract No. FH-116623 (30 April 1959).

Kinsbruner, P., *A Statistical Analysis of Vehicular Traffic.* M.S. thesis, Massachusetts Institute of Technology, Cambridge (1960).

Leutzbach, W., Bewegung als Function von Zeit und Weg. Transp. Res., 3(4): 421–428 (1969).

Mori, M., Takata, H., and Kisi, T., Fundamental considerations on the speed distributions of road traffic flow. Transp. Res., 2(1): 31–39 (1968).

Preston, E. J., Averaging of road traffic measurements. Traffic Eng. Control, 5(11): 656–658, 663 (1964).

Road Research Laboratory, *A Method of Measuring Saturation Flow at Traffic Signals.* Road Note 34. HM. Stationery Office, London (1963).

Tiplitz, C., Approximations to the mean and standard deviation of reciprocals of observations. Technometrics, 5(4): 522–524 (1963).

Tournerie, G., Sur la definition des grandeurs caracteristiques d'une circulation. Proc. Fourth Int. Symp. Theory Traffic Flow (held in Karlsruhe, June 1968), published as *Beitrage zur Theorie des Verkehrsflusses,*

Strassenbau und Strassenverkehrstechnik, 86: 241–244 (1969).

Wright, C. C., Some characteristics of traffic leaving a signalized intersection. Transp. Sci., 4(4): 331–346 (1970).

Gafarian, A. V., Lawrence, R. L., and Munjal, P. K., An experimental validation of various methods of obtaining relationships between traffic flow, concentration, and speed on multi-lane highways. Highw. Res. Rec. 349 (1971), pp. 13–30).

Haynes, J. J., Some considerations of vehicular density on urban freeways. Highw. Res. Rec. 99 (1965), pp. 58–80.

Pahl, J. The effect of discrete time measurements on speed data. Highw. Res. Rec. 349 (1971), pp. 1–12.

Wright, C. C., A second method of estimating traffic speeds from flows observed at the ends of a road link. Traffic Eng. Control, 15(9): 432–434 (1974).

2.8 PROBLEMS

1. A platoon of vehicles was observed over a distance of 1,200 ft entering the study area at Hickory Street and proceeding to Main Street 1,200 ft away. The following data were recorded:

Vehicle No.	Time at Hickory (sec)	Time at Main (sec)
1	0	41
2	2	46
3	3	49
4	5	50
5	10	53
6	13	55
7	15	75
8	20	77
9	25	79
10	35	80

(a) Plot the trajectories of these 10 vehicles on $8\frac{1}{2} \times 11$ graph paper with time on x-axis and distance on y-axis; (b) using the space–time domain, find volume, density, and velocity.

2. Two aerial photos were made over the three westbound lanes of the Connecticut Turnpike with a 10-sec interval. The following results were obtained:

Vehicle No.	Position First Photo [a]	Position Second Photo [a]
1	1500	2400
2	1450	2150
3	1300	1950
4	1250	2000
5	1100	1850
6	1000	1925
7	700	1360
8	300	1000
9	250	1050
10	0	900

[a] Expressed in terms of feet from a reference landmark.

(a) Plot these trajectories on $8\frac{1}{2} \times 11$ graph paper; (b) using the space–time domain, compute volume, density, and velocity.

3. The following list of speeds was observed on US 66 in Arizona between 8 AM and 2 PM. (Reported by Freund, J. F., *Modern Elementary Statistics,* Prentice-Hall (1960), p. 15).

52	47	50	54	59	63	67	57	55	48
54	35	53	58	47	53	52	61	54	53
49	46	56	43	57	45	53	28	42	55
60	56	61	55	55	51	56	37	53	51
55	47	51	52	48	60	44	49	57	62
40	53	57	46	57	62	53	57	47	45
58	48	50	53	54	44	52	50	55	50
55	61	47	63	49	59	54	59	46	56
51	54	63	53	53	47	54	38	41	49
57	48	30	42	56	49	46	56	60	55
45	47	44	52	54	59	56	49	58	43
60	52	58	55	61	51	50	48	50	54
56	51	46	46	58	38	52	55	51	52
52	42	53	60	45	48	56	50	46	53
54	51	47	56	54	54	52	57	53	43
59	55	62	50	47	59	66	53	49	74
53	53	56	51	49	53	58	44	55	64
65	65	45	57	52	46	52	57	48	58
56	55	48	53	54	51	56	64	68	54
44	53	54	58	49	61	55	50	47	55

(a) Compute the arithmetic mean and the variance about it. (b) Compute the harmonic mean directly from the arithmetic mean and the variance about it. (d) Estimate the variance of the "space mean speed". (*hint:* Combine Eqs. 2.7 and 2.8).

Chapter 3

STATISTICAL DISTRIBUTIONS OF TRAFFIC CHARACTERISTICS

3.1 INTRODUCTION

In designing new traffic facilities or new control plans it is necessary to predict the performance of traffic with respect to some particular characteristic, and it is often desirable to be able to make a prediction with a limited amount of data available or assumed. For instance, it may be necessary in designing a pedestrian control system to predict the frequency of headways of greater than 10 sec; in designing a left-turn pocket it may be necessary to predict how many times per hour the number of cars arriving during one signal cycle will exceed four. Statistical distribution models may enable the traffic engineer to make predictions such as these with a minimal amount of information.

Statistical distributions are useful in describing a wide variety of phenomena where there is a high element of randomness. In traffic the most important distributions are counting distributions—those useful in describing the occurrence of things that can be counted—and interval distributions—those useful in describing the occurrence of the (time) intervals between events. Distributions are also used, however, in describing such phenomena as speeds and gap acceptance.

In this chapter first elementary counting distributions, specifically the Poisson distribution, the binomial distribution, and the negative binomial distribution, are discussed and an elementary interval distribution for describing time headways is examined. Thereafter the inadequacy of elementary distributions in the general traffic case is demonstrated and some advanced distributions for traffic applications are described. Finally, distributions useful for speeds and gap acceptances are discussed. For each distribution discussed its form and a numerical example of fitting the distribution to field data are provided.

3.2 COUNTING DISTRIBUTIONS

Counting the number of cars arriving during an interval of time is the easiest and oldest measurement of traffic. When counts from a series of equal time intervals are compared, they appear to form a random series. This led early traffic engineers to investigate distributions as a means of describing the occurrence of vehicle arrivals during an interval.

3.2.1 Poisson Distribution

The appropriate distribution for describing the truly random occurrence of discrete events is the Poisson distribution. Thus it was natural for Kinzer [1] to discuss the possible use of the Poisson distribution for traffic in 1933, for Adams [2] to publish numerical examples in 1936, and for Greenshields and co-workers [3] to use the Poisson distribution in analysis of his classic work on traffic at intersections in 1947.

The Poisson distribution may be stated (for a derivation see Appendix B-1):

$$P(x)=\frac{m^x e^{-m}}{x!} \qquad x=0, 1, 2, \ldots \qquad (3.1)$$

or

$$P(x)=\frac{(\lambda t)^x e^{-\lambda t}}{x!} \qquad (3.2)$$

where $P(x)$ = probability that x vehicles will arrive during counting period of t; λ = average rate of arrival (vehicles/sec); t = duration of each counting period (sec); $m=\lambda t$ = average number of vehicles arriving during a period of duration t; and e = natural base of logarithms.

3.2.1.1 Fitting a Poisson Distribution to Observed Data. When a Poisson distribution is to be fitted to observed data, the parameter m is computed as

$$m = \frac{\text{Total occurrences (e.g., total cars observed)}}{\text{Total observations (e.g., total number of time periods)}} \quad (3.3)$$

The value e^{-m} is then obtained from tables, by slide rule calculation, or by a computer subroutine. The probabilities of various values of x are computed term by term. When calculation is to be performed by slide rule, by computer program, or by other direct computation process, the following relationships are helpful.

$$P(0) = e^{-m} \quad (3.4)$$

$$\frac{P(x)}{P(x-1)} = \frac{\dfrac{m^x}{x!}\exp(-m)}{\dfrac{m^{x-1}}{(x-1)!}\exp(-m)} = \frac{m}{x}$$

$$P(x) = \frac{m}{x}P(x-1) \qquad \text{for } x \geq 1 \quad (3.5)$$

Thus, from Eqs. 3.4 and 3.5 it follows that

$$P(0) = e^{-m}$$

$$P(1) = \frac{m}{1}P(0)$$

$$P(2) = \frac{m}{2}P(1)$$

$$P(3) = \frac{m}{3}P(2), \text{ etc.}$$

Equations of the type illustrated by Eq. 3.5 are known as "recursion formulas."

The theoretical number of observed intervals containing 0, 1, 2 cars, etc., is obtained by multiplying the total number of intervals observed by, respectively, the probabilities $P(0)$, $P(1)$, $P(2)$, etc.

3.2.1.2 Combination of Poisson Populations. It is shown in Appendix B-1 that if several Poisson populations, having the parameters m_A, $m_B \ldots m_N$, are combined, the result is still Poisson distributed with the parameter

$$m = \sum_{i=A}^{N} m_i$$

3.2.1.3 Numerical Example of Poisson Distribution. The first published numerical examples of the use of the Poisson distribution in traffic were those of Adams [2] (see Table 3.1).

3.2.1.4 Cumulative Poisson Distribution. The previous discussion has treated the probability of occurrence of a specific event (i.e., exactly x arrivals) using one term of a Poisson distribution. Frequently, in practice it is desirable to compute the probability of a range of events.

The terms of the Poisson distribution may be summed to give the probability of fewer than or more than x vehicles per period. If, for example, the traffic engineer desires to compute the probability that two or fewer cars will arrive during a given period, this is the sum of the probabilities that 0, 1, 2 cars arrive. This may be expressed as

$$P(\leq 2) = \sum_{i=0}^{2} \frac{m^i e^{-m}}{i!}$$

for which the general case may be stated

$$P(\leq x) = \sum_{i=0}^{x} \frac{m^i e^{-m}}{i!} \quad (3.6)$$

For the case of fewer than x, the statement becomes

$$P(< x) = \sum_{i=0}^{x-1} \frac{m^i e^{-m}}{i!} \quad (3.7)$$

For the case of more than x,

$$P(> x) = 1 - \sum_{i=0}^{x} \frac{m^i e^{-m}}{i!} \quad (3.8)$$

For the case of x or more,

$$P(\geq x) = 1 - \sum_{i=0}^{x-1} \frac{m^i e^{-m}}{i!} \quad (3.9)$$

or

$$P(\geq x) = \sum_{i=x}^{\infty} \frac{m^i e^{-m}}{i!} \quad (3.10)$$

For the case of at least x but not more than y,

$$P(x \leq i \leq y) = \sum_{i=x}^{y} \frac{m^i e^{-m}}{i!} \quad (3.11)$$

3.2.1.5 Limitations of the Poisson Distribution. In section 3.2.1 it was pointed out that the Poisson distribution is appropriate for describing discrete *random* events. When traffic is light and when there is no disturbing factor such as a traffic signal, the behavior of traffic may appear to be random, and the Poisson distribution will give satisfactory results. However, when traffic becomes congested or when there is some cyclic disturbance in the arrival rate of traffic (such as produced by a traffic signal), other elementary distributions give a better description of traffic behavior (i.e., they give a better fit between theoretical and observed data). (See Appendix B-7 for a discussion of goodness of fit, including χ^2 and Kolmogorov-Smirnoff (K-S) tests.)

It should be noted that for the Poisson distribution the mean and variance are equal. When the observed data present a ratio of variance/mean markedly different from 1.0, it is an indication that the Poisson distribution is not suitable. (See Appendix B-8 for a discussion of the variance of observed data.)

3.2.2 Binomial Distribution

For congested traffic (where the ratio of the observed variance/mean is substantially less than 1),* it has been found[4] that the binomial distribution can be used to describe the distribution of traffic arrivals.†

Stated in the form most useful for traffic purposes, the binomial distribution is:

$$P(x) = C_x^n\, p^x (1-p)^{n-x},\quad x = 0, 1, 2 \ldots, n \qquad (3.12)$$

where p is the probability that one car arrives and C_x^n is the combinations of n things taken x at a time $= \dfrac{n!}{x!(n-x)!}$.

For the binomial distribution, m is the mean, $= np$, and s^2 is the variance, $= np(1-p)$. If $\hat{p}$ is the estimated value of binomial parameter p, used in fitting, and $\hat{n}$ is the estimated value of binomial parameter n, used in fitting, these parameters may be estimated by the relationships[6]:

$$\hat{p} = (m - s^2)/m \qquad (3.13)$$

$$\hat{n} = m/p = m^2/(m - s^2) \qquad (3.14)$$

where m and s^2 are computed from the observed data. Fitting may be accomplished by direct computation or by the use of tables.[7]

* In congested traffic the opportunity for free movement is decreased, resulting in decreased variance.

† It has been shown[4,5] that as the binomial parameter n becomes very large and the parameter p becomes very small but the product pn is a constant, the Poisson distribution results.

TABLE 3.1 Poisson Arrival Frequencies, Compared with Observed Arrival Counts Measured on Vere Street, London

No. Vehicles/ Sec-10 Period [a]	Observed Frequency	Total Vehicles [b]	Theoretical Frequency [c]
0	94	0	97.0
1	63	63	59.9
2	21	42	18.5
3	2	6	3.8
>3	0	0	0.8
TOTAL	180	111	180.0

[a] Since there were 111 vehicles in 180 10-sec periods, the hourly volume was 222.

[b] Obtained by multiplying Col. 1 by Col. 2.

[c] (Theoretical frequency)=(total observed frequency) $\dfrac{m^x e^{-m}}{x!}$,

where $m = \dfrac{\text{(Total vehicles)}}{\text{(Total observed frequency)}} = \dfrac{111}{180}$; x = value in Col. 1.

3.2.2.1 Numerical Example of Binomial Distribution. Table 3.2 illustrates a binomial distribution fitted to data for congested freeway flow (where the variance/mean ratio equals 0.535). For comparison a Poisson distribution has also been fitted. It will be noted that the binomial distribution produces a much better fit.

3.2.3 Traffic Counts with High Variance (Negative Binomial Distribution)

When traffic counts extend over both a peak period and an offpeak period, combining the results into one distribution results in a high variance. A more common but less obvious situation occurs downstream from a traffic signal: During the early portion of the signal cycle traffic flow is high (usually at saturation level); during the later portion of the signal cycle there often will be very light traffic. If the counting period corresponds to the green portion of the signal cycle, or to the complete signal cycle, cyclic effects will be masked. However, if the counting period is short (say 10 sec), there will be periods of high flow and periods of low flow; there may even be periods of intermediate flow. Thus, combining all counting periods into one distribution will result in a very high variance.

Table 3.3 illustrates the phenomenon where the counting period is synchronized with the signal cycle; here, with 30-sec counting intervals the observed data can fit a Poisson distribution acceptably (at the 5 percent confidence level with a chi-square test). When the same data are analyzed in 10-sec intervals, the nonrandomness appears, and a Poisson distribution will not fit the data; however, a *negative binomial distribution* may fit acceptably, as will be shown.

The negative binomial distribution (sometimes called the Pascal distribution) may be stated [8]

$$P(x) = C_{k-1}^{x+k-1}\, p^k\, q^x \qquad x = 0, 1, 2, \ldots \tag{3.15}$$

Fitting is accomplished by using the sample mean and sample variance to estimate parameters $\hat{p}$ and $\hat{k}$. If m is the mean of observed data and s^2 is the variance of observed data,

$$\hat{p} = m/s^2 \tag{3.16}$$

$$\hat{k} = \frac{m^2}{s^2 - m} \tag{3.17}$$

and

$$\hat{q} = (1 - \hat{p}) \tag{3.18}$$

The various terms can then be obtained from tables [8] or by direct computation (e.g., using a computer program). The recursion equations for the negative binomial distribution are given by

$$P(0) = p^k \tag{3.19}$$

$$P(x) = \frac{x + k - 1}{x}\, q\, P(x-1) \qquad \text{for } x \geq 1 \tag{3.20}$$

TABLE 3.2 Comparison of Binomial and Poisson Distributions Fitted to Congested Traffic Arrivals [a]

Number of Cars per Interval	Observed Frequency	Theoretical Frequency by: Binomial Distribution [b]	Theoretical Frequency by: Poisson Distribution [b]
<3	0	0.3	1.3
3	3	1.0	2.5
4	0	2.9	4.7
5	8	6.2	7.1
6	10	9.8	8.8
7	11	12.3	9.4
8	10	12.1	8.8
9	11	9.4	7.3
10	9	5.8	5.4
11	1	2.8	3.7
12	1	1.0	2.3
>12	0	0.4	2.7
TOTAL	64	64.0	64.0

$m = 7.469$
$s^2 = 3.999$
$s^2/m = 0.535$

[a] Recorded at I-494 at 24th Avenue–Median Lane, during morning peak traffic; measured at 15-sec intervals.

[b] The fit of the binomial distribution is acceptable by a χ^2 test at the 5 percent significance level. The fit of the Poisson distribution is not acceptable. See Appendix B-7 for an illustration of the computations of this example.

3.2.3.1 Numerical Example of Negative Binomial Distribution. Table 3.4 illustrates the data for a cyclic situation downstream from a traffic signal. Although the negative binomial distribution fits the data acceptably, the Poisson distribution does not.

TABLE 3.3 Traffic Arrivals [a] at Durfee Avenue, Northbound; 30-Sec Intervals

Number of Cars per Interval	Frequency: Observed	Frequency: Theoretical, by Poisson Distr.[b]
0	9	5.6
1	16	17.2
2	30	26.3
3	22	26.9
4	19	20.6
5	10	12.6
6	3	6.5
7	7	2.8
8	3	1.1
≥9	1	0.4
TOTAL	120	120.0

$m = \frac{368}{120} = 3.067$

[a] Volume is 368 vehicles/hr.

[b] This fit is acceptable at the 5 percent significance level with a χ^2 test.

3.2.3.2 Some Notes Concerning the Negative Binomial Distribution. Buckley[9] points out that if a detector on a roadway spans several lanes, the resulting count is usually a negative binomial distribution. Kendall and Stuart[10] stated that if the Poisson parameter is independently distributed with a type III distribution, the resulting distribution is negative binomial.

TABLE 3.4 Comparison of Poisson and Negative Binomial Distributions for Durfee Avenue, Northbound; 10-Sec Intervals

Number of Cars per Interval	Observed Frequency	Theoretical Frequency: Poisson Distribution[a]	Theoretical Frequency: Negative Binomial Distribution[a]
0	139	129.6	140.4
1	128	132.4	122.0
2	55	67.7	62.2
3	25	23.1	24.2
4	10	5.9	8.0
5	3	1.2	2.3
>5	0	0.1	0.9
TOTAL	360	360.0	360.0

$m = 1.022$
$s^2 = 1.203$
$s^2/m = 1.177$

[a] The fit of the negative binomial distribution is acceptable by a χ^2 test at the 5 percent significance level; the fit of the Poisson distribution is not.

3.2.4 Summary of Elementary Counting Distributions

1. The Poisson distribution represents the random occurrence of discrete events.
2. In counts of light traffic where the observed data produce a ratio of variance/mean of approximately 1.0, the Poisson distribution may be fitted to the observed data.
3. In counts of congested traffic where the observed data produce a ratio of variance/mean of substantially less than 1.0, the binomial distribution may be fitted to the observed data.
4. In counts of traffic where there is a cyclic variation in the flow or where the mean flow is changing during the counting period, giving a ratio of variance/mean substantially greater than 1.0, the negative binomial distribution can be fitted to the observed data.

3.3 INTERVAL DISTRIBUTION

The previous analyses have developed the probability of discrete events occurring within a specific time interval. Another traffic characteristic of great importance is the time between events; i.e., the time headways between the arrival of vehicles. The class of distributions used for this purpose has been termed "interval distributions."

3.3.1 Negative Exponential Distribution

The elementary interval distribution is the negative exponential distribution, which may be derived as follows:

In Eq. 3.2 substitute $\lambda = V/3{,}600$ cars/sec, where V is the hourly volume; thus,

$$P(x) = \left(\frac{Vt}{3{,}600}\right)^x \frac{e^{-Vt/3{,}600}}{x!}$$

$$P(0) = e^{-Vt/3{,}600}$$

If there is no vehicle arrival in a particular interval of length t, there will be a headway of at least t sec between the last previous arrival and the next arrival. In other words, $P(0)$ is also the probability of a headway equal to or greater than t sec. This may be expressed:

$$P(h \geq t) = e^{-Vt/3{,}600} \tag{3.19}$$

From this relationship it may be seen that (under conditions of random flow) the number of headways greater than any given value will be distributed according to an exponential curve. (Though correctly a *negative* exponential, this is usually known simply as an exponential distribution.)

In the above equation m or $Vt/3{,}600$ is the mean of the arrival (counting) probability distribution. If we set $m = t/T$, T is the mean of the interval (headway) probability distribution $= 3{,}600/V$. Thus, the probability of a headway equal to or greater than t may be written:

$$P(h \geq t) = e^{-t/T} \tag{3.20}$$

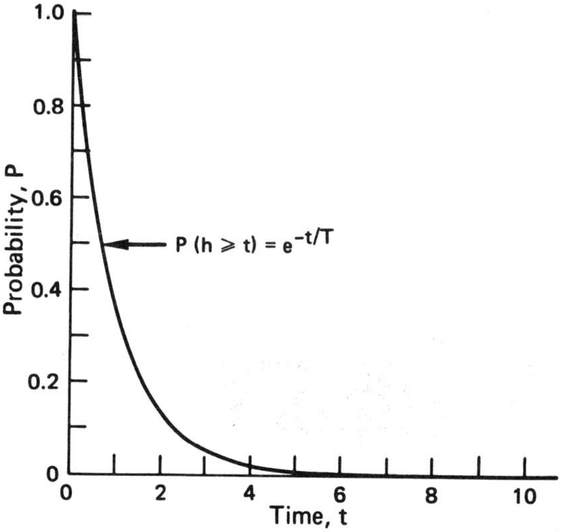

Figure 3.1 Probability of headways equal to or greater than *t*, as treated by the exponential distribution, with *T* = 1 sec.

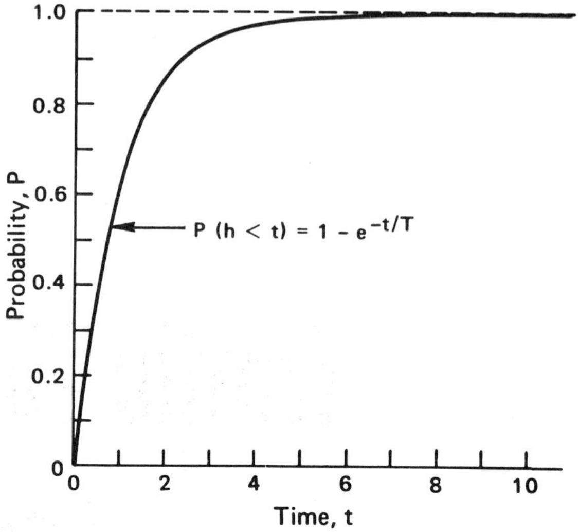

Figure 3.2 Probability of headways less than *t*, with *T* = 1 sec.

A sketch representing the general (negative) exponential distribution of Eq. 3.20 is shown in Figure 3.1.

For some purposes it is more convenient to use the complementary relationship:

$$P(h<t)=1-e^{-t/T} \qquad (3.21)$$

which is illustrated in Figure 3.2.

The exponential probability distribution is fitted by computing the mean interval T and then using tables [72] to obtain values of $e^{-t/T}$ for various values of t.

The variance of the exponential distribution is T^2. (Appendix B-3 presents a derivation.)

3.3.2 Numerical Example of Negative Exponential Distribution

The applicability of the above relationship may be illustrated by fitting an exponential distribution to data observed on the Arroyo Seco Freeway (now the Pasadena Freeway). The observations included 214 intervals, totaling 1,753 sec. Thus, $T=1{,}753/214=8.19$ sec, $m=t/T=t/8.19=0.122t$, $P(h\geq t)=e^{-0.122t}$, and H=expected number of headways$\geq t$, or

$$H=214e^{-0.122t}$$

Table 3.5 lists the results of the fitting. These results are also depicted graphically in Figure 3.3. It may be seen that agreement between the curve and the data is reasonable (acceptable by χ^2 test at the 0.05 level). It should be noted that traffic was light at the time of these observations: Volume=(3,600/1,753)214=439 vehicles/hr.

3.3.3 Alternate Viewpoint

The negative exponential probability models of Eqs. 3.20 and 3.21 can be viewed in an alternate way as follows: Consider the probability density of intervals (headways)

$$p(t)=\frac{dP}{dt}=\frac{1}{T}e^{-t/T} \qquad (3.22)$$

where T is the average of time intervals (headways) measured from the origin. Thus, the following probability relationships can be stated:

$$P(h<t)=\int_0^t p(h)\,dt=\int_0^t \frac{1}{T}e^{-t/T}dt=1-e^{-t/T}$$

$$P(h\geq t)=\int_t^\infty \frac{1}{T}e^{-t/T}dt=e^{-t/T}$$

3.4 INADEQUACY OF THE EXPONENTIAL DISTRIBUTION FOR INTERVALS

As discussed in sections 3.2 and 3.3, the Poisson distribution for vehicle counts and the negative exponential distribution for time headways are only applicable when traffic flows are light; i.e., when there is no interaction between vehicles, thus enabling them to move at random. As traffic becomes heavier, vehicles are restricted in their ability to pass at will, and interaction between vehicles increases. (The assumption of interaction between vehicles is fundamental to car-following analysis, as discussed in Chapter 4.) Vehicles then tend to operate in platoons where the minimum time headway is substantially greater than zero, which results from vehicle finite length and finite spacing between the rear of one vehicle and the front of the following vehicle. The exponential distribution, on the other hand, describes probabilities of headways extending down to very small values. (See Figure 3.1 or Figure 3.2.) When observations from several parallel lanes are combined, it is possible to have headways as low as zero.

Figures 3.1 and 3.3 illustrate the *cumulative* probability of headways t or greater, and Figure 3.2 illustrates the *cumulative* probability of headways less than t. The phenomenon of the scarcity of short headways can often be better illustrated by the frequency density or the probability density curve rather than the cumulative curve; i.e., Eq. 3.22 or the equivalent frequency curve obtained by multiplying the values from Eq. 3.22 by the total number of observations. Thus Figure 3.4 illustrates an example of headway observations on the Hollywood Freeway for flows of 33, 34, and 35 vehicles/min. It will be noted from visual examination that the fit of the exponential density distribution is very poor. (Buckley [11] estimates that "the probability of a worse fit is approximately 10^{-100}.")

TABLE 3.5 Intervehicle Headways, Arroyo Seco Freeway

Headway, t (sec)	Observed Cumulative Frequency $\geq t$	H, Expected Number of Headways[a] $\geq t$
0	214	214.0
1	185	191.3
2	171	171.0
3	149	153.0
4	136	136.7
5	125	122.2
6	111	109.4
7	95	97.8
8	84	87.3
9	72	78.1
10	61	69.8
11	52	62.5
12	40	55.8
13	34	49.9
14	32	44.5
15	29	39.8
16	26	35.7
17	19	31.9
18	16	28.5
19	14	25.5
20	11	22.7
21	10	20.3
22	9	18.2
23	8	16.3
24	8	14.6
25	7	13.1
26	7	11.6
27	6	10.5
28	4	9.2
29	3	8.3
30	1	7.5
31	0	6.6

[a] From negative exponential distribution.

3.5 ADVANCED HEADWAY DISTRIBUTIONS

Because of the poor agreement between the frequencies of headways observed in practice and the frequencies predicted by the negative exponential distribution, as well as theoretical considerations precluding very short headways, other distributions have been sought as a means of improving the predicted frequencies. For the purpose of this discussion these have been termed "advanced headway distributions."

3.5.1 Shifted Exponential Distribution

Section 3.4 emphasized one shortcoming of the exponential distribution: It predicts too many short headways. One approach to treating this situation is to introduce a minimum allowable headway; i.e., a region of the distribution in which headways are prohibited. (Some writers [13] have maintained that a deterministic prohibited period (i.e., a deterministic minimum headway) is philosophically unacceptable. They would rather have a period during which the probability of an arrival is

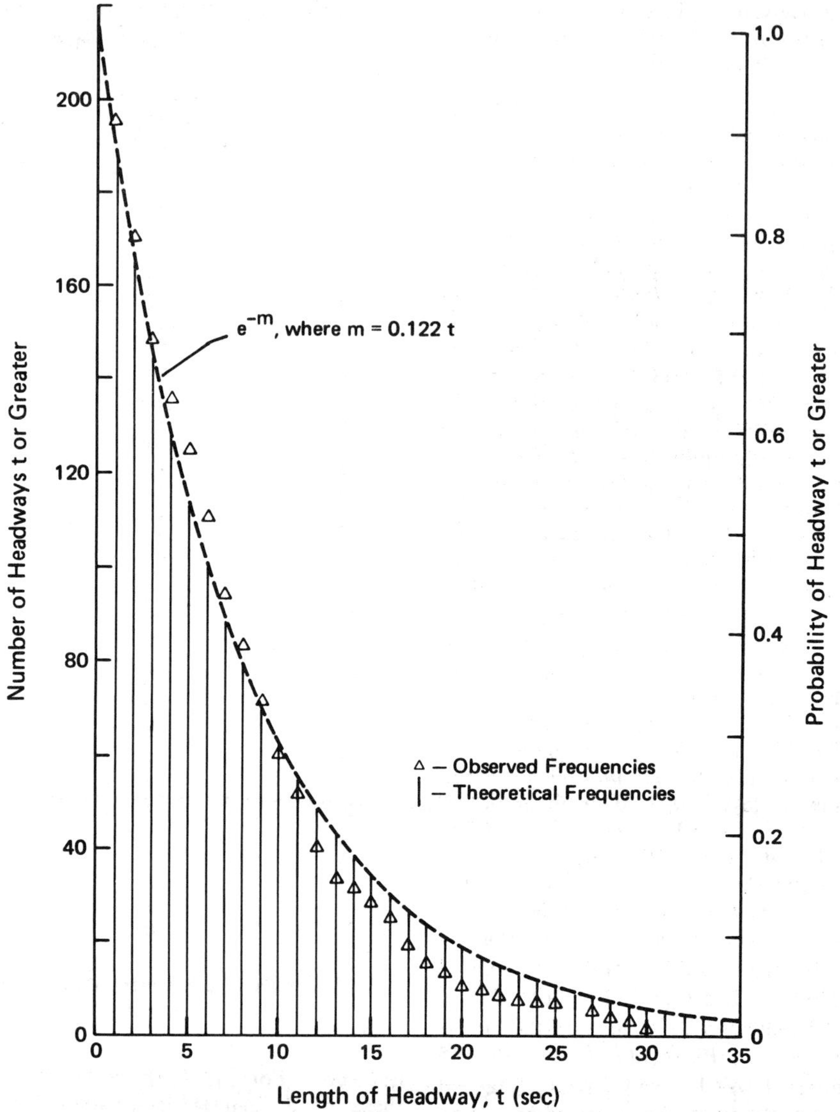

Figure 3.3 Plot of data [4] in Table 3.5 for lane 1 from 2 to 2:30 PM, October 1950. Note: Dashed curve applies only to probability scale.

very low but not zero. The authors wish to point out, however, the use of a deterministic "guarantee period" in reliability engineering.[14]) This may be visualized by taking the curve of Figure 3.2 and shifting it to the right by an amount τ equal to the minimum allowable headway. This is illustrated in Figure 3.5. In shifting, it is necessary to make appropriate adjustments to maintain the total area of unity under the probability curve, from which the following results [12]:

$$P(h \geq t) = e^{-(t-\tau)/(T-\tau)} \quad (3.22)$$

(Appendix B-3 derives the parameters of the shifted exponential distribution.) Note that the probability density is

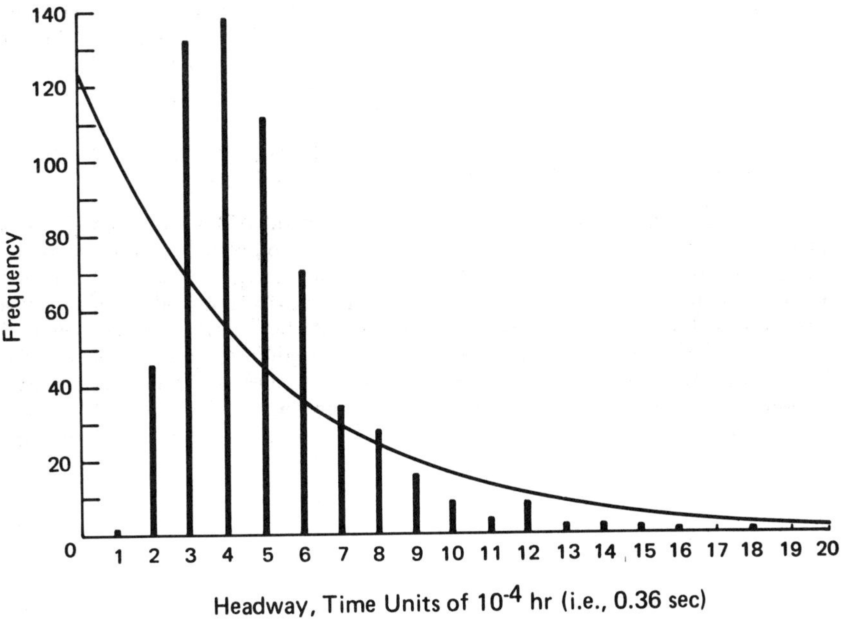

Figure 3.4 Example of negative exponential frequency curve fitted to Hollywood Freeway data.[11] Bars indicate observed data taken on sample size of 609.

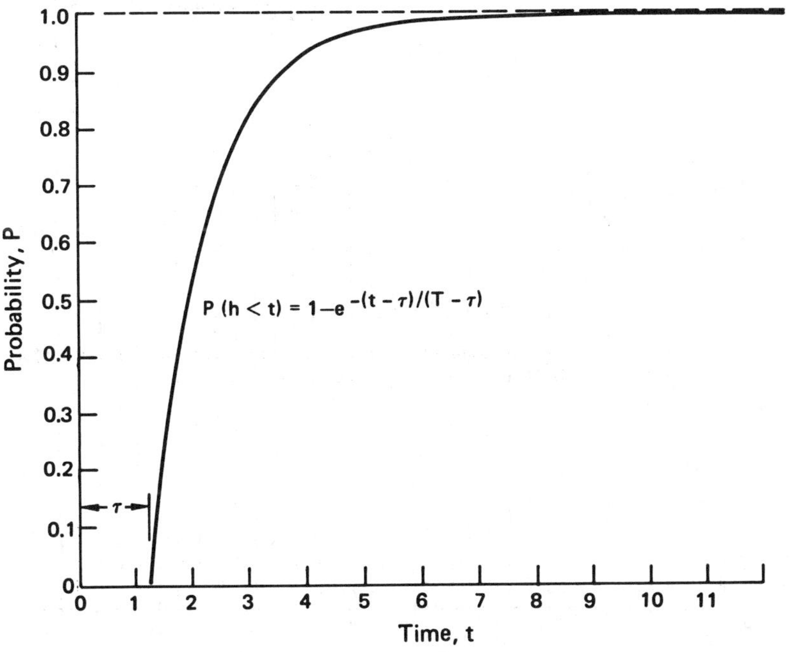

Figure 3.5 Shifted exponential distribution to represent the probability of headways less than *t* with a prohibition of headways less than τ. (Average of observed headways is *T*.)

$$p(t) = 0 \qquad \text{for } t < \tau$$

and

$$p(t) = \frac{1}{T - \tau} e^{-(t-\tau)/(T-\tau)} \qquad \text{for } t \geq \tau.$$

Fitting a shifted exponential distribution to field data requires estimation of the parameters $\hat{T}$ and $\hat{\tau}$, where $\hat{T}$ is the mean of measurements from the origin, $\hat{\tau}$ is the shift of curve with respect to the origin, and $(\hat{T} - \hat{\tau})^2$ is the variance about the origin.

An application of the shifted cumulative exponential distribution is shown in Figure 3.6. Figure 3.7 illustrates the relationship between a shifted exponential frequency density curve and Buckley's data for the Hollywood Freeway. Note that whereas the shifted exponential fits data for low flows, it is not suitable for high flows.

3.5.2 Other Pearson Type III Distributions

Appendix B-2 discusses the Pearson type III family of distributions and points out that both the negative exponential distribution and the shifted exponential distribution are special cases of this family. Other special cases have been found to be useful for describing traffic headways. One such special case is the Erlang distribution.

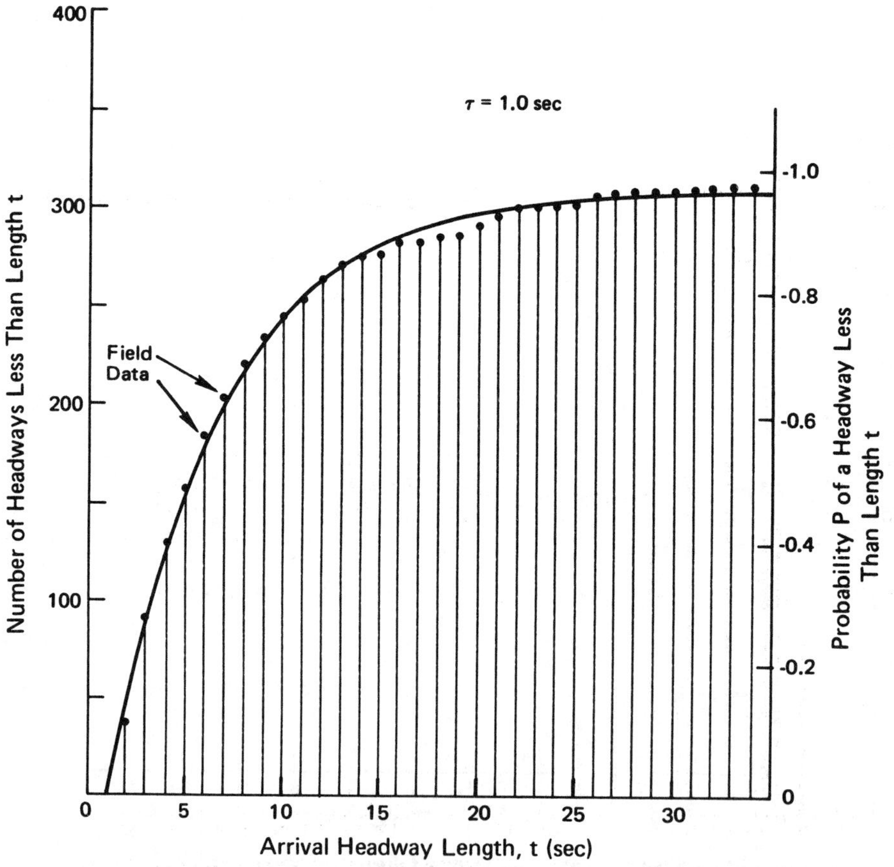

Figure 3.6 Shifted cumulative exponential distribution fitted to observations taken in Cambridge, Mass. Total time, 2,289 sec; total number of vehicles, 318; flow, 5,500 veh/hr.[15]

3.5.2.1 Erlang Distribution. The cumulative Erlang distribution may be stated:

$$P(h \geq t) = \sum_{i=0}^{k-1} \left(\frac{kt}{T}\right)^i \frac{e^{-kt/T}}{i!} \qquad (3.23)$$

For $k=1$, this reduces to the exponential distribution.

For $k=2$

$$P(h \geq t) = \left[1 + \left(\frac{kt}{T}\right)\right] e^{-kt/T}$$

For $k=3$

$$P(h \geq t) = \left[1 + \left(\frac{kt}{T}\right) + \left(\frac{kt}{T}\right)^2 \frac{1}{2!}\right] e^{-kt/T}$$

For $k=4$

$$P(h \geq t) = \left[1 + \left(\frac{kt}{T}\right) + \left(\frac{kt}{T}\right)^2 \frac{1}{2!} + \left(\frac{kt}{T}\right)^3 \frac{1}{3!}\right] e^{-kt/T}$$

Here the value of k, a parameter that determines the shape of the distribution, may be estimated from the mean and variance of the observed data[16]:

$$\hat{k} \approx \frac{T^2}{s^2} \qquad (3.24)$$

where T is the mean of the observed intervals and s^2 is the variance of the observed intervals. When the Erlang distribution is used, the value of $\hat{k}$ in Eq. 3.24 is rounded off to the nearest integer. It is possible to use the gamma distribution with noninteger values of k, but calculations for such cases are very involved, requiring use of tables of the gamma function[16,17] or a computer program of the gamma function.[18,19] The value of k may be a rough indication of the degree of nonrandomness. When $k=1$, the data appear to be random; as k increases, the degree of nonrandomness appears to increase.

3.5.2.2 Numerical Example of the Erlang Distribution. Tables 3.6 and 3.7 illustrate the fitting of an Erlang distribution to headway observations on a freeway; more specifically, Table 3.6 demonstrates the estimation of parameters from one-half of the data, and Table 3.7 the testing of the fit to the other half of the data by the Kolmogorov-Smirnoff (K-S) test (see Appendix B-7). This test is an alternative test for goodness of fit, especially suited to small sample sizes. The fit shown in Table 3.7 is acceptable at the 5 percent level.

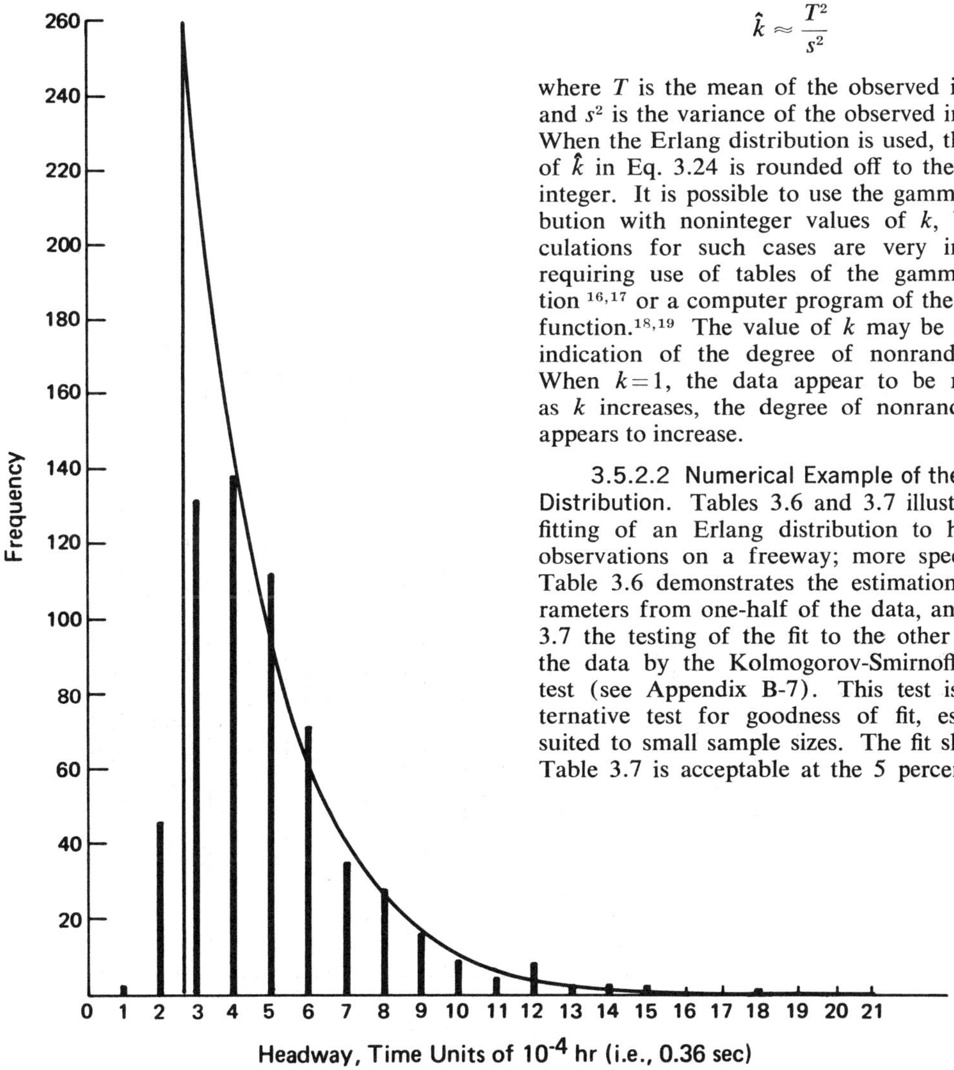

Figure 3.7 Buckley example of shifted exponential fitted to freeway data.[11]

(It should be noted that the data of Tables 3.6 and 3.7 are given in inverse cumulation form, analogous to Figure 3.3.)

Table 3.8 illustrates the changes in the theoretical distribution as the parameter k changes.

3.5.3 Lognormal Distribution of Headways

In the lognormal distribution the logarithm of the stochastic variable, rather than the variable itself, is distributed according to the normal distribution; Appendix B-4 describes this distribution and methods of fitting it to experimental data.

TABLE 3.6 Estimation of Parameters for Erlang Distribution [a]

Observed Headway [b]	Headway Squared
0.9	0.81
2.1	4.41
2.5	6.25
3.8	14.44
1.4	1.96
1.1	1.21
2.9	8.41
1.8	3.24
1.2	1.44
1.0	1.00
1.1	1.21
2.0	4.00
2.1	4.41
2.5	6.25
4.6	21.16
1.1	1.21
5.7	32.49
3.1	9.61
2.3	5.29
3.5	12.25
0.9	0.81
5.6	31.36
0.8	0.64
1.7	2.89
1.4	1.96
TOTAL 57.1	191.71

$T=57.1/25=2.204$

$$s^2=\frac{1}{24}\left[191.71-\frac{(57.1)^2}{25}\right]=2.554$$

$$\hat{k}=\frac{T^2}{s^2}=\frac{(2.204)^2}{2.554}=1.99\approx 2$$

[a] Data from Gerlough and Barnes.[4]

[b] Observations 2, 4, 6 . . . 48, 50 used.

Several writers, including Daou,[20,21] Greenberg,[22] May,[23] and Tolle,[24] have suggested the use of the lognormal distribution for headways, especially for traffic in platoons. One advantage is the ability to make a quick test of fitting by graphical means. The curve of a cumulative lognormal distribution on paper having a log scale on one axis and a normal probability scale on the other axis is a straight line. Figure 3.8 is such a plot for the data shown in Table 3.9.

3.6 COMPOSITE HEADWAY MODELS

Schuhl [25] has suggested that headways, especially where there are more lanes than one, appear to consist of two subpopulations: one of freely flowing cars and one of cars constrained by traffic ahead. The model he proposes is a composite of shifted and unshifted exponential distributions

$$p(h<t)=(1-\alpha)\left[1-\exp\left(\frac{-t}{T_1}\right)\right] +\alpha\left[1-\exp\left(-\frac{t-\tau}{T_2-\tau}\right)\right] \quad (3.25)$$

where

α = fraction of total flow made up of constrained vehicles;
T_1 = average headway of free-flowing vehicles;
T_2 = average headway (about origin) of constrained vehicles; and
τ = shift of curve (i.e., minimum headway) for constrained vehicles.

Figure 3.9 illustrates an example given by Schuhl.

Kell [26] has gone farther than Schuhl by including a minimum headway for the free-flowing vehicles. With four or five parameters to estimate, fitting of this model is sometimes difficult. Where the two populations can be readily identified, however, fitting is relatively straightforward; where this is not possible, other measures must be taken. Kell has developed some relationships from extensive field data.

Based on empirical results, Grecco and Sword [27] have developed a nomograph (Figure 3.10) by which it is possible to estimate probabilities of various headways from lane volume assuming a shifted exponential distribution. The equation represented by this nomograph is

TABLE 3.7 Fitting of Erlang Distribution and Testing by K–S Method [a]

t	Observed Gaps[b] $\geq t$	Observed Relative Frequency	Erlang [c] Probability	K–S Difference	Theoretical Gaps $\geq t$
0	25	1.00	1.000	0.000	25.0
1	21	0.84	0.720	0.120	18.0
2	11	0.44	0.462	0.022	11.6
3	7	0.28	0.246	0.034	6.2
4	4	0.16	0.125	0.035	3.1
5	2	0.08	0.061	0.019	1.5
6	1	0.04	0.026	0.014	0.7
7	0	0.00	0.015	0.035	0.4

Maximum difference: 0.12
$K\text{–}S_{.05.25}$: 0.27

[a] Data from Gerlough and Barnes.[4]
[b] Observations 1, 3, 5, . . . 47, 49 used.
[c] $P(h \geq t) = \left[1 + \frac{2t}{2.204}\right] e^{-2t/2.204}$; fit is acceptable at the 5 percent level.

TABLE 3.8 Effect of Varying Erlang Parameter k for Theoretical Frequency of Headways Greater Than t [a]

t	$k=1$	$k=2$	$k=3$	$k=4$	$k=5$	Observed Odd Values	Observed Even Values
0	25.0	25.0	25.0	25.0	25.0	25	25
1	15.9	19.2	21.1	22.2	23.0	21	22
2	10.1	11.5	12.2	12.7	13.1	11	13
3	6.4	6.1	5.7	5.2	4.8	7	6
4	4.1	3.1	2.3	1.7	1.3	4	3
5	2.6	1.5	0.9	0.5	0.3	2	2
6	1.6	0.7	0.3	0.1		1	0
7	1.0	0.3	0.1			0	
8	0.7	0.1					
9	0.4						
10	0.3						

[a] Data from Gerlough and Barnes[4] (as corrected by the authors Mar. 8, 1974).

$$P(h \leq t) = \frac{115\,V\,e^{-(t-1)/2.5}}{100{,}000} + \left(1 - \frac{115V}{100{,}000}\right) e^{-t/(24-0.0122V)} \qquad t \geq 1.0 \text{ sec} \tag{3.26}$$

where V is the hourly volume.

Shifting of distributions and use of multiparameter distributions and mixed distributions to represent mixtures of subpopulations have been exploited by various investigators in various ways. The works of Dawson and of Buckley are cited as examples (without derivation).

Dawson[28,29] has combined the improved shape of the Erlang distribution and the feature of shifting the distribution to the right to account for the minimum headway. His final model, which he calls the hyper-Erlang ("hyperlang") distribution, may be expressed

$$P(h \geq t) = \alpha_1 e^{-(t-\delta_1)/(\gamma_1-\delta_1)} + \alpha_2 e^{-k(t-\delta_2)/(\gamma_2-\delta_2)} \sum_{x=0}^{k-1} \frac{k\left[\frac{(t-\delta_2)}{(\gamma_2-\delta_2)}\right]^x}{x!} \tag{3.27}$$

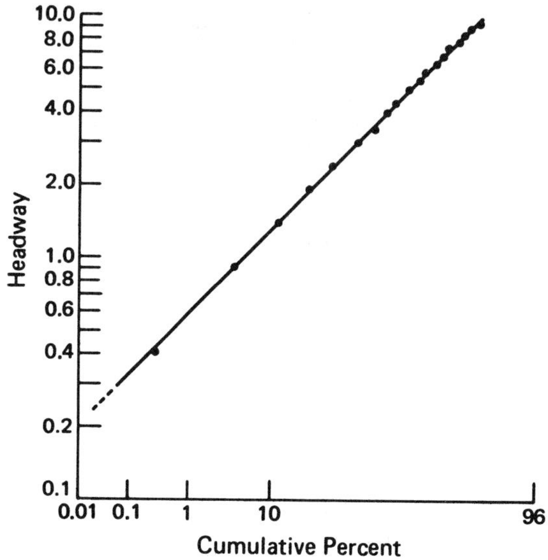

Figure 3.8 Lognormal plot of data in Table 3.9 with visual fit.

TABLE 3.9 Headways on Eisenhower Freeway at First Avenue, 10–14 Vehicles/Minute [a]

Headway	Frequency	Cumulative Percent
0.4	5	0.3
0.9	63	4.1
1.4	122	11.4
1.9	163	21.2
2.4	152	30.3
2.9	198	42.2
3.4	139	50.5
3.9	98	56.4
4.4	85	61.5
4.9	92	67.0
5.4	75	71.5
5.9	55	74.8
6.4	58	78.3
6.9	38	80.6
7.4	42	83.1
7.9	55	86.4
8.4	27	88.0
8.9	23	89.3
9.4	27	91.0
>9.5	150	100.0

[a] Data from May.[23]

where t=any time duration; δ_1=the minimum "free" headway; γ_1=the average "free" headway; δ_2=the minimum headway in the constrained headway distribution; γ_2=the average headway in the constrained distribution; and k=an index that indicates the degree of nonrandomness in the constrained headway distribution. (See section 3.5.2.1.)

Table 3.10 lists the values for the various parameters as computed for data observed by investigators at Purdue University. Figure 3.11 plots the distributions from these parameters.

Buckley [11] has proposed a "semirandom" model or a distribution representing two subpopulations, one having type 1 headways and the other having type 2 headways. Type 1 headways result from a vehicle being placed exactly at the rear limit of a zone of emptiness, where vehicles never enter; its length is measured in time units and is normally distributed. Type 2 headways result when a vehicle occurs

TABLE 3.10 Hyperlang Model Parameters [a]

Flow Rate: Monitored	Flow Rate: Computed	R^2	α_1	γ_1	δ_1	k	α_2	γ_2	δ_2
158	184	0.9959	0.86	22.34	0.69	1	0.14	2.88	1.65
251	219	0.9977	0.70	22.09	0.35	1	0.30	2.90	1.44
353	311	0.9996	0.61	16.75	0.74	1	0.39	3.35	1.12
450	492	0.9996	0.64	9.81	0.61	2	0.36	2.81	0.70
547	489	0.9997	0.56	11.05	0.70	2	0.44	2.73	0.90
651	567	0.9995	0.43	11.06	0.79	2	0.57	2.81	0.71
746	710	0.9996	0.40	8.35	0.88	3	0.60	2.92	0.57
836	740	0.9998	0.20	11.57	0.95	3	0.80	3.23	0.52
957	971	0.9997	0.53	4.58	1.06	6	0.47	2.71	0.72

[a] Data from Purdue University.

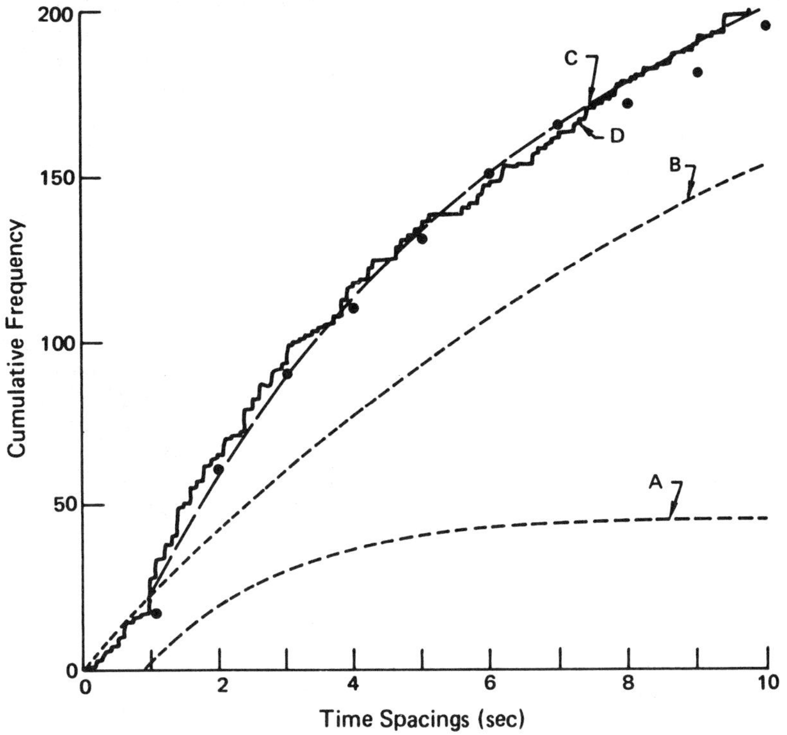

Figure 3.9 Composite exponential arrivals:[10] curve A computed for constrained vehicles; curve B, for unconstrained vehicles; curve C, for composite flow; curve D, from field data (taken from Schuhl). Points in circles are results of computer generation.

at some position to the rear of the extreme rear limit of a zone of emptiness; i.e., vehicles with type 2 headways are presumed to be not influenced by the front vehicle.

Using 609 headway observations from the Hollywood Freeway at volumes of 33, 34, and 35 vehicles/min, Buckley plotted comparisons of data against several distributions. These plots appear as Figures 3.4, 3.7, and 3.12. By inspection, the semirandom distribution appears to provide the best fit.

3.7 SELECTION OF HEADWAY DISTRIBUTION

As in many engineering selection processes, selection of a suitable headway distribution represents a compromise between economic considerations and faithfulness of the model. Greater faithfulness is often obtained by using a model with a greater number of parameters; such a model, on the other hand, results in a more complex computational procedure. In some cases the intended use of the model can help in the selection procedure. For instance, Newell[30] has shown that delays are relatively insensitive to the form of the distribution of the arriving traffic. Thus, if the objective is simply the computation of delays, the simplest (i.e., the negative exponential) distribution should be used. If, however, the objective is the determination of gaps for, say, crossing purposes, a more faithful distribution may be needed.

3.8 ADVANCED COUNTING DISTRIBUTIONS

In section 3.2 traffic counts (and headways) for low traffic flows (i.e., where cars can maneuver with relative ease) appear to be random. In such cases the Poisson distribution can serve as a model of traffic counts. When the flow becomes high, however, freedom to maneuver is diminished and counts per unit time become more uniform. In this situation the binomial distribution provides a simple means of modeling the counting process. On the other hand, when there is some disturbing factor, such as a traffic signal or a rapidly changing average flow, the variance of the counting process becomes large. The simple model for such situations is the negative binomial distribution.

In section 3.4 simple headway distributions are not deemed adequate from a theo-

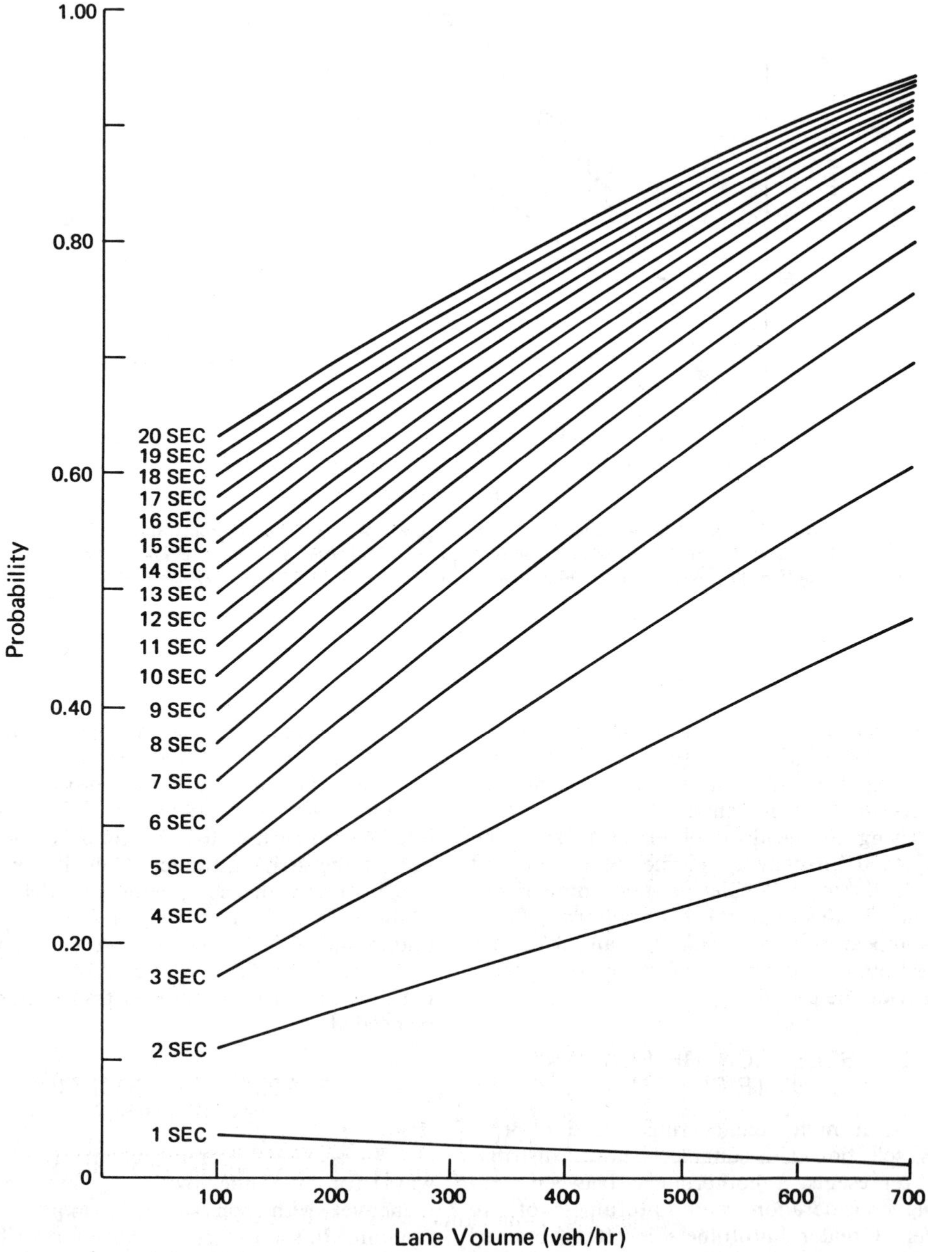

Figure 3.10 Probability of a headway less than x sec.[27]

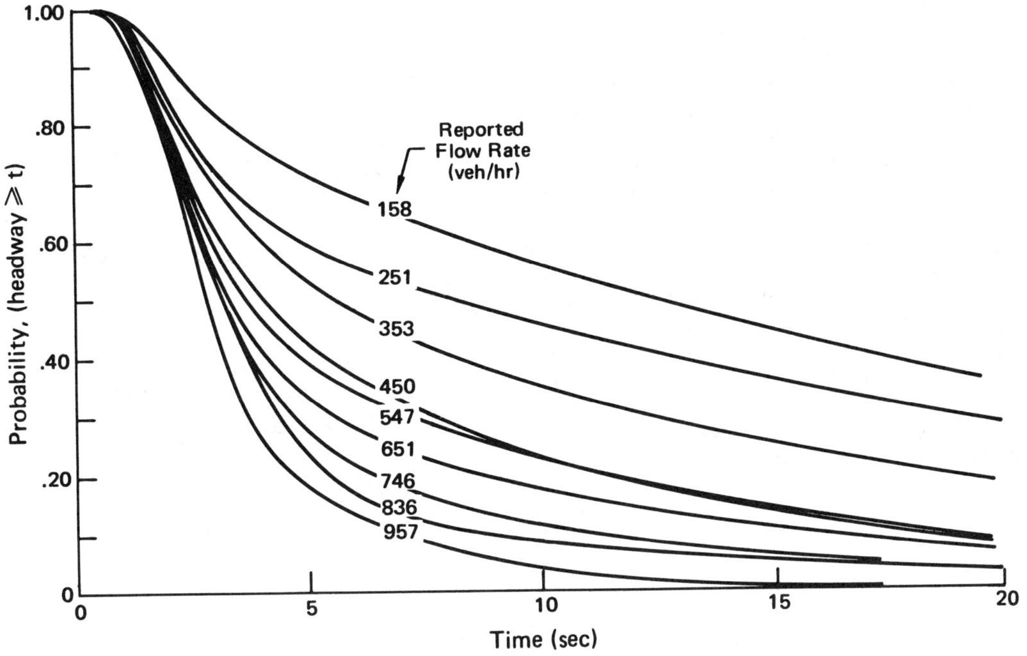

Figure 3.11 Hyperlang headway distributions for Purdue research project data.[28]

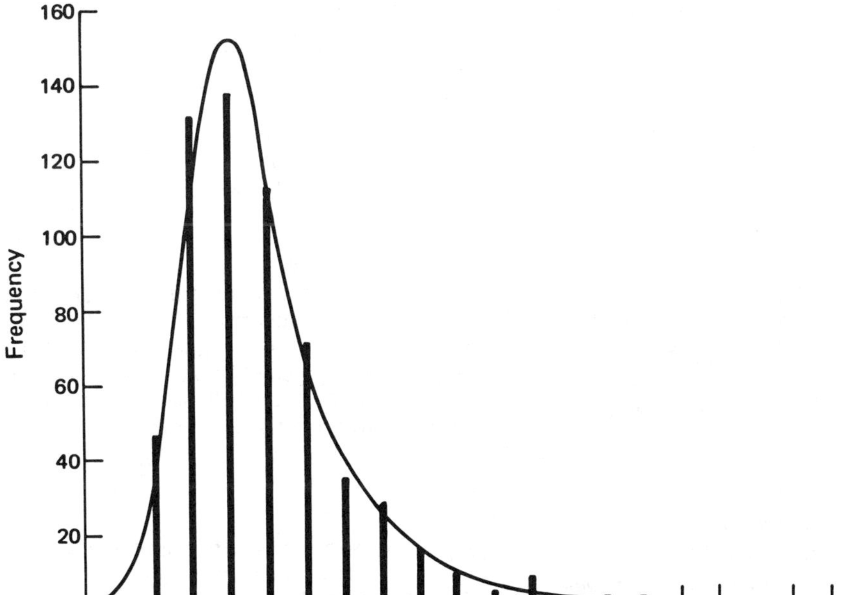

Figure 3.12 Fit of Buckley's semirandom distribution to Hollywood Freeway data.[11] Bars are observed data and curve is best fit for the semirandom distribution; sample size = 609.

retical standpoint, and in sections 3.5 and 3.6 several advanced distributions of headways were discussed.

In section 3.1.1 it was pointed out that the negative exponential distribution for headways can be easily derived from the Poisson distribution of counts. The derivation process could have been reversed, taking the negative exponential distribution of headways and deriving the Poisson distribution by the mathematical operation known as convolution. This procedure, even for the simple case of the negative exponential headway distribution, is very tedious. For more complex headway distributions the convolution process can be intractable even though theoretically possible.

In summary, the following pertain: (1) Although simple distributions will often work for traffic data, they are not necessarily correct; (2) for each headway distribution there is a related counting distribution, even though such distributions cannot be stated explicitly.

By use of the approaches taken in developing advanced headway distributions we can search for advanced counting distributions directly (i.e., independent of the related headway distribution).

3.8.1 Synchronous and Asynchronous Counting

Before proceeding it is necessary to clarify a particular point. Jewell [31] has cited two types of counting—one in which a new period is started immediately on the completion of the previous period, and the other in which the start of a counting period is delayed until the instant a vehicle passes. Haight [32] has termed these cases asynchronous and synchronous, respectively (Figure 3.13). Because synchronous counting has relatively minor application in traffic engineering, only the asynchronous form (with counting intervals forming a continuum) is covered in this monograph.

3.8.2 Generalized Poisson Distribution

In dealing with headways of dense traffic it was found beneficial to introduce a shape parameter k, which as an integer led to the Erlang distribution. Similarly, for counting car arrivals it is beneficial to introduce a shape parameter k. This results in the generalized Poisson distribution,[13] which, when k is an integer, takes the form:

$$P(x) = \sum_{j=kx}^{k(x+1)-1} \frac{\lambda^j e^{-\lambda}}{j!} \qquad x=0, 1, 2, \ldots$$

or

$$P(x) = \sum_{i=1}^{k} \frac{e^{-\lambda} (\lambda)^{xk+i-1}}{(xk+i-1)!} \qquad x=0, 1, 2, \ldots \tag{3.28}$$

When $k=1$, the simple Poisson distribution results.

When $k=2$

$$P(0) = e^{-\lambda} + \lambda e^{-\lambda}$$

$$P(1) = \frac{\lambda^2 e^{-\lambda}}{2!} + \frac{\lambda^3 e^{-\lambda}}{3!}$$

$$P(2) = \frac{\lambda^4 e^{-\lambda}}{4!} + \frac{\lambda^5 e^{-\lambda}}{5!}, \text{ etc.}$$

When $k=3$

$$P(0) = e^{-\lambda} + \lambda e^{-\lambda} + \frac{\lambda^2 e^{-\lambda}}{2!}$$

$$P(1) = \frac{\lambda^3 e^{-\lambda}}{3!} + \frac{\lambda^4 e^{-\lambda}}{4!} + \frac{\lambda^5 e^{-\lambda}}{5!}$$

That is, each term of the generalized Poisson distribution consists of the sum of k Poisson terms. It should be quickly pointed out that the parameter λ in the generalized Poisson distribution is not equal to the Poisson parameter m (except when $k=1$). Furthermore, estimation of the parameter k is not a straightforward process. To aid this situation, Haight et al.[33] have provided a nomograph and an estimating equation. To fit a generalized Poisson distribution to observed data, the best way of determining k is by means of the nomograph of Figure 3.14. This nomograph is entered with the mean m and the variance s^2 of the observed data, from which the closest value of k is read. Although it is possible to read λ from Figure 3.14, a more reliable way is to compute its value from the estimating equation:

$$\lambda = k\,m + \frac{1}{2}(k-1). \tag{3.29}$$

Poisson terms with parameter λ are then computed or read from tables,[34,35] and successive groups of k-terms are summed to obtain the generalized-Poisson terms, to comply with Eq. 3.28. It should be noted that the generalized Poisson distribution describes the counting of the same population whose headways are described by the Erlang distribution.

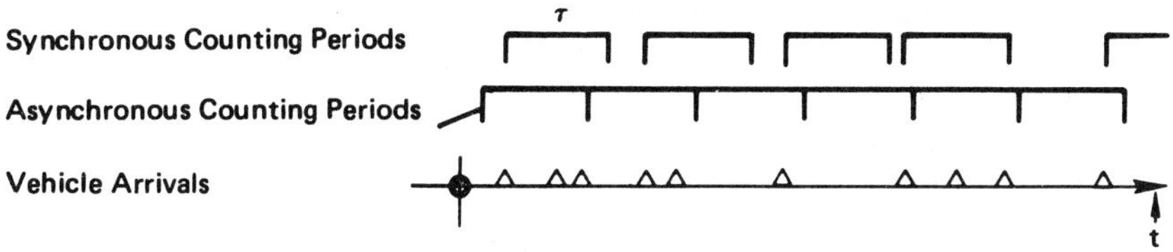

Figure 3.13 Schematic illustrating synchronous and asynchronous counting.[32]

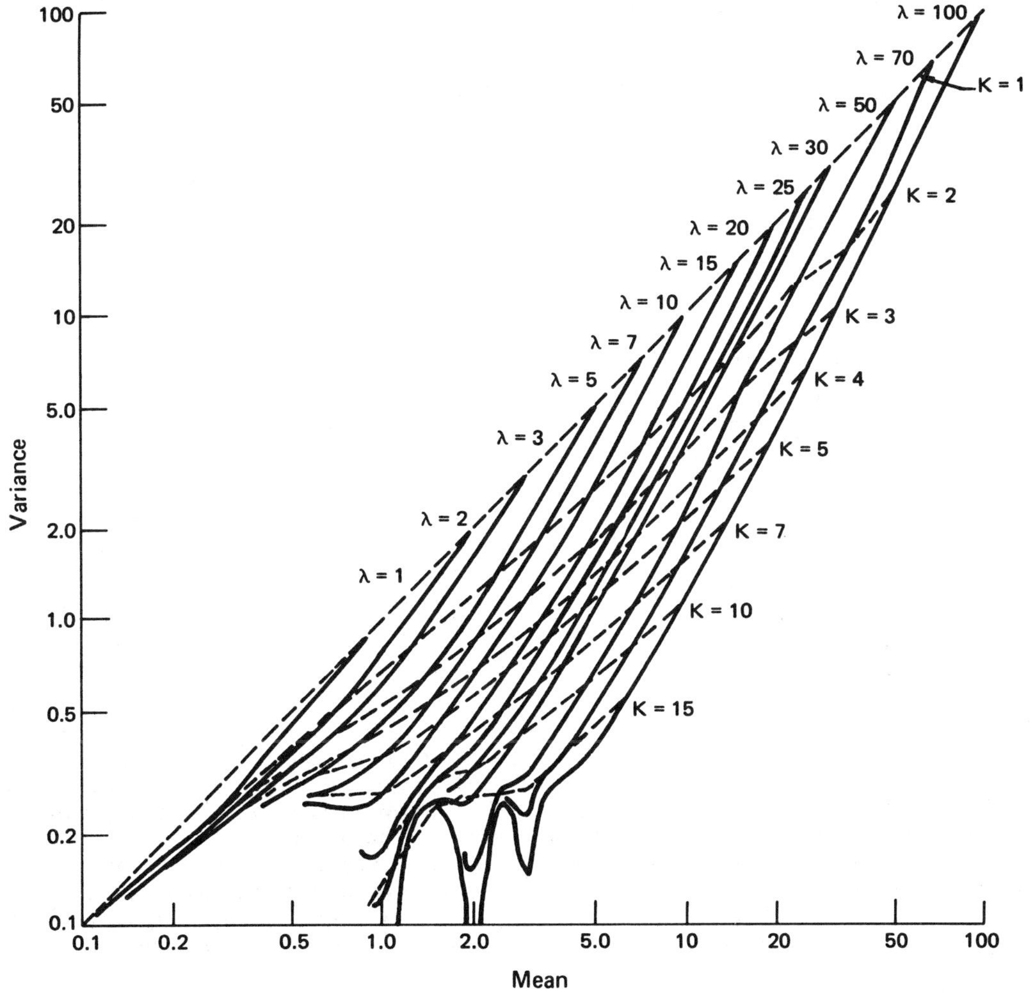

Figure 3.14 Nomograph for estimating parameter *k* for generalized Poisson distribution.[33]

3.8.2.1 Numerical Example of Generalized Poisson Distribution. Table 3.11 exemplifies a generalized Poisson distribution fitted to freeway data. It will be noted that the fit is very good as measured by the χ^2 test.

3.8.2.2 Generalized Poisson Distribution with Nonintegral *k*. When nonintegral values of *k* are to be used, it is necessary to make use of the "normalized" incomplete gamma function [13,17,18]:

TABLE 3.11 Vehicle Arrivals for Median Lane at Peak Morning Flow, I-494 at 24th Avenue [a]

Number of Cars per Interval [b] (x_i)	Observed Frequency (f_i)	Total Cars Observed ($f_i x_i$)	$p(2x_i)$	$p(2x_i+1)$	Probability [c] $P(x_i)$	Theoretical Frequency [d] $64\,P(x_i)$	$f(x_i)/F(xi)$
0			0.0000	0.0000	0.0000	0.0	
1			0.0000+	0.0000+	0.0001	0.0	
2			0.0005	0.0015	0.0020	0.1	
3	3	9	0.0039	0.0084	0.0123	0.8	11.9
4	0	0	0.0162	0.0275	0.0437	2.8	
5	8	40	0.0423	0.0590	0.1013	6.5	
6	10	60	0.0755	0.0894	0.1649	10.6	9.4
7	11	77	0.0983	0.1010	0.1993	12.8	9.5
8	10	80	0.0974	0.0885	0.1859	11.9	8.4
9	11	99	0.0760	0.0619+	0.1380	8.8	13.8
10	9	90	0.0480	0.0354	0.0834	5.3	12.5
11	1	11			0.0691	10> 4.4	
12	1	12					
TOTAL	64	478			1.0000	64.0	65.5

(Theoretical frequencies for 0–5 are grouped, giving 11.9; those for 10 and 10> are grouped, giving 12.5.)

$$m = \frac{478}{64} = 7.469$$

$$s^2 = \frac{3822 - \dfrac{(478)^2}{64}}{63} = 3.999$$

$$s^2/m = 0.535$$

[a] Data from Minnesota Department of Highways.
[b] Fifteen-second intervals.
[c] Col. 4 + Col. 5; $\chi_{0.05}{}^2 = 7.81$.
[d] For the generalized Poisson distribution.
[e] From nomograph of Figure 3.14.
$k \approx 2$

$$\lambda = km + \frac{1}{2}(k-1) = 2\,(7.469) + 0.5 = 15.438\ ^{e}$$

$$\chi^2 = 65.5 - 64.0 = 1.5$$

$$\therefore \nu = 6 - 3 = 3$$

$$I(u,\,p) = \frac{\displaystyle\int_0^{\lambda} e^{-\lambda}\ \lambda^p\ d\lambda}{\displaystyle\int_0^{\infty} e^{-\lambda}\ \lambda^p\ d\lambda}$$

where $p=(x+1)k-1$ and $u=\lambda/(p+1)^{1/2}$

Appendix B-6 describes this procedure.

For most traffic engineering applications the relatively modest improvement achieved by use of noninteger values of k does not justify the extra effort involved.[33]

3.8.3 Other Counting Distributions

Although other counting distributions have been proposed,[9,36,37] most are beyond the scope of the present discussions. Particularly worthy of note, however, is the distribution proposed by Oliver and Thibault.[38] This distribution makes use of a composite headway distribution such as that discussed by Kell.[26] Oliver states the cumulative probability of x or fewer in an interval t (in the notation of this monograph, $P(\leq x)$) as:

$$P(\leq x) = \sum_{i=0}^{x} C_i^{x+1}(1-\alpha)^{x+1-i}\alpha^i \frac{\gamma[(x+1-i),y]}{(x-i)!} + \alpha^{x+1} \qquad (3.30)$$

where

$\alpha =$ proportion of restrained vehicles;
$y = \mu(1-\alpha)[t-(x+1)\tau]$ for
$\mu =$ mean flow rate for unrestrained vehicles and
$\tau =$ minimum headway; and
$\gamma(u,\,z) =$ incomplete gamma function of $(u,\,z)$

$$= \int_0^z t^{x-1}e^{-t}dt$$

3.8.4 Selection of Counting Distribution

The selection of a counting distribution for a particular application should, of course, balance the requirement for faithfulness against the degree of complexity required for a given faithfulness. In performing such a selection, it is useful to consider the statement of Underwood [39]:

While less accurate than the more sophisticated distributions that have been proposed by various workers, the virtue of the Poisson distribution is that it is relatively easy to use. Bearing in mind the various other assumptions and approximations that often must be used in road traffic calculations, it is believed that the Poisson approximation is sufficiently accurate to use for many practical problems, provided it is realized that errors increase as the volume increases.

Where traffic conditions are such that traffic is clearly not random, Table 3.12 provides a guide to the selection of an appropriate distribution where minimum complexity is desired.

3.9 DISTRIBUTION MODELS FOR SPEEDS

Inasmuch as speeds are measured on a continuous scale and the mean speed (whether space mean or time mean) will usually be a substantial distance from zero, one might think of the normal distribution as a model for speeds. Although the normal distribution is sometimes appropriate, at times the lognormal distribution is more useful.

3.9.1 Normal Distributions of Speeds

Numerous investigators have used normal distributions to represent speeds. An excellent example of the use of the normal distribution for speeds is contained in the work of Leong.[42] He reports the results of radar speedmeter measurements of free-flow speeds at 31 rural locations in Australia during three different years. When the data for passenger cars on level roads are assembled, the speed distributions are found to be normal, with the standard deviations equal to 0.17 times the respective arithmetic means. Figure 3.15 shows cumulative speed distributions at four locations; Figure 3.16 shows the same four cumulative distributions after each has been normalized * by dividing each observation by the arithmetic mean for that distribution. When thus normalized, these four sets of data appear to form one generalized distribution with the standard deviation equal to 0.17 times the mean.

Several investigators,[43–45] however, have found speed distributions to be quite skewed when a fit of the normal distribution is attempted.

* In this context "normalized" is used in the engineering sense of dividing all values by the maximum or by the expected value.

TABLE 3.12 Guide to Selecting Counting Distribution

Range of Variance: Mean Ratio [a]	Situations Where Condition Prevails	Suggested Distribution
>1	Variation in mean value; cyclic fluctuation	Negative binomial
≈ 1	Essentially random behavior	Poisson
<1	Congested flow [b]	Generalized Poisson; binomial

[a] At present it is impossible to give definitive values for the ratios of variance to mean at which the Poisson distribution fails to fit. However, Figure 3.14 provides a basis for selecting the appropriate value of k in the generalized Poisson distribution. Miller [71] points out that the variance:mean ratio increases as the length of the counting period.

[b] Pak-Poy [40] suggests that the critical volume is approximately the practical capacity as computed by the 1950 *Highway Capacity Manual*.[41]

3.9.2 Lognormal Model of Speeds

Haight and Mosher [46] have pointed out that the lognormal distribution may be an appropriate model for speeds. (The details of the lognormal distribution are discussed in Appendix B-4.) Table 3.13 summarizes field observations of spot speeds at a freeway location, along with the computed theoretical frequencies from the fitted lognormal distribution. Figure 3.17 is a plot of the observed relative frequency on logarithm-probability paper.

3.10 GAP (AND LAG) ACCEPTANCE DISTRIBUTIONS

The distribution of gap acceptance at intersections and freeway ramps and by pedestrians constitutes an important consideration in the computation of delays and in the design of control systems. Early investigators [47] presented curves of percent or number accepted and percent or number rejected as rectilinear plots; the point of intersection of the acceptance and rejection curves was then termed the

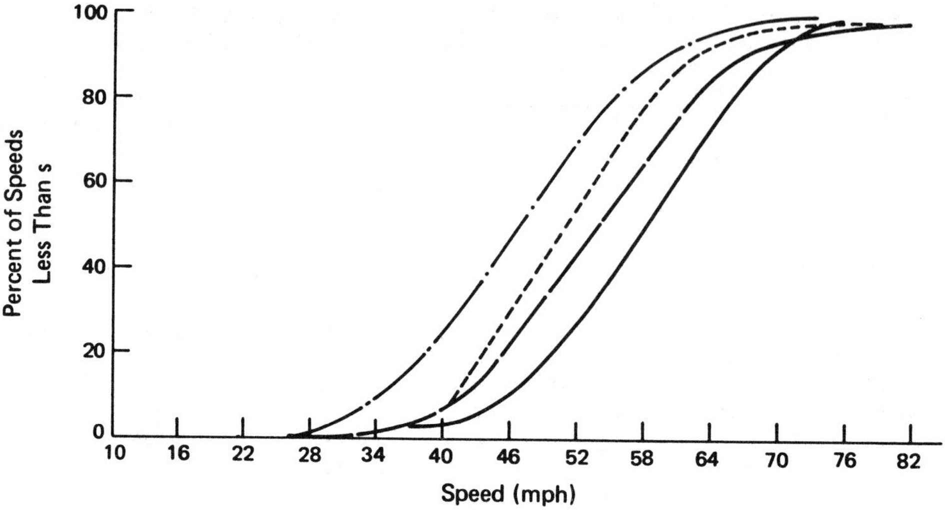

Figure 3.15 Cumulative (normal) distributions of speeds at four locations.[42]

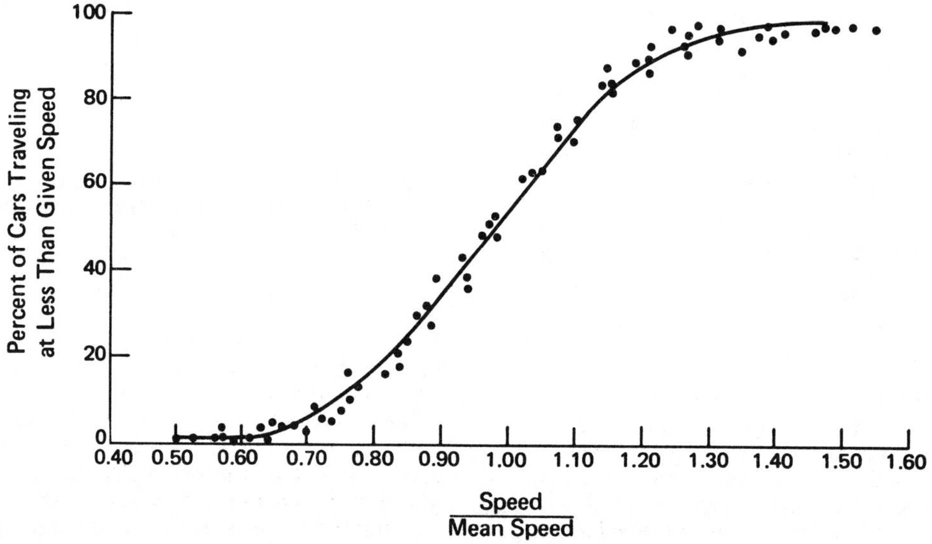

Figure 3.16 Same data as Figure 3.15 but with each distribution normalized.[42]

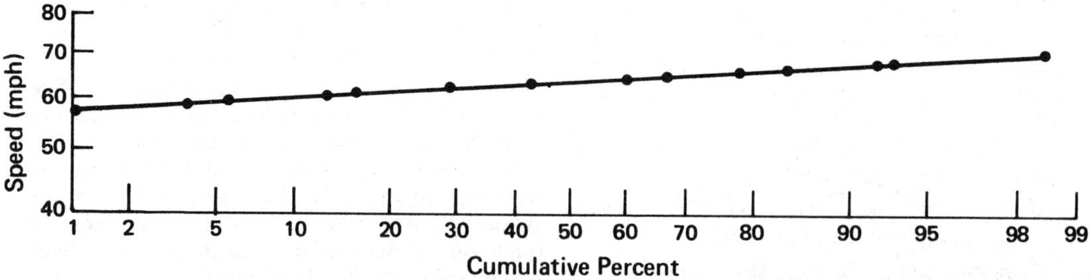

Figure 3.17 Lognormal plot of freeway spot speeds.

TABLE 3.13 Spot Speeds on Interstate 94 at Prior Street Bridge, 8:30–9:00 AM, August 20, 1970

	Observed Data			Theoretical Data for Lognormal Distribution	
	Frequency				
mph	No.	Cumul.	Cumul. %	Cumul. Prob.	Cumul. Freq.
57	2	2	1.1	0.02	3.7
58	7	9	3.7	0.04	7.4
59	1	10	5.4	0.09	16.6
60	14	24	13.0	0.15	27.6
61	5	29	15.8	0.25	46.0
62	25	54	29.4	0.36	66.2
63	24	78	43.0	0.49	90.1
64	32	110	60.0	0.61	112.2
65	13	123	67.0	0.72	132.5
66	22	145	78.0	0.82	150.9
67	9	154	84.0	0.89	163.8
68	13	167	92.0	0.93	171.1
69	4	171	93.0	0.96	176.6
70	10	181	98.5	0.98	180.3
71	0	181	98.5	0.99	182.2
72	2	183	99.5	1.00	184.0
73	1	184	100.0	1.00	184.0

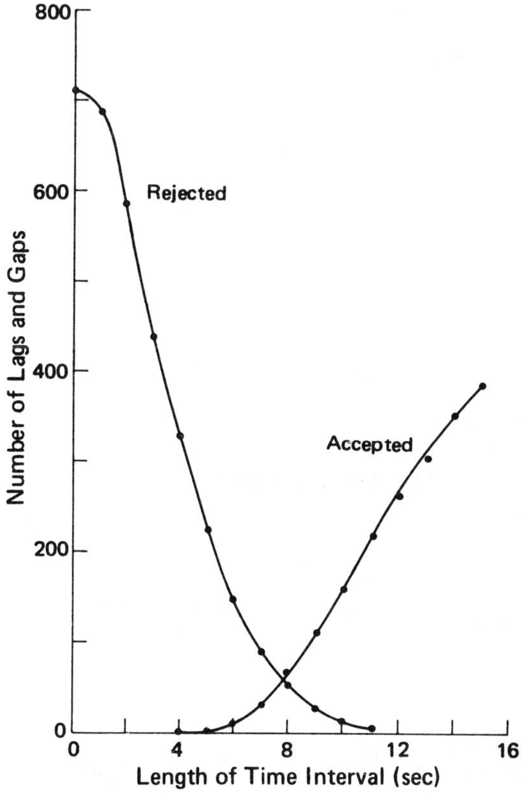

Figure 3.18 Distribution of accepted and rejected lags and gaps at intersection left turns.[51]

"critical gap" or the "critical lag." Such a plot is shown in Figure 3.18. Blunden et al.[48] have attempted to fit Erlang distributions to rejection curves (Figure 3.19). Other studies [49,51,54] have indicated that the lognormal distribution constitutes a good representation of both acceptance and rejection curves (Figure 3.20).

Still others have found probit analysis (discussed in Appendix B-5) convenient for analyzing gap acceptance data (see, for instance, Robinson [50]). Figure 3.21 demonstrates how probits can be combined with a logarithmic plot to display the distribution of gap acceptances.

In making field observations of gap acceptances and rejections sample size difficulties arise inasmuch as each driver can reject many gaps but can accept only one gap each time he makes an acceptance and acts upon his selection. Blunden and co-workers [48] propose two methods of treating this situation for sample size considerations.

Ashworth [52] has demonstrated that normal field practice in observing acceptance data, as well as Blunden's method of adjustment, results in biases to the data. In recognition of these biases he proposes the following correction: The unbiased estimate of the mean of the critical gap distribution may be obtained by subtracting s^2q from the 50 percentile acceptance gap,

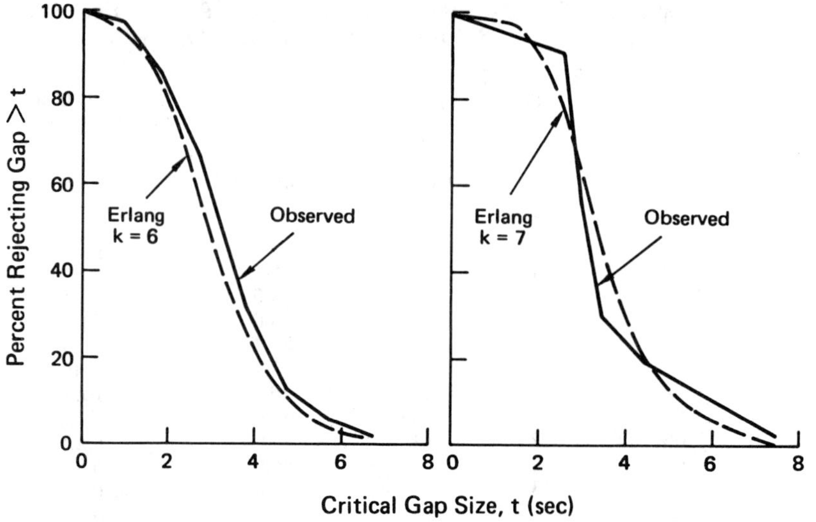

Figure 3.19 Comparison of observed and theoretical distributions of rejected gaps.[48]

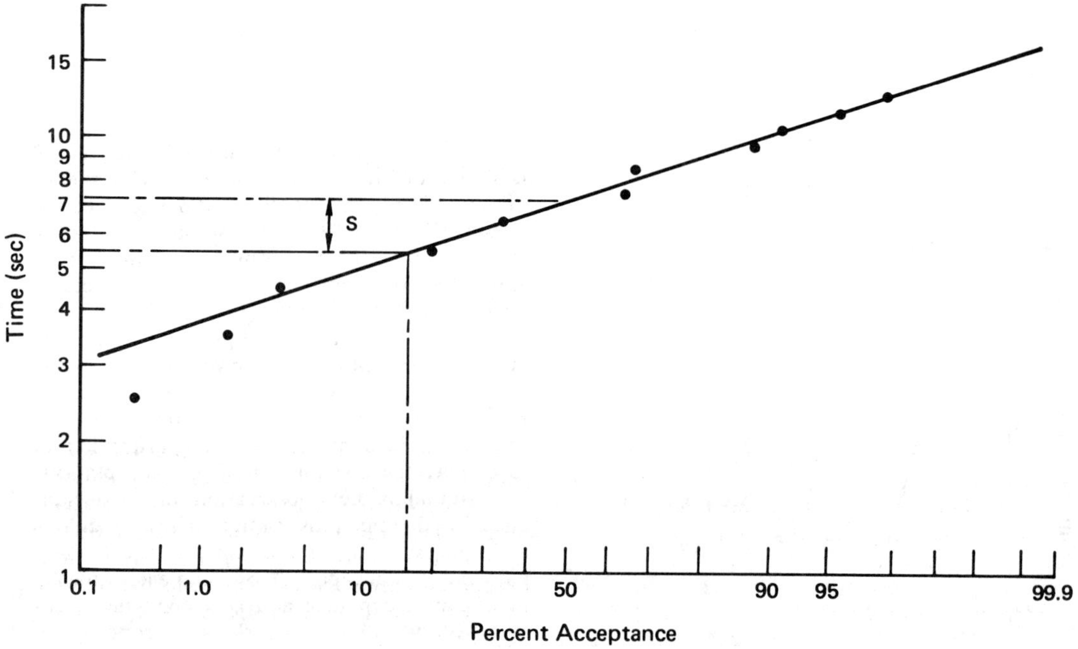

Figure 3.20 Lag and gap distribution for through movements in Lafayette and Indianapolis.[51]

where s^2 is the variance of the normal curve fitted to the observed data and q is the flow in vehicles/sec.

Tsongos and Wiener [53] have found different distributions of acceptance and rejection for night than for day.

3.11 Summary

The statistical distributions of various traffic characteristics deal with two types of quantities. First is the *vehicle counts* or flows that mainly have discrete distributions such as Pois-

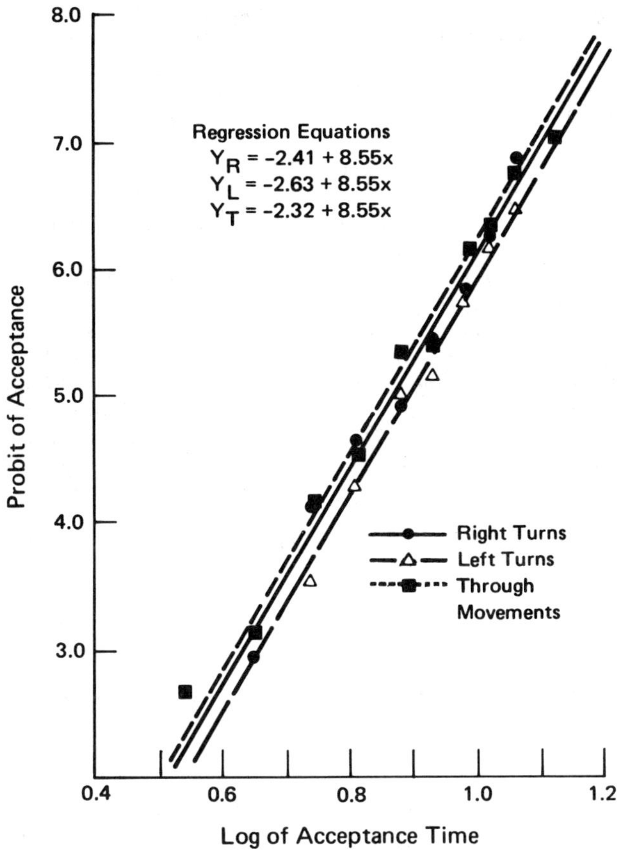

Figure 3.21 Probit regression lines for estimating difference in acceptance time for various movements at intersections in Lafayette and Indianapolis (intersections pooled; lags and gaps pooled.[51]

son (for low-density traffic), negative binomial distribution (for varying flows), and the generalized Poisson as well as the binomial (for congested flow). The second type are those that obey continuous distributions such as *headways* (exponential, shifted exponential, composite exponential, Erlang, hyper-Erlang, semirandom, and normal) and *speeds* (normal, lognormal, and gamma). The choice of distribution depends on how much complexity is desired as well as the behavior of the traffic.

In several of the above cases it was shown that if the vehicle count (flow) obeys a given discrete distribution the headway will obey a unique corresponding distribution. For example, if flow is purely random (Poisson), headways are exponentially distributed. Similarly, if the headways have an Erlang distribution of order k (positive integer), the corresponding flow is itself a discrete event variable consisting of every kth event of a Poisson series. This unique correspondence, although true, is difficult to obtain under real conditions except for special cases; that is, to the discrete flow process there corresponds a continuous headway process.

Gap acceptances were found to obey either the Erlang or the lognormal distributions. Probits were useful for expressing the normal scales.

3.12 REFERENCES

1. Kinzer, J. P., *Application of the Theory of Probability to Problems of Highway Traffic*. B.C.E. thesis, Polytechnic In-

stitute of Brooklyn (June 1, 1933); also in Proc. Inst. Traffic Eng., 5:118–124 (1934). (Abstr.)

2. Adams, W. F., Road traffic considered as a random series. J. Inst. Civil Eng., 4: 121–130 (1936).
3. Greenshields, B. D., Shapiro, D., and Ericksen, E. L., *Traffic Performance at Urban Street Intersections.* Tech. Rep. No. 1, Yale Bureau of Highway Traffic, New Haven, Conn. (1947).
4. Gerlough, D. L., and Barnes, F. C. The Poisson and other probability distributions in highway traffic. In *Poisson and Other Distributions in Traffic.* Eno Foundation (1971).
5. Gerlough, D. L., Use of Poisson distribution in highway traffic, pp. 1–58. In *Poisson and Traffic,* Eno Foundation (1955).
6. Drew, D. R., *Traffic Flow Theory and Control.* McGraw-Hill, New York (1968), p. 130.
7. Harvard University Computation Laboratory, *Tables of the Cumulative Binomial Probability Distribution.* Harvard University Press, Cambridge, Mass. (1955); Weintraub, S., *Tables of the Cumulative Binomial Probability Distribution for Small Values of p.* Collier-Macmillan Ltd., London (1963).
8. Williamson, E., and Bretherton, M. H., *Tables of the Negative Binomial Probability Distribution.* Wiley, New York (1963), pp. 9–11.
9. Buckley, D. J., Road traffic counting distributions. Transp. Res., 1 (2):105–116 (1967).
10. Kendall, M. G., and Stuart, A., *The Advanced Theory of Statistics.* Hafner Publ. (1958), Vol. 1, p. 147.
11. Buckley, D. J., Road traffic headway distributions. Proc. Austr. Road Res. Board, 1962:153–187.
12. Gerlough, D. L., Traffic inputs for simulation on a digital computer. Proc. Highw. Res. Board, 38:480–492 (1959).
13. Haight, F. A., The generalized Poisson distribution. *Ann. Inst. Stat. Math.* (Tokyo), 11:101–105 (1959).
14. Epstein, B., Estimation of the parameters of two parameter exponential distributions from censored samples. Technometrics, 2(3):403–406 (1960).
15. Wohl, M., and Martin, B. V., *Traffic System Analysis for Engineers and Planners.* McGraw-Hill, New York (1967), p. 506.
16. Burgin, T. A., and Wild, A. R., Stock control—experience and usable theory. Oper. Res. Q., 18(1):35–52 (1967).
17. Pearson, K., *Tables of the Incomplete Gamma Function.* Cambridge University Press, Cambridge, England (1965).
18. Whittlesey, J. R. B., Incomplete gamma functions for evaluating Erlang process probabilities. Math. Comput., 17 (81): 11–17 (1963).
19. Whittlesey, J. R. B., *Normalized Incomplete Gamma Function with Poisson Term—Double Precision.* SHARE General Program Library UR GAM 2 (DO) 3138, February 28, 1964.
20. Daou, A., On flow within platoons. *Austr. Road Res.* 2 (7):4–13 (1966).
21. Daou, A., The distribution of headways in a platoon. Oper. Res., 12 (2):360–361 (1964).
22. Greenberg, I., The log-normal distribution of headways. Austr. Road Res., 2 (7): 14–18(1966).
23. May, A. D., Jr., Gap availability studies. Highw. Res. Rec. 72 (1965), pp. 101–136.
24. Tolle, J. E., The lognormal headway distribution model. Traffic Eng. Control, 13 (1):22–24 (1971).
25. Schuhl, A., The probability theory applied to distribution of vehicles on two-lane highways, pp. 59–75. In *Poisson and Traffic.* Eno Foundation (1955).
26. Kell, J. H., Analyzing vehicular delay at intersections through simulation. Highw. Res. Board Bull. 356 (1962), pp. 28–39.
27. Grecco, W. L., and Sword, E. C., Prediction of parameters for Schuhl's headway distribution. Traffic Eng., 38 (5): 36–38 (1968).
28. Dawson, R. F., and Chimini, L. A., The hyperlang probability distribution—a generalized traffic headway model. Highw. Res. Rec. 230 (1968), pp. 1–14.
29. Dawson, R. F., The hyperlang probability distribution—a generalized traffic headway model. Proceedings of Fourth International Symposium on Theory of Traffic Flow, held in Karlsruhe, Germany, June 1968. Published as *Beitrage zur Theorie des Verkehrsflusses, Strassenbau und Strassenverkehrstechnik* (1969), Vol. 86, pp. 30–36.

30. Newell, G. F., Statistical analysis of the flow of highway traffic through a signalized intersection. Q. Appl. Math., 13 (4):353–369 (1956).
31. Jewell, W. S., The properties of recurrent-event processes. Oper. Res., 8 (4):446–472 (1960).
32. Haight, F. A., *Mathematical Theories of Traffic Flow*. Academic Press, New York (1963), pp. 97–98.
33. Haight, F. A., Whisler, B. F., and Mosher, W. W., Jr., New statistical method for describing highway distribution of cars. Proc. Highw. Res. Board, 40:557–563 (1961).
34. Molina, E. C., *Poisson's Exponential Binomial Limit*. Von Nostrand, Princeton, New Jersey (1945).
35. General Electric Co., Defense Systems Department, *Tables of Individual and Cumulative Terms of the Poisson Distribution*. Von Nostrand, Princeton, New Jersey (1962).
36. Buckley, D. J., A semi-Poisson model of traffic flow. Transp. Sci., 2 (2):107–133 (1968).
37. Serfling, R. J., Non-Poisson models for traffic flow, Transp. Res., 3 (3):299–306 (1969).
38. Oliver, R. M., and Thibault, B., A high-flow traffic counting distribution. Highw. Res. Board Bull. 356 (1962), pp. 15–27.
39. Underwood, R. T., Discussion of Pak-Poy paper. Proc. Austr. Road Res. Board, 2:243 (1964).
40. Pak-Poy, P. G., The use and limitation of the Poisson distribution in road traffic. Proc. Austr. Road Res. Board, 2:223–247 (1964).
41. Highway Research Board, *Highway Capacity Manual*. Washington, D.C. (1950).
42. Leong, H. J. W., The distribution and trend of free speeds on two-lane rural highways in New South Wales. Proc. Austr. Road Res. Board, 4: 791–814 (1968).
43. Bleyl, R. L., Skewness of spot speed studies. Traffic Eng., 40(10): 36–37, 40 (1970).
44. Berry, D. S., and Belmont, D. M., Distribution of vehicle speeds and travel times. Proc. Second Berkeley Symp. Math. Stat. Probab. (1951), pp. 589–602.
45. Taylor, W. C., Speed zoning—a theory and its proof. Proc. Inst. Traffic Eng., 34: 11–23 (1964).
46. Haight, F. A., and Mosher, W. W., Jr., A practical method for improving the accuracy of vehicular speed distribution measurements. Highw. Res. Board Bull. 341 (1962), pp. 92–116.
47. Raff, M. A., *A Volume Warrant for Urban Stop Signs*. Eno Foundation (1950).
48. Blunden, W. R., Clissold, C. M., and Fisher, R. B., Distribution of acceptance gaps for crossing and turning maneuvers. Proc. Austr. Road Res. Board, 1: 188–205 (1962).
49. Bissell, H. H., *Traffic Gap Acceptance From a Stop Sign*. Research Report submitted as partial fulfillment of the requirement for the degree of Master of Engineering, University of California, Berkeley (May 1960).
50. Robinson, C. C., Pedestrian interval acceptance. Proc. Inst. Traffic Eng., 22: 144–150 (1951).
51. Solberg, P., and Oppenlander, J. C., Lag and gap acceptances at stop-controlled intersections. Highw. Res. Rec. 118 (1966), pp. 48–67.
52. Ashworth, R., The analysis and interpretation of gap acceptance data. Transp. Sci., 4(3): 270–280 (1970).
53. Tsongos, N. G., and Wiener, S., Comparison of day and night gap-acceptance probabilities. Public Roads, 35(7): 157–165 (1969).
54. Wagner, F. A., Jr., An evaluation of fundamental driver decisions and reactions at an intersection. Highw. Res. Rec. 118 (1966), pp. 68–84.
55. Arley, N., and Buch, K. R., *Introduction to the Theory of Probability and Statistics*. Wiley, New York (1950), pp. 30–31.
56. Feller, W., *An Introduction to Probability and Its Applications,* 1st ed. Wiley, New York (1950), Vol. 1, pp. 364–367.
57. Fry, T. C., *Probability and Its Engineering Uses*. Van Nostrand, Princeton, New Jersey (1928), pp. 220–227.
58. Ford, L. R., *Differential Equations*. McGraw-Hill, New York (1933), p. 172.
59. Elderton, W. P., *Frequency Curves and Correlation,* 4th ed. Harren Press (1953).

60. Kendall, M. G., and Stuart, A., *The Advanced Theory of Statistics.* Hafner Publ. (1958), Vol. 1, pp. 148–152.
61. Aitchison, J., and Brown, J. A. C., *The Lognormal Distribution.* Cambridge University Press, Cambridge, England (1963).
62. Hald, A., *Statistical Theory with Engineering Applications.* Wiley, New York (1952), pp. 160–174.
63. Kendall, M. G., and Stuart, A., *The Advanced Theory of Statistics.* Hafner, Publ. (1958), Vol. 1, pp. 168–173.
64. Finney, D. J., *Probit Analysis.* Cambridge Univ. Press (1947).
65. Dixon, Wilfred J., and Massey, Frank J., Jr., *Introduction to Statistical Analysis.* McGraw-Hill, New York (1957), pp. 226–227.
66. Yule, G. Udny, and Kendall, M. G., *An Introduction to the Theory of Statistics,* 14th ed. Hafner Publ. (1950), p. 475, 477.
67. Cramer, H., *Mathematical Methods of Statistics.* Princeton University Press, Princeton, New Jersey (1946), p. 435.
68. Massey, F. J., Jr., The Kolmogorov-Smirnov test for goodness of fit. J. Am. Stat. Assoc., 46: 68-78 (1951).
69. Walker, H. M., and Lev, J., *Statistical Inference.* Henry Holt and Co., New York (1953), p. 443.
70. Neville, A. M., and Kennedy, J. B., *Basic Statistical Methods for Engineers and Scientists.* International Textbook Co. (1964), pp. 289–290.
71. Miller, A. M., An empirical model for multilane road traffic. Transp. Sci., 4(2): 164–186 (1970).
72. *Handbook of Chemistry and Physics* or *Standard Mathematical Tables.* Chemi-Rubber Co., various editions; also, U.S. National Bureau of Standards, *Tables of the Exponential Function* e^x, Applied Mathematics Series, No. 14, (June 29, 1951).

3.13 RELATED LITERATURE

Counting Distributions

Haight, F. A., Counting distributions for renewal processes. Biometrika, 52(3,4): 395–403 (1965).

Haight, F. A., *Handbook of the Poisson Distribution.* Wiley, New York (1967).

Newell, G. F., Stochastic properties of peak short-time traffic counts. Transp. Sci., 1(3): 167–183 (1967).

Oliver, R. M., A traffic counting distribution. Oper. Res., 9(6): 807–810 (1961).

Oliver, R. M., Note on a traffic counting distribution. Oper. Res. Q., 13(2): 171–178 (1962).

Shumway, R., and Gurland, J., *A Fitting Procedure for Some Generalized Poisson Distributions.* University of Wisconsin, Madison (Feb. 1960); NTIS No. PB 167 087.

Vaughan, R. J., The distribution of traffic volumes. Transp. Sci., 4(1): 97–112 (1970).

Whittlesey, J. R. B., and Haight, F. A., Counting distributions for an Erlang process. Ann. Stat. Math. (Tokyo), 13(2) (1961).

Interval Distributions—Headways, Gaps, etc.

Allan, R. R., Extension of the binomial model of traffic to the continuous case. Proc. Austr. Road Res. Board, 3: 276–316 (1966).

Breiman, L., Gafarian, A. V., Lichtenstein, R., and Murthy, V. K., An experimental analysis of single-lane headways in freely flowing traffic. Proceedings of Fourth International Symposium on Theory of Traffic Flow, held in Karlsruhe, Germany, June 1968. Published as *Beitrage zur Theorie des Verkehrsflusses, Strassenbau und Strassenverkehrstechnik* (1969), Vol. 86, pp. 22–29.

Dunne, M. C., Rothery, R. W., and Potts, R. B., A discrete Markov model of vehicular traffic. Transp. Sci., 2(3): 233–251 (1968).

Kell, J. H., A theory of traffic flow on urban streets. Proc. West. Sec. Inst. Traffic Eng., 1960: 66–70.

Leutzbach, W., Distribution of time gaps between successive vehicles. Int. Road Safety Traffic Rev., 5 (3): 31–36 (1957).

Lewis, R. M., A proposed headway distribution for traffic simulation studies. Traffic Eng., 33(5): 16–19, 48 (1963).

May, A. D., Jr., and Wagner, F. A., Jr., Headway characteristics and interrelationships of the fundamental characteristics of traffic flow. Proc. Highw. Res. Board, 39: 524–547 (1960).

McNeil, D. R., and Morgan, J. H. T., Estimating minimum gap acceptances for merging motorists. Transp. Sci., 2(3): 265–277 (1968).

Miller, A. J., Nine estimates of gap-acceptance parameters. Presented at Fifth International

Symposium on Theory of Traffic Flow and Transportation, Berkeley, Calif.

Mitchell, H. R., Freeway accidents and slow-moving vehicles. Traffic Eng., 37(2): 22–25 (1966).

Moskowitz, K., Waiting for a gap in a traffic stream. Proc. Highw. Res. Board, 33: 385–394 (1954).

Oliver, R. M., Distribution of blocks and gaps in a traffic stream. Oper. Res., 10(2): 197–217 (1962).

Raff, M. S., *A Volume Warrant for Urban Stop Signs*. Eno Foundation (1950).

Raff, M. S., The distribution of blocks in an uncongested stream of automobile traffic. J. Am. Stat. Assoc., 46: 114–123 (1951).

Schuhl, A., Distribution of gaps between successive vehicles. (Letter to Editor) Int. Road Safety Traffic Rev., 41 (1958).

Suzuki, T., Some results in road traffic distributions. Oper. Res. Soc. Jap., 9(1): 16–25 (1966).

Weiss, G. H., The intersection delay problem with gap acceptance function depending on speed and time. Transp. Res., 1(4):367–371 (1967).

Widermuth, B., Average vehicle headways at signalized intersections. Traffic Eng., 33(2): 14–16 (1962).

Platoon Phenomena

Athol, P., Headway groupings. Highw. Res. Rec. 72 (1965), pp. 137–155.

Buckley, D. J., Tomlin, J. A., and Minson, W. G., Delays to traffic platoons. Proc. Austr. Road Res. Board, 3: 364–376 (1966).

Edie, L. C., and Foote, R. S., Analysis of single lane traffic flow. Traffic Eng., 34(4): 21–27 (1963).

Gerlough, D. L., Some problems in intersection traffic control, pp. 10–27. In R. Herman (ed.) *Proceedings of [First] International Symposium on the Theory of Traffic Flow*. Elsevier, New York (1961).

Grace, M. J., and Potts, R. B., Diffusion of traffic platoons. Proc. Austr. Road Res. Board, 1: 260–267 (1962).

Grace, M. J., and Potts, R. B., A theory of the diffusion of traffic platoons. Oper. Res., 12(2): 255–275 (1964).

Herman, R., Potts, R. B., and Rothery, R. W., Behaviour of traffic leaving a signalized intersection. Traffic Eng. Control, 5(9): 539–533 (1964).

Lewis, B. J., Platoon movement of traffic from an isolated signalized intersection. Highw. Res. Board Bull. 178 (1958), pp. 1–11.

Nemeth, Z. A., and Vecellio, R. L., Investigation of the dynamics of platoon dispersion. Paper presented at Annual Meeting of Highway Research Board, Washington, D.C. (January 1970).

Pacey, G. M., *The Progress of a Bunch of Vehicles Released from a Traffic Signal*. Res. Note No. RN/2665/GMP. Road Research Laboratory (Jan. 1956), mimeo.

Underwood, R. T., Traffic flow and bunching. *Austr. Road Res.*, 1(6): 8–24 (1963).

Wright, C. C., Some characteristics of traffic leaving a signalized intersection. Transp. Sci., 4(4): 331–346 (1970).

Speed Distributions, etc.

Duncan, N. C., *A Method of Estimating the Distribution of Speeds of Cars on a Motorway*. Transport and Road Research Laboratory Rep. 598 (1973).

Lefeve, B. A., Speed habits observed on a rural highway. Proc. Highw. Res. Board, 33: 409–428 (1954).

Mori, M., Takata, H., and Kisi, T., Fundamental considerations on speed distributions of road traffic flow. Transp. Res., 2(1): 31–39 (1968).

Ogawa, T., Fisher, E. S., and Oppenlander, J. C., Driver behavior study—influence of speed limits on spot speed characteristics in a series of contiguous rural and urban areas. Highw. Res. Board Bull. 341 (1962), pp. 18–29.

Oppenlander, J. C., Bunte, W. F., and Kadakia, P. L., Sample size requirements for vehicular speed studies. Highw. Res. Board Bull. 281 (1961), pp. 68–86.

Oppenlander, J. E., A theory on vehicular speed regulation. Highw. Res. Board Bull. 341 (1962), pp. 77–91.

Pahl, J., The effect of discrete time measurement on speed data. Paper presented at Annual Meeting of Highway Research Board, Washington, D.C. (January 1971).

Rowan, N. J., and Keese, C. J., A study of factors influencing traffic speeds. Highw. Res. Board Bull. 341 (1962), pp. 30–76.

Schumate, R. P., and Crowther, R. F., Variability of fixed-point speed measurements. Highw. Res. Board Bull. 281 (1961), pp. 87–96.

Wagner, F. A., Jr., and May, A. D., Jr., Volume and speed characteristics at seven study locations. Highw. Res. Board Bull. 281 (1961), pp. 48–67.

Wortman, R. H., A multivariate analysis of vehicular speeds on four-lane rural highways. Highw. Res. Rec. 72 (1965), pp. 1–18.

Gap Acceptance

Ashworth, R., and Green, B. D., Gap acceptance at an uncontrolled intersection. Traffic Eng. Control, 7: 676–678 (1966).

Ashworth, R., The capacity of priority-type intersections with a non-uniform distribution of critical acceptance gaps. Transp. Res., 3(2): 273–278 (1969).

Bissell, H. H., *Traffic Gap Acceptance from a Stop Sign.* Unpublished graduate research report. University of California, Berkeley (1960).

Dart, O. K., Jr., Left-turn characteristics at signalized intersections on four-lane arterial streets. Highw. Res. Rec. 230 (1968), pp. 45–59.

Drew, D. R., Gap acceptance characteristics for ramp-freeway surveillance and control. Highw. Res. Rec. 157 (1967), pp. 108–135.

Hawkes, A. G., Gap-acceptance in road traffic. J. Appl. Probab., 5(1): 84–92 (1968).

Exponential Distributions

Rider, P. R., The method of moments applied to a mixture of two exponential distributions. Ann. Math. Stat., 32(1): 143–147 (1961); reprinted by NTIS as AD 701 962.

Tallis, G. M., and Light, R., The use of fractional moments for estimating the parameters of a mixed exponential distribution. Technometrics, 10(1): 161–175 (1968).

Gamma Distribution and Related Distributions

Bain, L. J., and Antle, C. E., Estimation of parameters in the Weibull distribution. Technometrics, 9(4): 621–627 (1967).

Choi, S. C., and Wette, R., Maximum likelihood estimation of the parameters of the gamma distribution and their bias. Technometrics, 11(4): 683–690 (1969).

Clarke, A. B., and Disney, R. L., *Probability and Random Processes for Engineers and Scientists.* Wiley, New York (1970), pp. 186–189.

Dubey, S. D., On some permissible estimators of the location parameter of the Weibull and certain other distributions. Technometrics, 9(2): 293–307 (1967).

Greenwood, J. A., and Durand, D., Aids for fitting the gamma distribution by maximum likelihood. Technometrics, 2(1): 55–65 (1960).

Harter, H. L., Maximum-likelihood estimation of the parameters of a four-parameter generalized gamma population from complete and censored samples. Technometrics, 9(1): 159–165 (1967).

Krishnaiah, P. R., and Rizvi, M. H., A note on moments of gamma order statistics. Technometrics, 9(2): 315–318 (1967.

Sasieni, M., Yaspan, A., and Friedman, L., *Operations Research—Methods and Problems.* Wiley, New York (1959), pp. 32, 56–57.

Normal Distribution

Cohen, A. C., Jr., Estimation in mixtures of two normal distributions. Technometrics, 9(1): 15–28 (1967).

Lognormal Distribution

Joffe, A. D., and Sichel, H. S., A chart for sequentially testing observed arithmetic means from lognormal populations against a given standard. Technometrics, 10(3): 605–612 (1968).

Distributions—General

Cohen, A. C., Jr., Estimation of the Poisson parameter from truncated samples and from censored samples. J. Am. Stat. Assoc., 158–168 (Mar. 1954).

Falls, L. W., Estimation of parameters in compound Weibull distributions. Technometrics, 12(2): 399–407 (1970).

Haight, F. A., Index to the distributions of mathematical statistics. J. Res. Natl. Bur. Stand. B. Math. Math. Phys., 65B(1): 23–60 (1961).

Hammersley, J. M., and Morton, K. W., The estimation of location and scale parameters from grouped data. Biometrika, 41: 296–301 (1954).

Mann, N., *Estimation of Location and Scale Parameters Under Various Models of Censoring and Truncation,* Rocketdyne Div., North American Rockwell Corp. (February 1970); NTIS No. AD 707 868.

Miscellaneous Applications

Allen, R. R., Special methods in probability for traffic engineers. Proc. Austr. Road Res. Board, 3: 317–333 (1966).

Gaver, D. P., Jr., A problem arising in reliability and traffic studies. Oper. Res., 12(4): 534–542 (1964).

Kisi, T., and Hiyoshi, K., The effect of a traffic light on a Poisson flow. J. Oper. Res. Soc. Jap., 5(2): 67–73 (1962).

Newell, G. F., Theories of instability in dense highway traffic. J. Oper. Res. Soc. Jap., 5(1): 9–54 (1962).

Newell, G. F., Equilibrium probability distributions for low density highway traffic. J. Appl. Probab., 3: 247–260 (1966).

Oliver, R. M., and Newman, L., Effect of trucks on freeway flows. Highw. Res. Rec. 15 (1963), pp. 67–72.

Rallis, T., The application of Erlang's theory to the calculation of road capacity. Presented at Third International Symposium on Theory of Traffic Flow, June 1965.

Schuhl, A., The probability theory applied to distribution of vehicles on two-lane highways. *Poisson and Traffic Revisited.* Eno Foundation (1971).

3.14 PROBLEMS

1. Given the following data, determine the most appropriate distribution and fit it to the data. Check the goodness-of-fit with a χ^2 test.

Arrival on Interstate 94 at Arlington During Evening Peak Traffic

Number of Arrivals per Interval	Observed Frequency
≤ 2	0
3	1
4	3
5	8
6	7
7	6
8	16
9	15
10	5
11	3
12	4
≥ 13	0

2. A sample of 100 over-all travel speeds was observed on a four-lane, undivided arterial street. The sample mean was 35.45 mph, and the standard deviation of the sample was 5.20 mph. Assume that the sample of travel speeds was taken from a normal population. (a) Find the percentages of vehicles in the population traveling between 33 and 41 mph. (b) Between what limits may the mean of the population be expected to fall for a 95 percent level of confidence? (c) Between what limits may the variance of the population be expected to fall for a 90 percent level of confidence? (d) Between what limits may the standard deviation of the population be expected to fall for a 99 percent level of confidence?

3. The driver acceptances of time gaps in the main street were observed for the traffic on the minor street controlled by a "stop" sign. For the following frequency distribution, test the hypothesis that these data represent a random sample from a normal population.

Midvalue (sec)	Observed Frequency
1.0	0
2.0	6
3.0	34
4.0	132
5.0	179
6.0	218
7.0	183
8.0	146
9.0	69
10.0	30
11.0	3
12.0	0
TOTAL	1,000

4. Let x represent the number of vehicles observed per 15-sec increment. Three case studies show these results:

Case 1	Case 2	Case 3
$\bar{x}=3.0$ veh/int	$\bar{x}=3.0$ veh/int	$\bar{x}=3.0$ veh/int
$s^2=2.4$	$s^2=3.0$	$s^2=4.0$

Use the Poisson, binomial, and negative binomial distributions as appropriate for the above data. Plot the three frequency functions on a single chart (use color to discriminate among functions).

Chapter 4

TRAFFIC STREAM MODELS

4.1 INTRODUCTION

Chapter 3 presented the statistical variations of various traffic characteristics about their mean values. The use of these distributions for prediction implies constant mean values. For certain types of traffic design it is desirable to make predictions based on changing mean values. In such situations it is often helpful to use a functional relationship between the mean value of the characteristic under study and some other characteristic. For instance, in the design of freeway surveillance and control systems, a curve relating flow and concentration is most helpful. Furthermore, such a curve is quite useful in defining the capacity of a facility.

The relationship among the three variables u, k, and q is called a traffic stream model. A typical model is shown in Figure 4.1. The model must be on the three-dimensional surface $u=q/k$. Two specific points on the model can also be established: (1) As concentration approaches zero (light traffic), mean speed approaches the mean free-flow speed u_f and the flow approaches zero; (2) as concentration approaches its maximum value, called jam density or k_j, speed approaches zero and flow again approaches zero.

It is usually more convenient to show the model of Figure 4.1 as one or more of the three separate relationships in two dimensions shown in Figure 4.2. These relationships are the orthographic projections of Figure 4.1. Such a presentation also shows another important feature that was not immediately obvious on Figure 4.1: At some concentration k_m and corresponding speed u_m, flow passes through a maximum value q_m.

4.2 SPEED–CONCENTRATION MODELS

It is an observable fact that drivers decrease their speeds as the number of cars around them increases. Some of the reasons for this are discussed in Chapter 5. Because of this close interaction between concentration (cars on the highway) and speed, and knowing both concentration and speed, from which flow can be computed, it is not surprising that early investigators explored relationships between speed and concentration. The simplest (and perhaps the most obvious) of such relationships is a linear relationship, as proposed by Greenshields.

4.2.1 Greenshields' Linear Speed–Concentration Model

Greenshields,[1] as one of the early investigators of traffic characteristics, proposed a linear relationship between flow and concentration that is usually expressed

$$u=u_f(1-k/k_j) \qquad (4.1)$$

where u_f is the free-flow speed and k_j is the jam density. This model is simple to use and several investigators have found good correlation between the model and field data. (See, for instance, Figure 4.3, which shows results found by Huber.[2]) For various theoretical and practical reasons, however, other models have found greater acceptance.

4.2.2 Logarithmic Speed–Concentration Models

Greenberg,[3] using a theoretical background (discussed in Chapter 7), has postulated a speed–concentration model of the form

$$u=u_m \ln(k_j/k) \qquad (4.2)$$

where u_m is a constant that will be shown (section 4.3) to be the speed at maximum flow. Greenberg found good agreement between this model and field data for congested flows (Figure 4.4). This model, however, breaks down at low concentrations, as may be seen by letting $k=0$ in Eq. 4.2.

Underwood [4] has demonstrated a model of the form

$$u=u_f\, e^{-k/k_m} \qquad (4.3)$$

at low concentrations (Figure 4.5). In Eq. 4.3, k_m is the concentration at maximum flow.

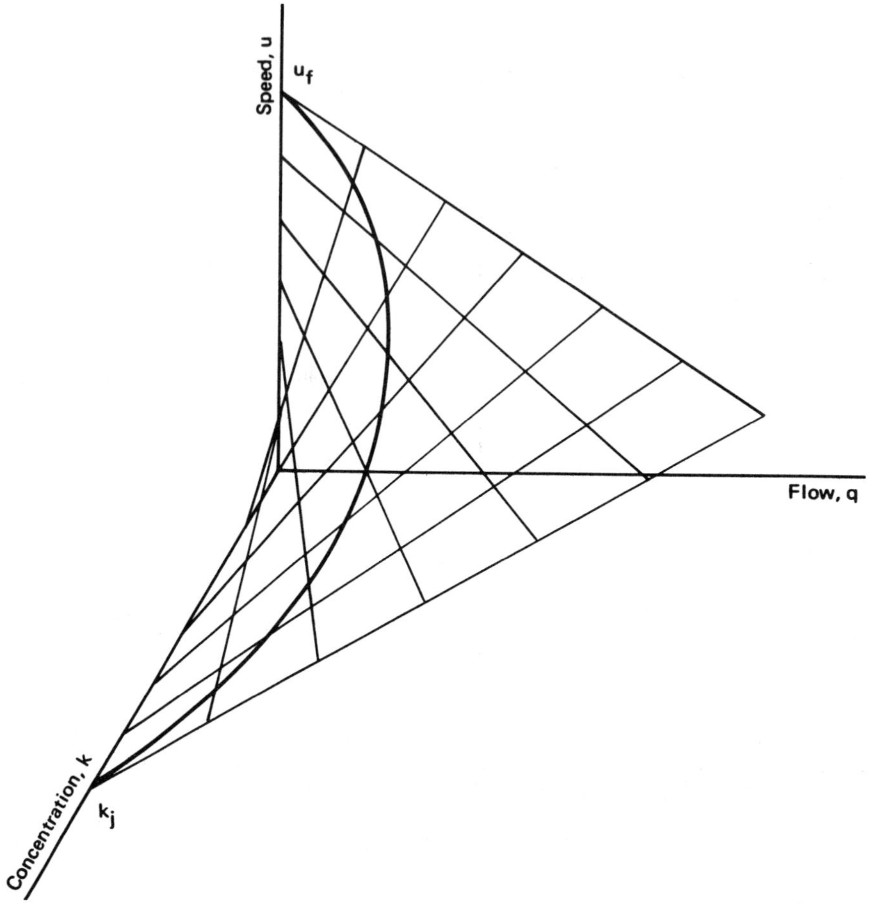

Figure 4.1 A traffic stream model where $u = q/k$ represents the surface of admissible traffic stream models.

This model has shortcomings in that it does not represent zero speed at high concentrations. In sections 4.3.2 and 4.3.2.1 it is shown that for Eq. 4.2 $k_m = k_j/e$, whereas for Eq. 4.3 $u_m = u_f/e$.

4.2.3 Generalized Single-Regime Speed–Concentration Models

Recently, Pipes and Munjal [5,6] have described a general family of speed–concentration models of which the linear model is a special case. Drew [7] has described a family of models of which Greenberg's logarithmic model is a special case. Other families result from car-following analyses.

4.2.3.1 Pipes–Munjal Models. Pipes and Munjal [5,6] have proposed a family of models of the form

$$u = u_f[1 - k/k_j]^n \qquad (4.4)$$

where n is a real number greater than zero. Three conditions of this model ($n<1$, $n=1$, $n>1$) are illustrated in Figure 4.6. It will be seen that for $n=1$ the relationship reduces to Greenshields' model.

4.2.3.2 Drew Models. Drew [7] has proposed a family of models of the form

$$\frac{du}{dk} = u_m\, k^{(n-1)/2} \qquad (4.5)$$

where n is a real number. When $n = -1$, Eq. 4.5 can be solved to yield Greenberg's model.

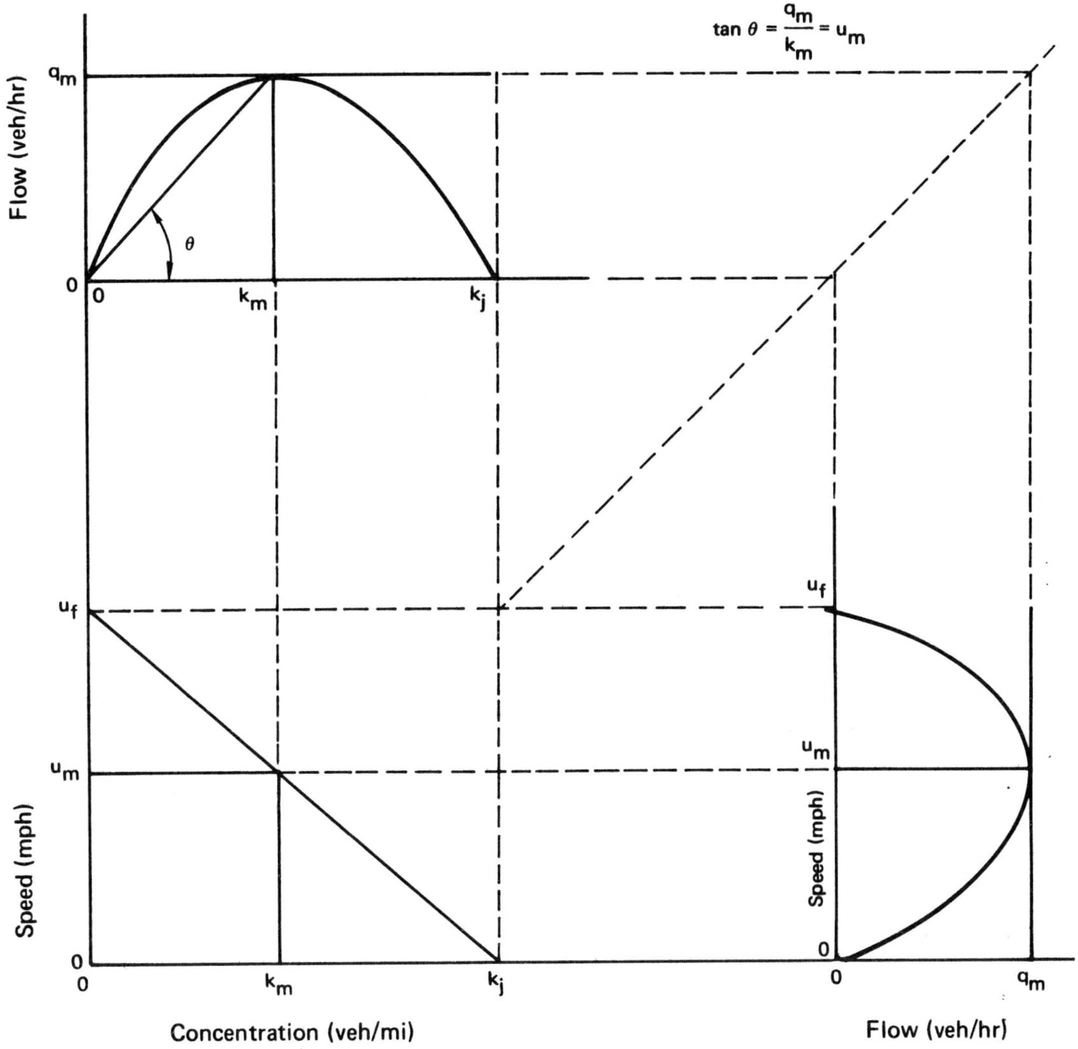

Figure 4.2 Flow-concentration, speed-flow, and speed-concentration curves (assuming single-regime linear speed-concentration model).

Figure 4.7 illustrates the results for three values of n $(-1, 0, +1)$.

4.2.3.3 Car-Following Models. In Chapter 6 the methods of car-following analysis are introduced, including parameters m and ℓ. Figure 4.8 shows how different speed–concentration models can be obtained by manipulating these parameters. Figure 4.9 lists the more important models of this family.

4.2.3.4 Bell-Shaped Curve Model. Drake et al.[9] have proposed use of the bell-shaped or normal curve as a model of speed–concentration using the form

$$u = u_f\, e^{-1/2)(k/k_m)^2} \quad (4.6)$$

4.2.4 Multiregime Speed–Concentration Models

As noted in Section 4.2.2, Greenberg's model is useful for high concentrations but not for low concentrations; conversely, Underwood's model is useful for low concentrations but not for high concentrations. Additional weight is added to this argument by noting that the high-congestion portion of the model usually represents cars accelerating from a jam condition and taking into account Forbes'[10–12] finding that cars have a different speed and acceleration after a slowdown caused by congestion.

4.2.4.1 Edie's Model. Edie[13] has described a model that is a composite of Eqs. 4.2 and 4.3, where Eq. 4.3 is at low concentrations and Eq. 4.2 at high concentrations. When normalized speed is plotted against normalized concentration, the two models become tangent in the midrange of concentration (Figure 4.10).

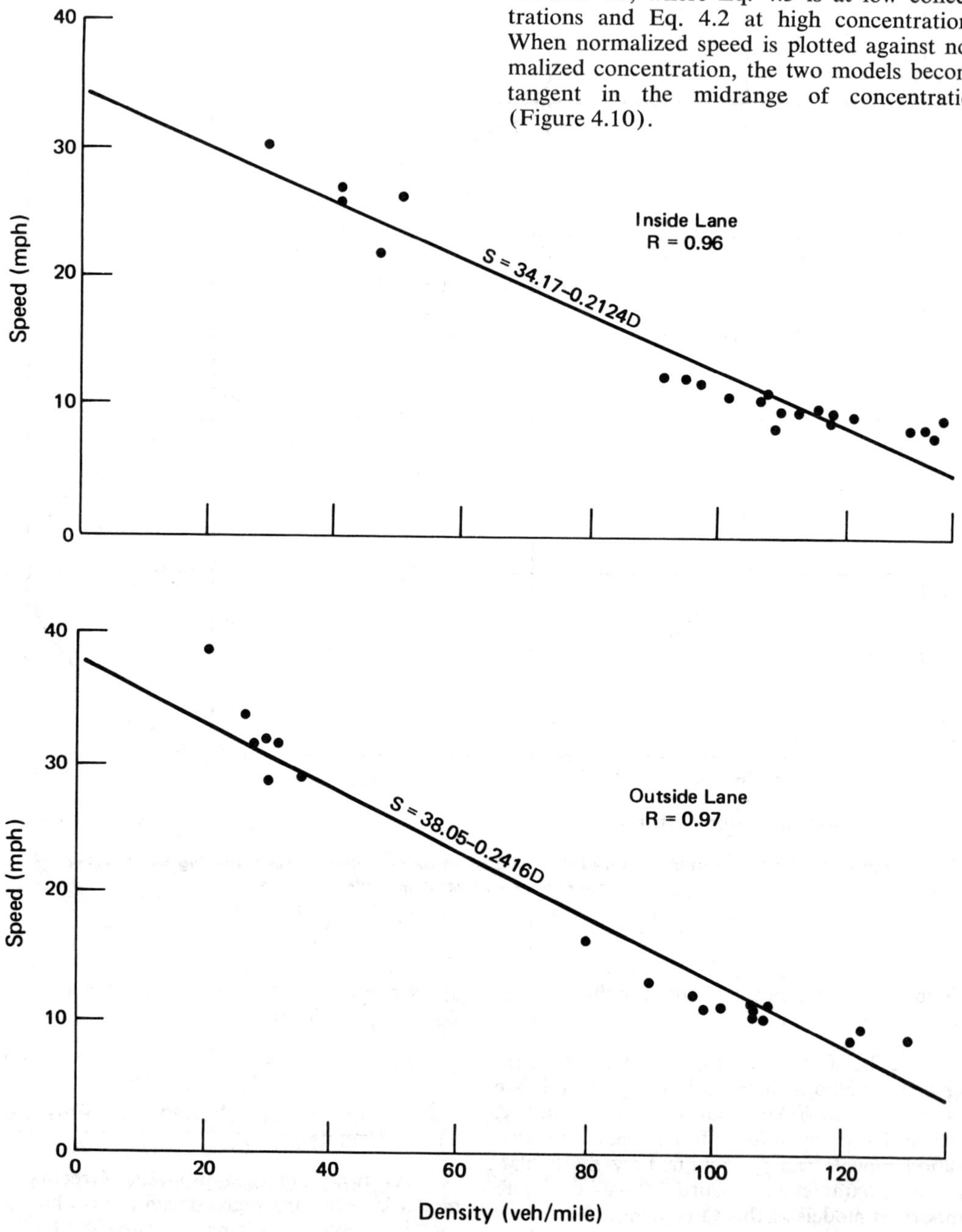

Figure 4.3 Study showing high correlation coefficient between field data and linear model for speed vs. density.[2]

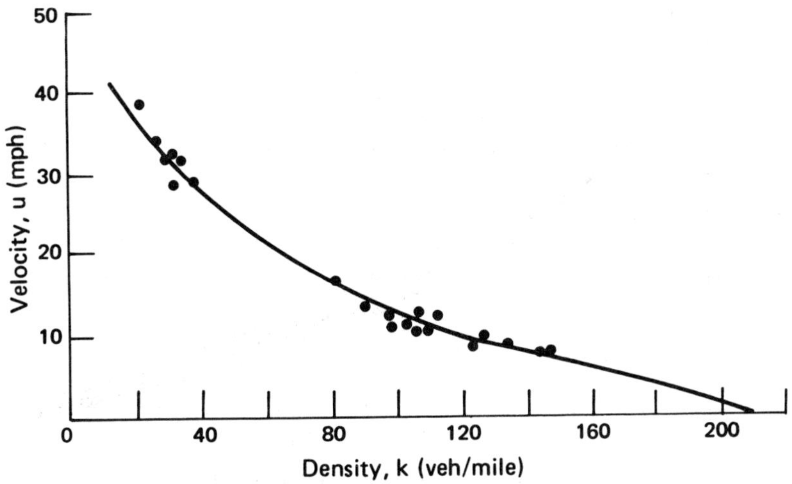

Figure 4.4 Greenberg's[3] speed-concentration model for Merritt Parkway.

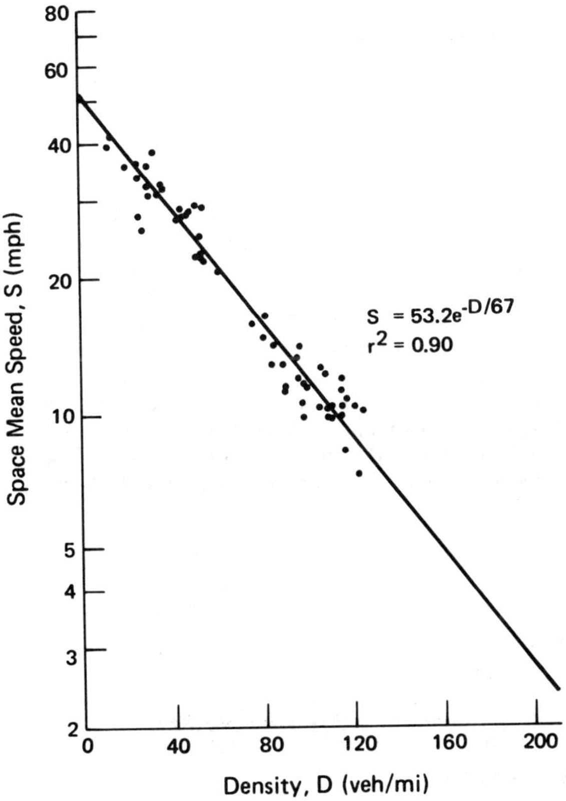

Figure 4.5 Underwood's[4] speed-concentration plot for Merritt Parkway.

Whereas Edie combined two theoretical models at a point of tangency, several other investigators started with one theoretical model and added relatively arbitrary modifications.

4.3.4.2 Underwood's Two-Regime Model. Underwood adapted Eq. 4.3 by making the modification shown in Figure 4.11.

4.2.4.3 Dick's Model. In developing a model to represent urban traffic, Dick[14] as-

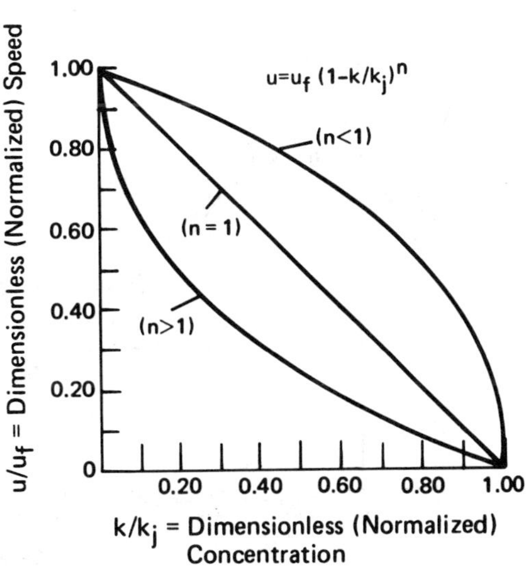

Figure 4.6 Family of speed-concentration models proposed by Munjal and Pipes.[6]

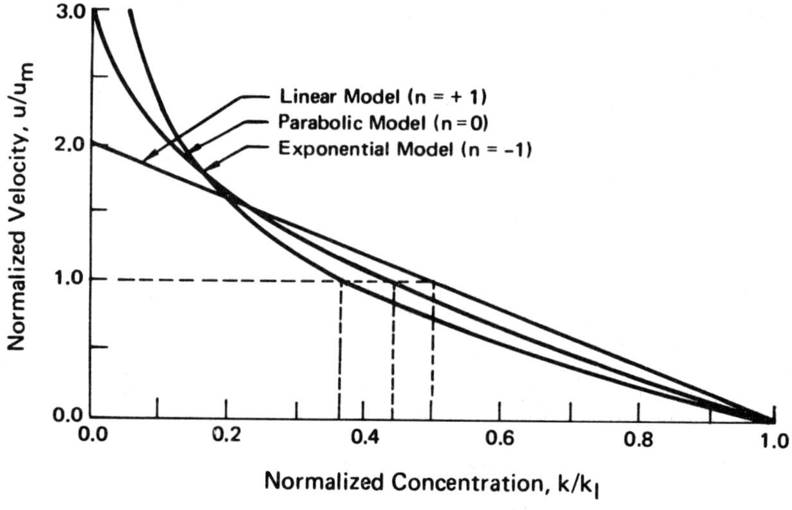

Figure 4.7 Illustration of Drew's[7] family of speed-concentration models.

sumed that there is a fixed upper limit to speed. He combined this assumption with Greenberg's model to produce the result shown in Figure 4.12.

4.2.4.4 Fitting Multiregime Models. May and Keller[15] have discussed methods for fitting multiregime models.

4.2.5 Summary

Figure 4.13 summarizes several types of speed–concentration models.

4.2.6 Field Measurements

Most investigations of the characteristics of traffic have been performed in field situations where the investigator could select only the site and the time at which observations were made. Wardrop,[16] however, has reported results from controlled experiments conducted at the (British) Road Research Laboratory. These tests have shown that the relationship between speed and concentration depends on the radius of the road section. Figure 4.14 shows a plot for straight sections (infinite radius).

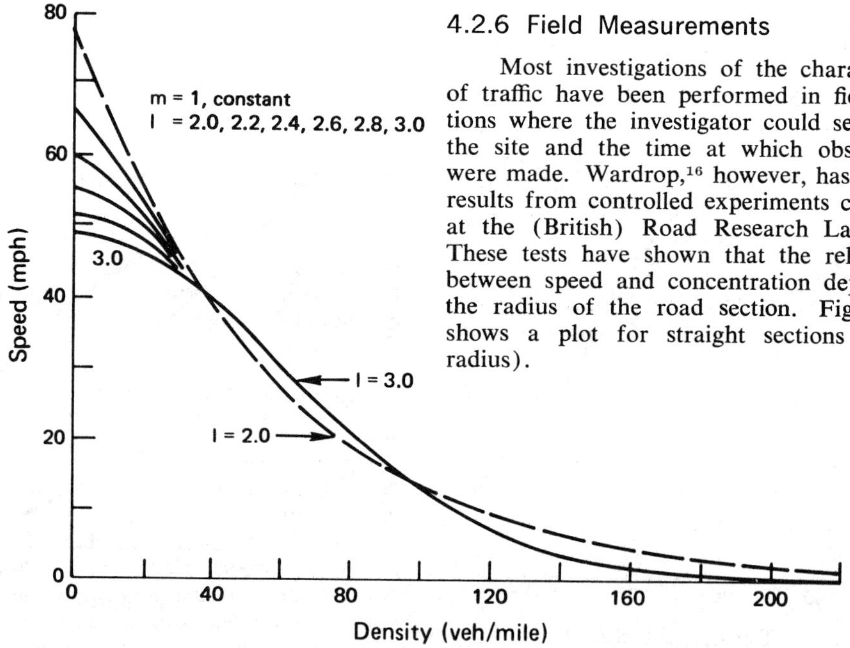

Figure 4.8 Example of car-following models[8] of speed-concentration, showing effect of varying parameter *l*.

l \ m	m = 0	m = 1
l = 0	$u = \frac{1}{s}(\frac{1}{k} - \frac{1}{k_j})$ s = a constant	—
l = 1	$u = u_m \ln(\frac{k_j}{k})$	—
l = 1.5	$u = u_f [1 - (\frac{k}{k_j})^{1/2}]$	—
l = 2	$u = u_f [1 - (\frac{k}{k_j})]$	$u = u_f e^{-(k/k_m)}$
l = 3	—	$u = u_f e^{-\frac{1}{2}(k/k_m)^2}$

Figure 4.9 Matrix of speed-concentration models from car-following analysis.[3]

4.3 FLOW–CONCENTRATION RELATIONSHIPS

Early studies of highway capacity followed two principal approaches. Some investigators examined speed–flow relationships at low concentrations; others discussed headway phenomena at high concentrations. Lighthill and Whitham [17] have proposed use of the flow–concentration curve as a means of unifying these two approaches. Because of this unifying feature, and because of the great usefulness of the flow-concentration curve in traffic control situations (such as metering a freeway), Haight [18,19] has termed the flow–concentration curve "the basic diagram of traffic."

Some important features of the flow–concentration (q–k) diagram may be summarized as follows:

1. In the absence of concentration there can be no flow; thus the curve must pass through the origin. Furthermore, if space mean speed is taken as the ratio q/k, the slope with which the curve leaves the origin is the free-flow speed. (Note that this is the maximum slope of the curve.)

2. It is an observable fact that it is possible to have high concentrations with no flow where the leader of a stream has stopped and followers have thus been forced to stop. This may be seen in queues at traffic signals; under certain situations it can also be seen on freeways; and although it occurs at many other situations, the two cited examples are best

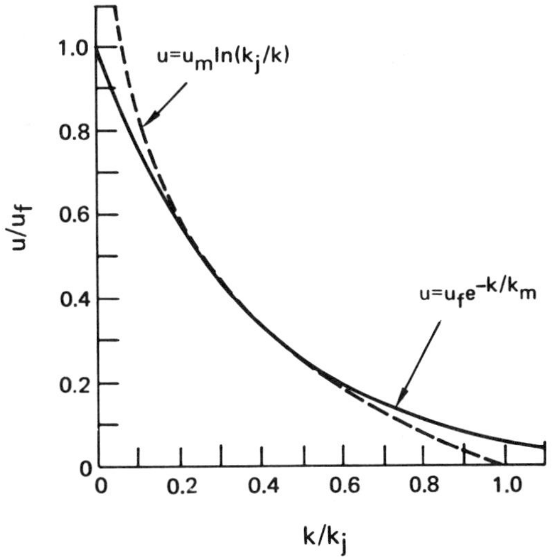

Figure 4.10 Normalized speed vs. normalized concentration for Eqs. 4.2 and 4.3.[13]

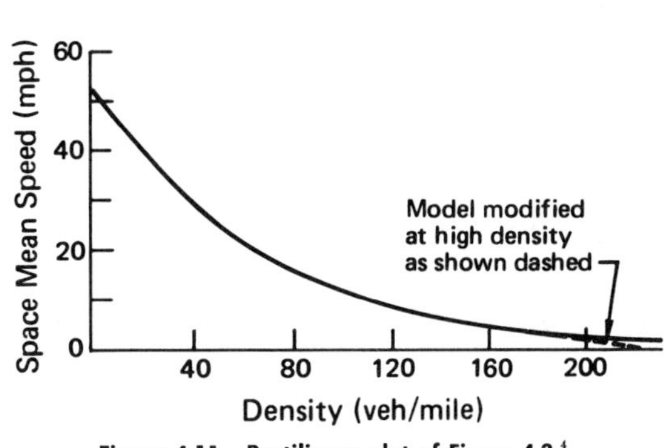

Figure 4.11 Rectilinear plot of Figure 4.3.[4]

known. Thus, the curve must have a point representing maximum (jam) concentration with zero flow.

3. Inasmuch as there are observable flows at intermediate concentrations, there must be one or more points of maximum flow between the two zero points.

4. It is not necessary for the q–k curve to be continuous.

Lighthill and Whitham [17] have discussed phenomena of shock waves and other topics related to flow–concentration relationships, which are covered in Chapter 7. Several of the characteristics of the flow–concentration curve (as defined by Lighthill and Whitham) however, are useful at this point, and have conveniently been summarized by Edie and Foote [20,21] (Figure 4.15). Figure 4.15 illustrates the characteristic measurements representing vehicles traveling with an average speed of 25 mph, at a flow rate of 1,200 vehicles/hr and a concentration of 48 vehicles/mile/lane. For any point on the curve the slope of the radius vector represents the average speed, and the slope of the tangent represents the velocity of propagation of small changes of flow. In Figure 4.15 the jam density (concentration) is shown as 279 vehicles/mile, the maximum flow as 1,400 vehicles/hr. The scale on the right shows average headway in seconds as computed by the relationship

$$h=\frac{1}{q/3{,}600} \tag{4.7}$$

Edie and Foote emphasize the importance of the flow–concentration curve as a means for showing headway and speed in addition to flow and concentration.

4.3.1 Parabolic Flow–Concentration Model

If Greenshields' (linear) speed–concentration model is adopted, a parabolic flow–concentration model results (Figure 4.2). The characteristics of this model can be developed as follows:

$$q=k\,u=k\,u_f(1-k/k_j)=u_fk-\frac{u_fk^2}{k_j} \tag{4.8}$$

Differentiating Eq. 4.8, setting $\frac{dq}{dk}=0$ to obtain conditions for maximum flow, and defining q_m=maximum flow, k_m=concentration at maximum flow, and u_m=speed at maximum flow,

$$k_m=k_j/2$$
$$u_m=u_f/2$$
$$q_m=u_f\,k_j/4=u_m\,k_j/2$$

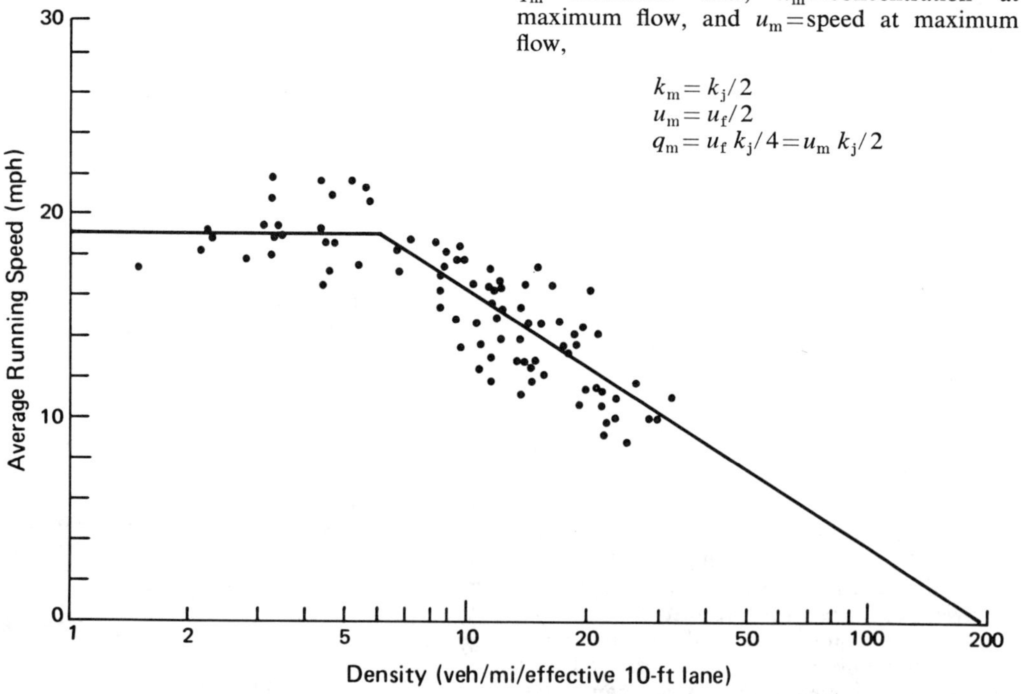

Figure 4.12 Dick's [14] speed-concentration model for urban traffic. (Note logarithmic scale for density.)

4.3.2 Logarithmic Flow–Concentration Model

If Eq. 4.2—the Greenberg logarithmic model of speed–concentration—is used, the flow–concentration model becomes

$$q = k\,u = k\,u_m\,\ell n(k_j/k) \qquad (4.9)$$

Again using differentiation to obtain conditions for maximum flow,

$$k_m = k_j/e$$
$$u_m = u_m$$
$$q_m = u_m\,k_j/e$$

Note that in this model u_m is a parameter; that is, u_m is specified and determines the other characteristics. Figure 4.16 shows such a model fitted to field data. (Measurements that result in a curve including the portion for high concentrations must be taken at a bottleneck.[20])

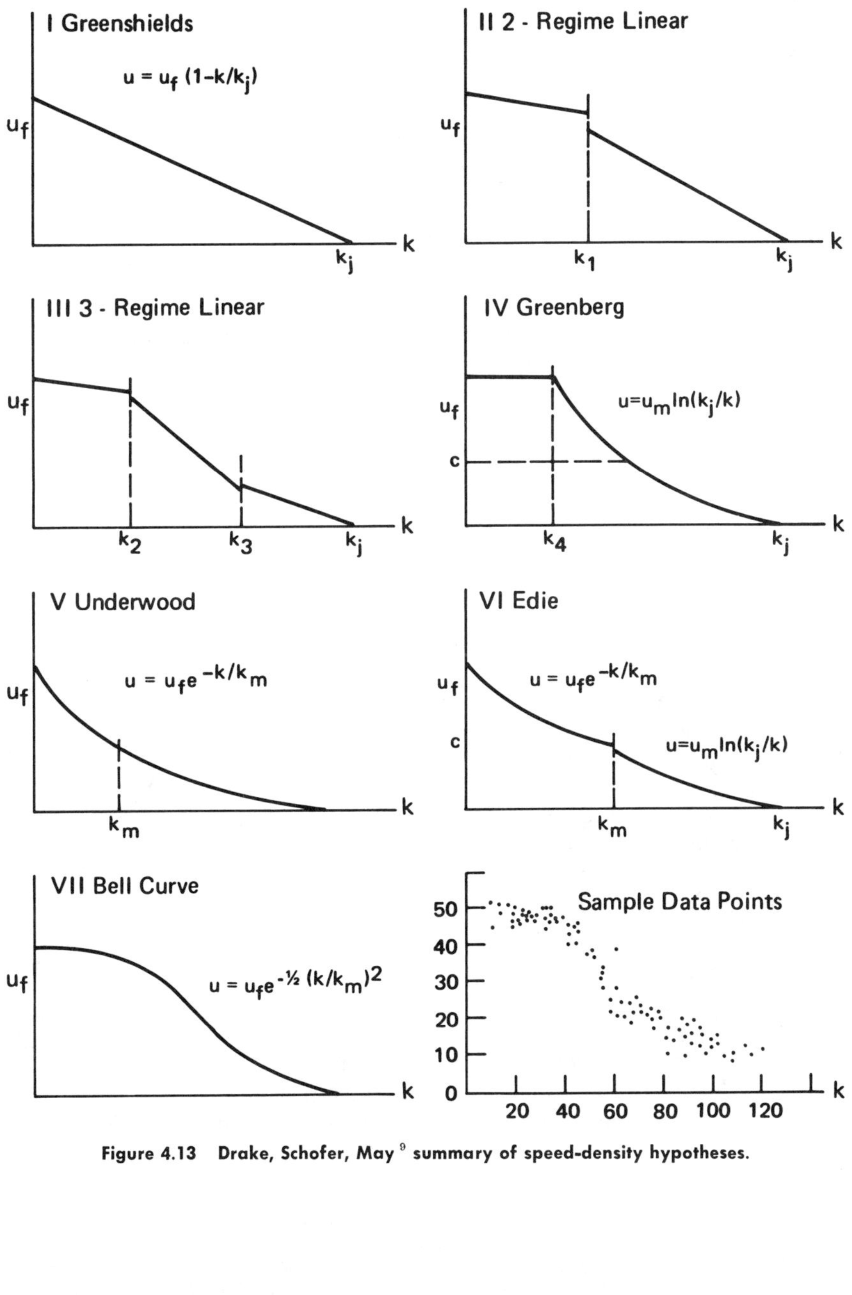

Figure 4.13 Drake, Schofer, May[9] summary of speed-density hypotheses.

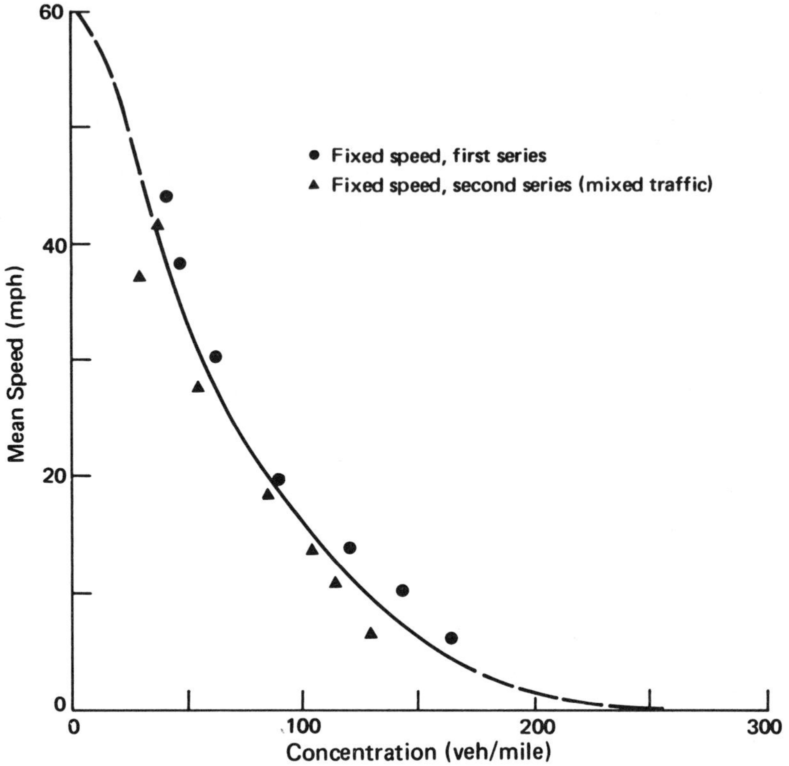

Figure 4.14 Speed-concentration curve for straight track at British Road Research Laboratory.[16]

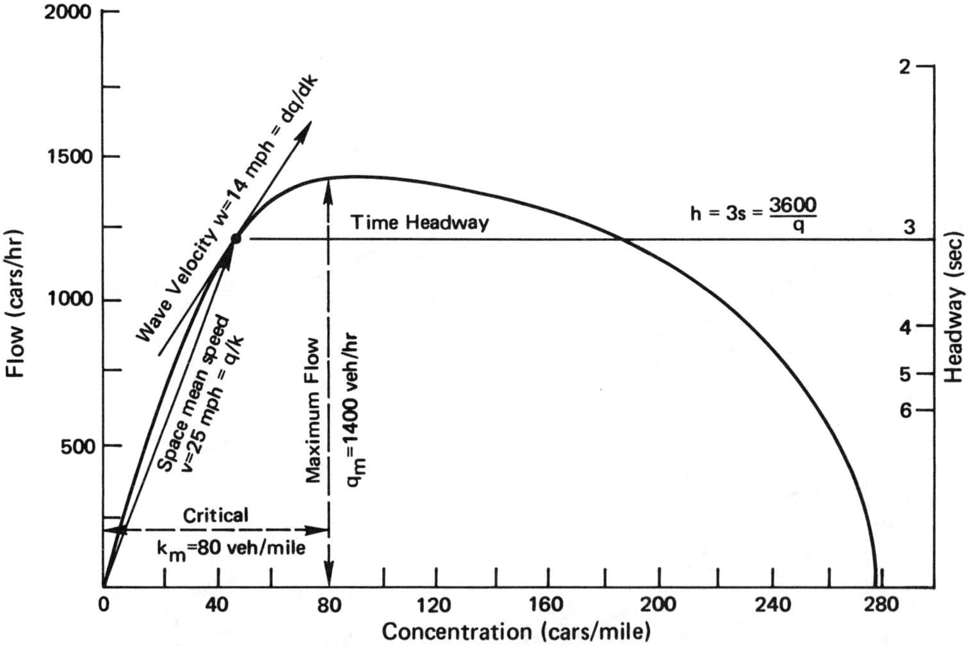

Figure 4.15 Characteristics measurable from the flow-concentration diagram assuming a continuous relationship with one maximum.[20]

Figure 4.17 shows a normalized logarithmic flow–concentration diagram in which all concentration values have been divided by the jam concentration, and all flow values have been divided by $u_m k_j$ or the product of the jam concentration and the speed at maximum flow. The normalized maximum flow value is 1/e and the concentration at the time of maximum flow is 1/e.

In Underwood's logarithmic model for Eq. 4.3,

$$q = k\, u_f \exp(-k/k_m) \qquad (4.10)$$

where $q_m = k_m u_f/e$, $u_m = u_f/e$, and $k_m = k_m$. Here, k_m is a parameter.

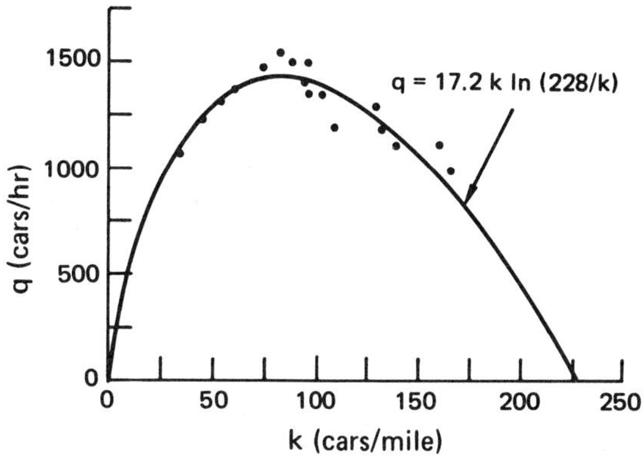

Figure 4.16 Logarithmic flow-concentration diagram.[13] Data points taken in Lincoln Tunnel, where $u_m = 17.2$ mph and $k_j = 228$ cars/mile.

4.3.3 Discontinuous Flow–Concentration Models

Edie [13] has pointed out that traffic behavior appears to be different at high concentrations and at low concentrations, and has introduced the idea of two speed–concentration models. Two speed–concentration curves can lead to two flow–concentration curves as shown in Figure 4.18. There are ample experimental data to indicate that there may in fact be two types of behavior of traffic—one upstream from the bottleneck (or before the bottleneck reaches capacity) and another when the capacity of the bottleneck has been exceeded. (See section 4.3.5.1.)

4.3.4 Special Flow–Concentration Models

Whereas most stream flow models are used to describe one-lane flows, it is possible to develop models describing the total flow on one roadway of a freeway. Figure 4.19 shows a flow–concentration model describing three lanes of a freeway.

4.3.5 Applying Flow–Concentration Models

The flow–concentration representation of the traffic stream is frequently used in studying

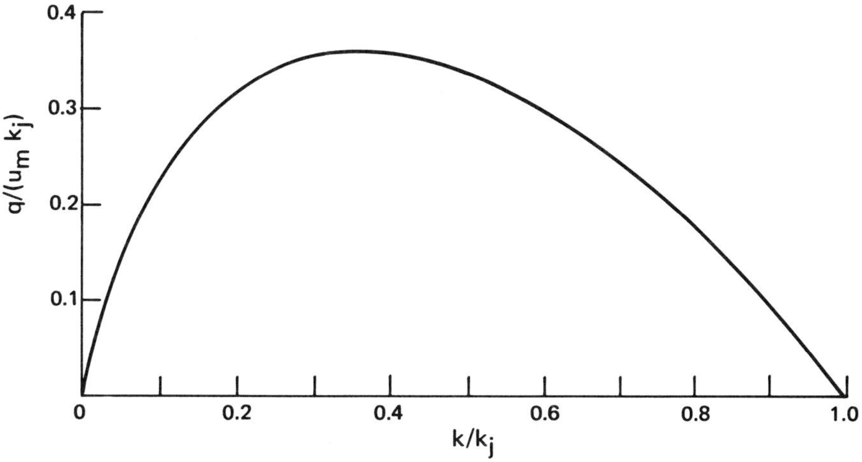

Figure 4.17 Normalized logarithmic flow-concentration diagram.[3]

capacity and in controlling flows on freeways. More details are given in Chapter 7. The following paragraphs give two examples of such applications.

4.3.5.1 Traffic Flow at a Bottleneck. A bottleneck is a section of roadway having a capacity that is less than that of the section of road leading up to it. Figure 4.20 shows flow–concentration curves for the highway and for the bottleneck. The capacity of the highway may be taken as the point of maximum flow on the curve for the highway; the capacity of the bottleneck is indicated by point 1. As the highway flow approaches the capacity of the bottleneck, operation switches to the right-hand side (point 2) of the highway curve. Any slight increase in arrival flow above the capacity of the bottleneck causes the formation of a queue, and a wave of increasing density is transmitted rearward with a speed $\Delta q/\Delta k$.

4.3.5.2 Freeway Control. In several parts of the United States it has been found important to establish control systems in order to operate freeways at maximum efficiency. The tactics of such control systems are to sense critical situations in the freeway flow and then to control the actions of cars seeking to enter the freeway via nearby entrance ramps. The

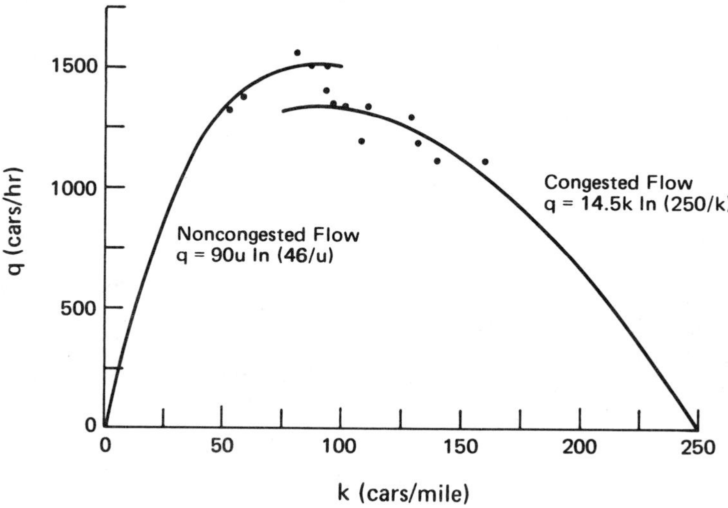

Figure 4.18 Discontinuous flow-concentration curve.[13]

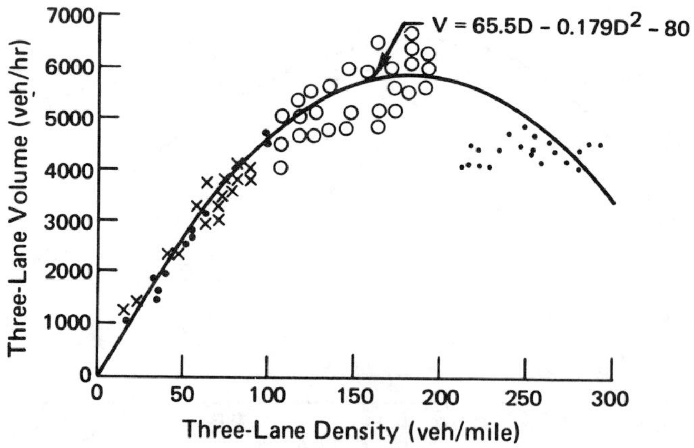

Figure 4.19 Flow-concentartion (volume-density) model for three-lane section of freeway.[23]

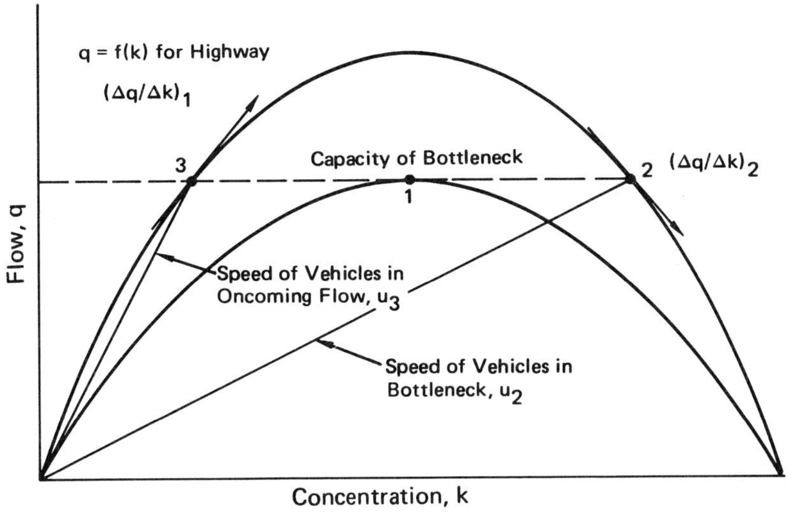

Figure 4.20 Flow-concentration curves for main highway and bottleneck.[24]

most basic of such control systems limit the entrance rate when the freeway flow is near capacity. Concentration has been found to be a good estimator of the flow/capacity ratio. This is partly because concentration increases in a monotonic manner during periods of increasing flow; that is, concentration does not decrease until after the end of the peak period. To control the facility represented by Figure 4.19, it would be necessary to keep the combined concentration in the three lanes at or below about 150 vehicles/mile. At present, there is no good on-line instrument to measure freeway concentration (density) directly (although in certain tunnels it is possible to provide direct concentration measurement [43]). Thus, lane occupancy is used as an estimate of lane density or concentration. References 37–42 provide advanced methods of estimation.

4.4 SPEED–FLOW MODELS

As pointed out in section 4.2, once a speed–concentration model has been determined, a speed–flow model can be determined from it. In all realistic speed–concentration models, the free-flow speed at zero concentration is the maximum attainable speed. (See Figure 4.13.) Thus, the highest point on the speed–flow curve will be the point at free-flow speed and zero flow. Inasmuch as the flow values are the products of the corresponding speed and concentration values, there will be a second point of zero flow, corresponding to zero speed (maximum concentration). Thus, regardless of the shape of the speed–concentration curve, the speed–flow curve will have one point at the origin and one point on the speed axis at the maximum value of speed. Between zero and maximum speeds, the diagram will form some type of loop toward maximum flow. If the speed–concentration curve is a straight line, as suggested by Greenshields, the resulting speed–flow curve is a parabola (Figure 4.2). Other shapes are associated with other speed–concentration curves. (Creighton [25] has presented a qualitative description of the way in which various portions of the speed–flow diagram come about.)

Some early investigators (e.g., Walker [26]) postulated a linear relationship between flow and speed out to maximum flow, with a curvilinear segment between maximum flow and the origin (Figure 4.21). An extreme case is the model developed by the (British) Road Research Laboratory (Figure 4.22). Here the speed is taken as constant for a substantial range of flow that finally breaks to a linear decrease of speed with increasing flow. Here road width was an important parameter.

In the experimental speed–flow curves of Figures 4.23, 4.24, 4.25 roadway radius of curvature is shown to be an important parameter.

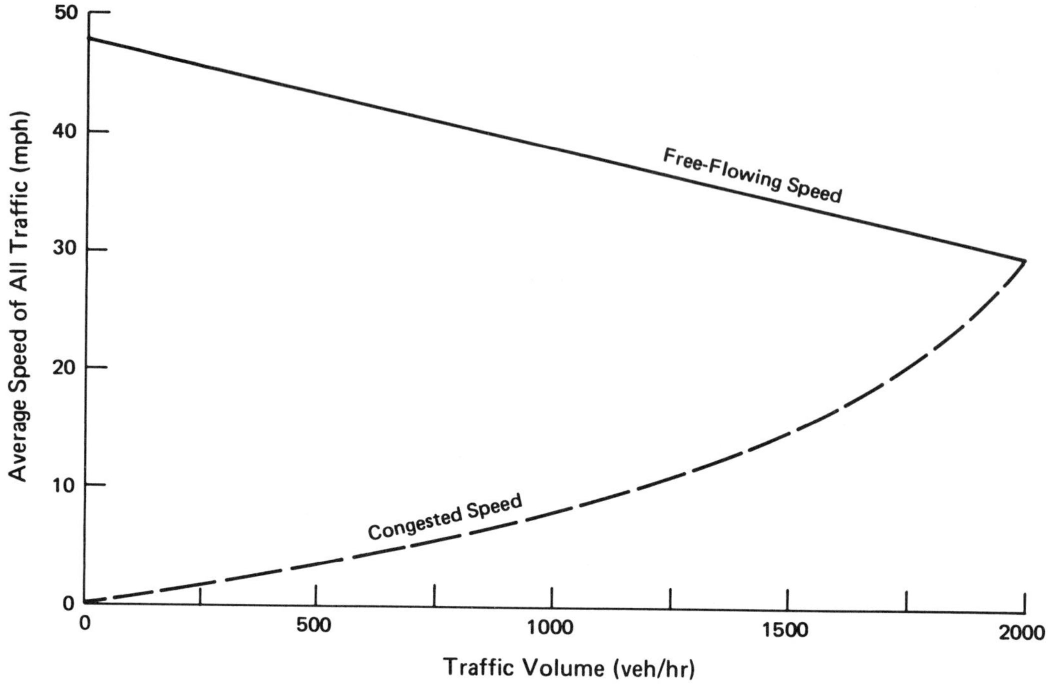

Figure 4.21 Speed-flow model adopted by U.S. Bureau of Public Roads.[26] Relation between average speed of traffic and traffic volume on a two-lane highway having a possible capacity of 2,000 vehicles per hour under favorable operating conditions.

The *Highway Capacity Manual*[44] defines "level of service" and suggests its use in the design and evaluation of facilities. May[28] has proposed the use of speed–flow curves for establishing levels of service for traffic flow (Figure 4.26).

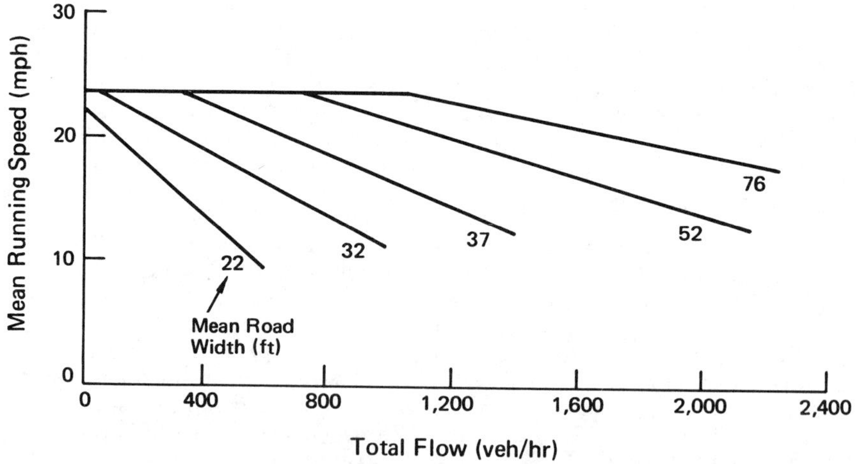

Figure 4.22 Speed-flow model developed by British Road Research Laboratory.[27]

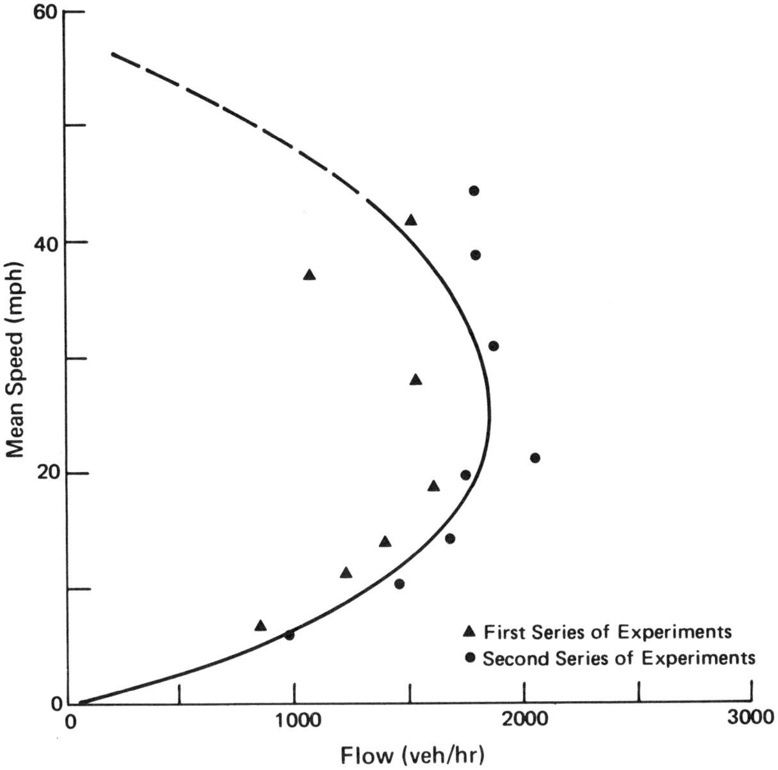

Figure 4.23 Experimental speed-flow relationship on test track with straight roadway section.[16]

4.5 TRAVEL TIME RELATIONSHIPS

Although much attention has been given by various investigators to speed–concentration, speed–flow, and flow–concentration relationships, relatively little attention has been devoted to travel time–flow relationships. Rothrock and Keefer [29] empirically demonstrated the form of travel time–flow relationships (Figure 4.27). Guerin [30] has proposed empirical boundary curves representing travel time–volume (flow) and travel time–density (concentration), including percentage contours, as shown in Figure 4.28. Weinberg et al.[31] and Greenberg and Crowley [32] have also discussed travel time relationships.

4.5.1 Travel Time Models

Haase [33,34] has proposed freeway travel time models having the form

$$t_i = n_i\left[\frac{1}{q_1}-\frac{1}{q_0}\right]+\frac{d}{u}+n_i\left[\frac{1}{q_2}-\frac{1}{q_1}\right] \qquad (4.11)$$

where t_i = total trip time for the ith car;
n_i = the ith car to arrive on the on-ramp queue;
q_0 = the average arrival rate at the on-ramp queue;
q_1 = the average departure rate from the on-ramp queue;
q_2 = the average departure rate from the off-ramp queue;
u = effective steady-state velocity of the N cars on the freeway; and
d = distance traveled on the freeway.

Haase summarizes all trips for the facility and produces curves such as Figure 4.29.

4.5.2 Other Models Involving Travel Time

Smeed [35] has discussed a special situation in which drivers delay starting their trips in order to minimize travel time. He has developed a model that includes the fraction of

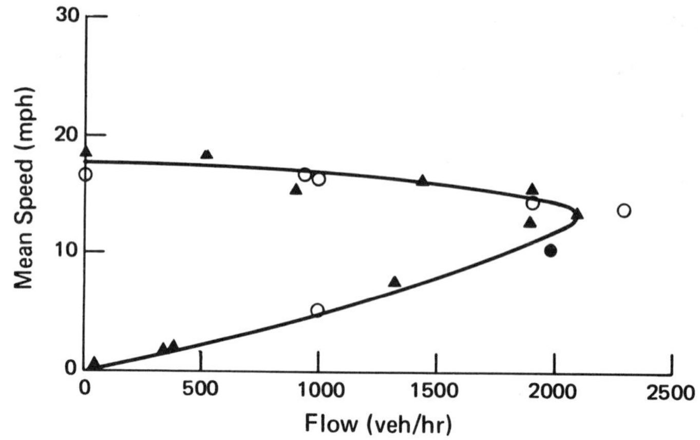

Figure 4.24 Experimental speed-flow relationship on test track with 56-ft radius.[16]

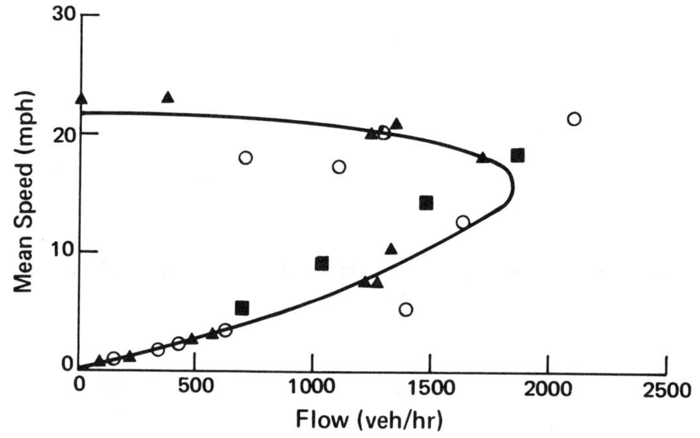

Figure 4.25 Experimental speed-flow relationship on test track with 106-ft radius.[16]

the central business district (CBD) area devoted to streets:

$$\overline{T}=t/2+(7.409/10^6)\,A^{1/2}/[1-n/33tfA^{1/2}]^{1/3} \qquad (4.12)$$

where $\overline{T}$ = average journey time measured from the time the first vehicle enters the CBD;
t = period over which entries to the CBD are spread;
n = number of vehicles entering CBD during period t;
A = area of CBD in square feet; and
f = fraction of CBD area devoted to streets.

A plot of this model is given in Figure 4.30.

Wardrop,[27] in his classic paper, postulated two principles concerning travel time by regular users over several alternate routes. In such situations regular users will distribute themselves over the various routes so that:

(1) The journey times on all routes actually used are equal, and less than those which would be experienced by a single vehicle on any unused route.

(2) The average journey time is a minimum.

In transportation planning, the expression

$$T=T_{\min}\left[1+0.15\left(\frac{q}{q_m}\right)^4\right] \qquad (4.13)$$

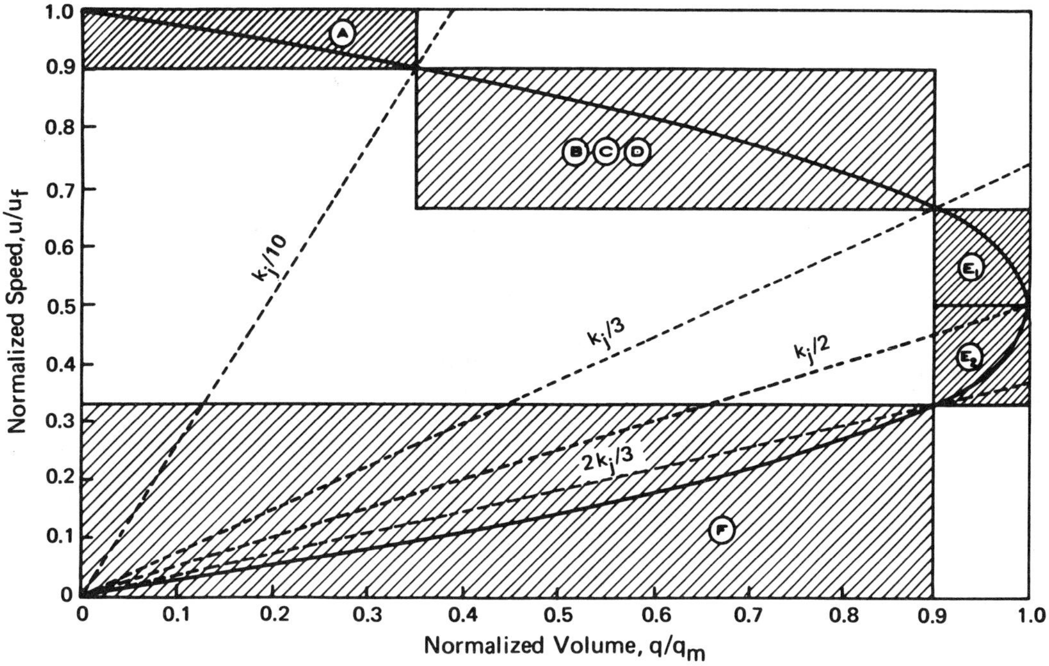

Figure 4.26 Use of speed-flow (speed-volume) curve to establish levels of service.[28]

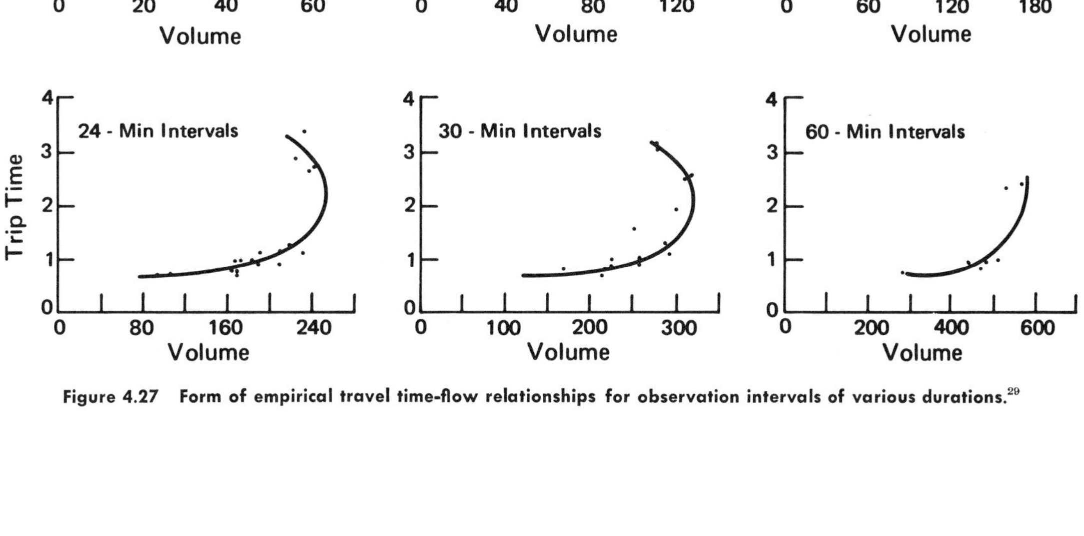

Figure 4.27 Form of empirical travel time-flow relationships for observation intervals of various durations.[29]

is sometimes used when estimating travel time as a function of volume,[36]

where T = the travel time for a particular flow, q;
q_{m} = the maximum flow or capacity of the facility; and
T_{min} = the minimum travel time.

4.6 SUMMARY

Although the speed–concentration–flow relationship is three-dimensional in reality, it is often treated by using one or more of its two-dimensional orthographic projections. From a causation viewpoint, the speed–concentration relationship appears to be most fundamental in that drivers appear to adjust their speeds according to the concentration of the traffic around them. The flow–concentration relationship is generally the most useful because it unifies various theoretical ideas, and it provides relationships for traffic control activities.

A variety of speed–concentration, flow–concentration, and speed–flow models has been presented. The model(s) selected for a specific application must depend on the particular situation and purpose of the analysis. In the absence of a strong reason to the contrary, the simplest appropriate model should be considered. In Chapter 7 an example is presented of traffic behavior at a signal in which solutions are obtained using both linear and logarithmic speed–concentration models. Which result is better depends on local characteristics and judgment.

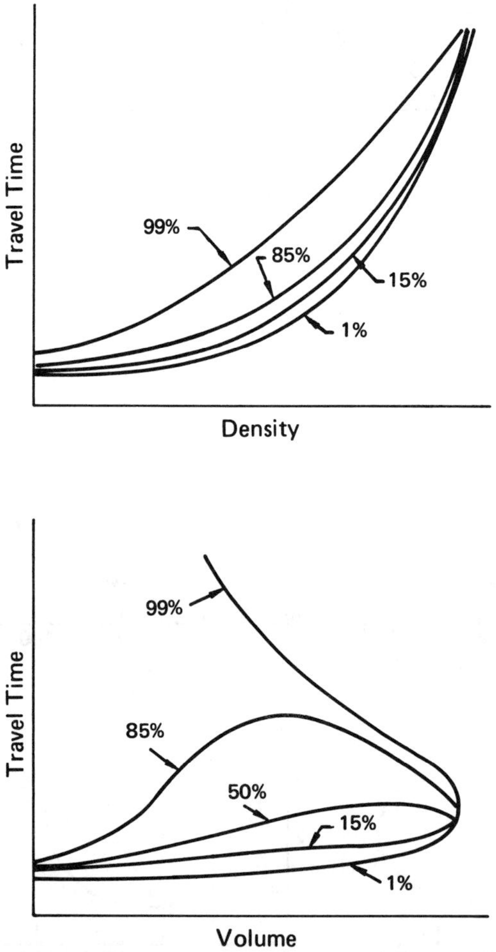

Figure 4.28 Guerin's travel time relationships.[30] Percentage indicated is percentage of observations of travel time that are expected to be below the curve.

4.7 REFERENCES

1. Greenshields, B. D., A study of traffic capacity. Proc. Highw. Res. Board, 14:448–477 (1934).
2. Huber, M. J., Effect of temporary bridge on parkway performance. Highw. Res. Board Bull. 167 (1957), pp. 63–74.
3. Greenberg, H., An analysis of traffic flow. Oper. Res., 7 (1):79–85 (1959).
4. Underwood, R. T., Speed, volume and density relationships, pp. 141–187. *In Quality and Theory of Traffic Flow.* Bureau of Highway Traffic, Yale University, New Haven, Conn. (1961).
5. Pipes, L. A., Car following models and the fundamental diagram of road traffic. Transp. Res., 1 (1):21–29 (1967).
6. Munjal, P. K., and Pipes, L. A., Propagation of on-ramp density perturbations on unidirectional and two- and three-lane freeways. Transp. Res., 5 (4):241–255 (1971).
7. Drew, D. R., Deterministic aspects of freeway operations and control. Highw. Res. Rec. 99 (1965), pp. 48–58.
8. May, A. D., Jr., and Keller, E. M., Non-integer car-following models. Highw. Res. Rec. 199 (1967), pp. 19–32.
9. Drake, J., Scofer, J., and May, A. D., Jr., A statistical analysis of speed–density hypotheses. Proceedings of Third International Symposium on Theory of Traffic Flow (New York, June 1965). Published as *Vehicular Traffic Science,*

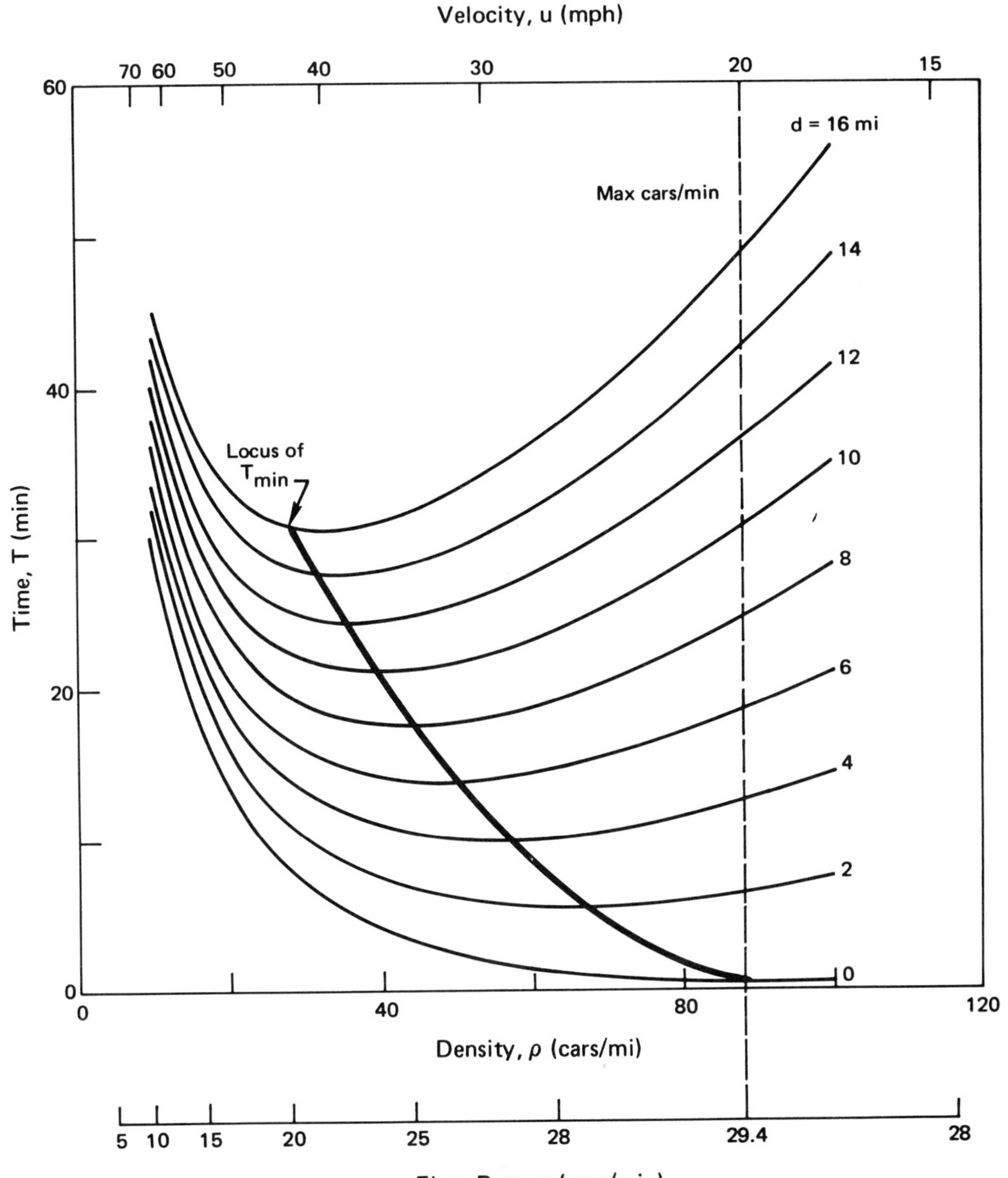

Figure 4.29 Minimum average total travel time for 1,000 cars as a function of distance when $q_0 = 30$ cars/min.[33]

American Elsevier, New York (1967), pp. 112–117.

10. Forbes, T. W., Zagorski, H. J., Holshouser, E. L., and Deterline, W. A., Measurement of driver reactions to tunnel conditions. Proc. Highw. Res. Board, 37:345–357 (1958).

11. Forbes, T. W., Human factor considerations in traffic flow theory. Highw. Res. Rec. 15 (1963), pp. 60–66.

12. Forbes, T. W., and Simpson, M. E., Driver-and-vehicle response in freeway deceleration waves. Transp. Sci., 2 (1): 77–104 (1968).

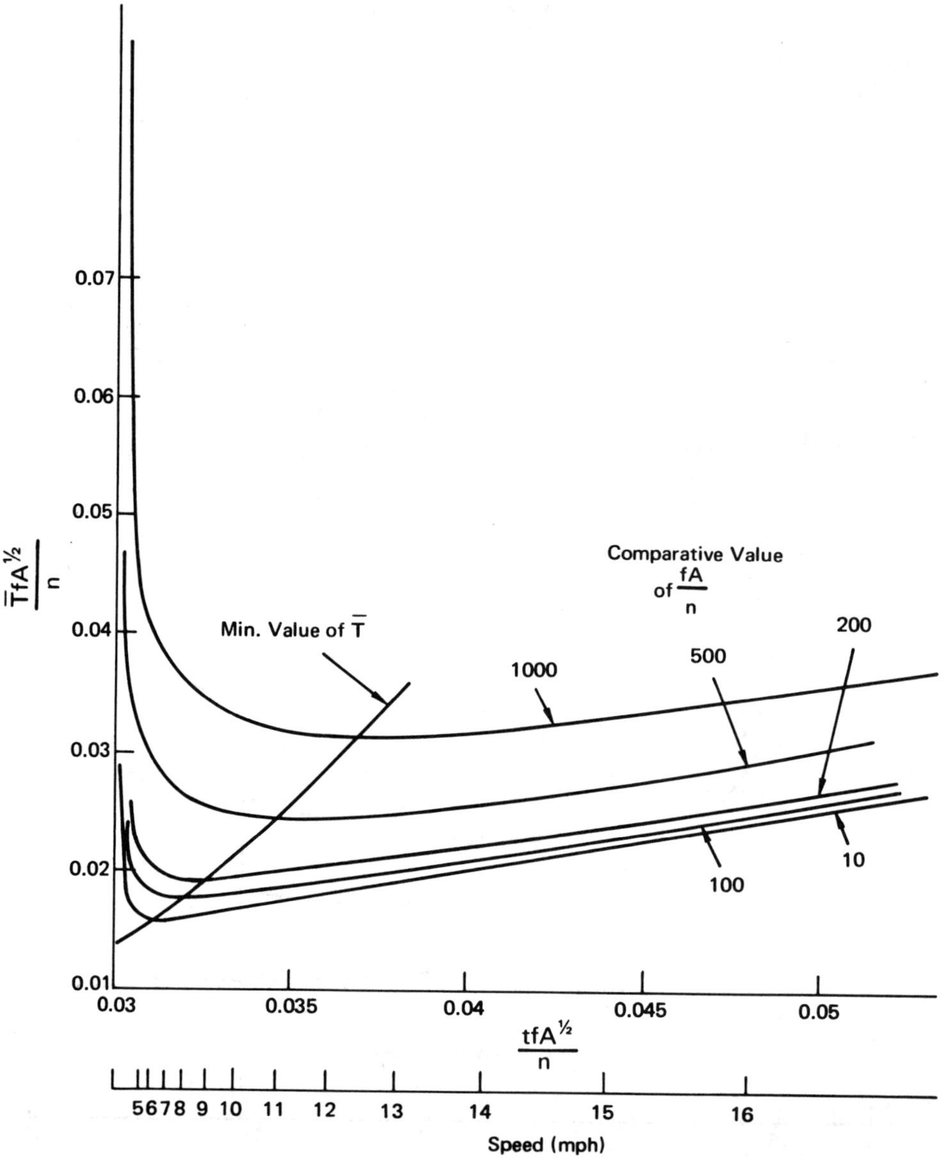

Figure 4.30 Plot of Smeed's[35] model relating minimization of travel time to spread of period during which travel takes place.

13. Edie, L. C., Car-following and steady-state theory for noncongested traffic. Oper. Res., 9 (1):66–76 (1961).
14. Dick, A. C., Speed/flow relationships within an urban area. Traffic Eng. Control, 8 (6):393–396 (1966).
15. May, A. D., Jr., and Keller, H. E. M., Evaluation of single and multi-regime traffic flow models. Proceedings of Fourth International Symposium on Theory of Traffic Flow (Karlsruhe, Germany, June 1968). Published as *Beitrage zur Theorie des Verkehrsflusses, Strassenbau und Strassenverkehrstechnik* (1969), vol. 86, pp. 37–48.

16. Wardrop, J. G., Experimental speed/flow relations in a single lane. *Proceedings of the Second International Symposium on the Theory of Traffic Flow* (London, 1963). OECD (1965), pp. 104–119.
17. Lighthill, M. J., and Whitham, G. B., On kinematic waves: II. A theory of traffic flow on long crowded roads. Proc. R. Soc., Lond., Ser. A., 229 (1178):317–345 (1955); reprinted in *An Introduction to Traffic Flow Theory*. Special Report 79, Highway Research Board, Washington, D.C. (1964), pp. 7–35.
18. Haight, F. A., The volume, density relation in the theory of road traffic. Oper. Res., 8 (4):572–573 (1960).
19. Haight, F. A., *Mathematical Theories of Traffic Flow*. Academic Press, New York (1963), pp. 67–95.
20. Edie, L. C., and Foote, R. S., Traffic flow in tunnels. Proc. Highw. Res. Board, 37:334–344 (1958).
21. Edie, L. C., and Foote, R. S., Experiments on single-lane flow in tunnels. Proceedings of [First] [International] Symposium on the Theory of Traffic Flow (Warren, Michigan, 1959). Elsevier, New York (1961), pp. 175–192.
22. Gazis, D. C., Herman, R., and Potts, R. B., Car-following theory of steady-state traffic flow. Oper. Res., 7 (4):499–505 (1959).
23. Haynes, J. J., Some considerations of vehicular density on urban freeways. Highw. Res. Rec. 99 (1965), pp. 59–80.
24. Drew, D. R., and Keese, C. J., Freeway level of service as influenced by volume and capacity characteristics. Highw. Res. Rec. 99 (1965), pp. 1–39.
25. Creighton, R. L., Deriving analytically the relationships between speed and volume as observed on signalized roads in urban areas. C.A.T.S. Res. News, 1 (11): 6–11 (1957).
26. Walker, W. P., Speed and travel time measurements in urban areas. Highw. Res. Board Bull. 156 (1957), pp. 27–44.
27. Wardrop, J. G., Some theoretical aspects of road traffic research. Proc. Inst. Civil Eng., Part II, 1 (2):325–362 (1952).
28. May, A. D., Jr., Discussion to paper by Drew and Keese. Highw. Res. Rec. 99 (1965), pp. 39–43.
29. Rothrock, C. A., and Keefer, L. A., Measurement of urban traffic congestion. Highw. Res. Board Bull. 156 (1957), pp. 1–13.
30. Guerin, N. W., Travel time relationships, pp. 69–103. In *Quality and Theory of Traffic Flow*. Bureau of Highway Traffic, Yale University, New Haven, Conn. (1961).
31. Weinberg, M. I., Deleys, N. J., and Schneberger, R. F., *Surveillance and Ways and Means of Communicating with Drivers*. NCHRP Report 28, Highway Research Board, Washington, D.C. (1966), p. 37.
32. Greenberg, I. and Crowley, K. W., Holland Tunnel study aids efficient increase of tube's use. Traffic Eng., 35(6):20–22, 38–41 (1965).
33. Haase, R. H., *Decreasing Travel Time for Freeway Users*. The Rand Corporation, Memorandum RM-3099-FF, October 1962.
34. Haase, R. H., Increasing freeway utilization for motorists. Paper presented at 34th national meeting of Operations Research Society of America, November 6–9, 1968, Philadelphia.
35. Smeed, R. J., Some circumstances in which vehicles will reach their destinations earlier by starting later. Transp. Sci., 1 (4):308–317 (1967).
36. Bureau of Public Roads, Federal Highway Administration, U.S. D.O.T., *Urban Transportation Planning, General Information and Introduction to System 360* (1970).
37. Gazis, D. C., and Knapp, C. H., On-line estimation of traffic densities from time-series of flow and speed data. Transp. Sci., 5 (3):283–301 (1971).
38. Gazis, D. C., and Szeto, M. W., Estimation of traffic densities at the Lincoln Tunnel from time series of flow and speed data, pp. 151–165. In P. Lewis (ed.) *Stochastic Point Processes*. John Wiley & Sons, New York (1972).
39. Szeto, M. W., and Gazis, D. C., Application of Kalman filtering to the surveillance and control of traffic systems. Transp. Sci., 6 (4):419–439 (1972).
40. Gazis, D. C., and Foote, R. S., Surveillance and control of tunnel traffic by an on-line digital computer. Transp. Sci., 3 (3):255–275 (1969).
41. Gazis, D. C., and Szeto, M. W., On the design of density measuring systems for roadways. Paper presented at Highway

Research Board Annual Meeting, Washington, D.C. (January 1974).
42. Nahi, N. E., and Trivedi, A. N., Recursive estimation of traffic variables: section density and average speed. Transp. Sci., 7 (3):269–286 (1973).
43. Foote, R. S., Instrumentation development and contributions from traffic flow theory. Highw. Res. Rec. 368 (1971), pp. 1–8.
44. Highway Research Board, *Highway Capacity Manual.* Special Report 87. Washington, D.C. (1965), pp. 7–8, 78–87.

4.8 RELATED LITERATURE

Campbell, E. W., Speed–volume relationship determined by probability of signal failure. Chicago Area Transportation Study. C.A. T.S. Res. News, 1 (16):12–16 (1957).

Carstens, R. L., and Ring, S. L., Pedestrian capacities of shelter entrances. Traffic Eng., 41 (3):38–43 (1970).

Edie, L. C., Foote, R. S., Herman, R., and Rothery, R., Analysis of single lane traffic flow. Traffic Eng., 33 (5):21–27 (1963).

Forbes, T. W., Speed, headway, and volume relationships on a freeway. Proc. Inst. Traffic Eng., 22:103–126.

Forbes, T. W., Mullin, J. J., and Simpson, M. E., Interchange spacing and driver behavior effects on freeway operation. Proceedings of Third International Symposium on Theory of Traffic Flow (New York, June 1965). Published as *Vehicular Traffic Science*. American Elsevier, New York (1967), pp. 97–108.

Foster, J., An investigation of the hydrodynamic model for traffic flow with particular reference to the effect of various speed-density relationships. Proc. Austr. Road Res. Board, 1:229–257 (1962).

Franklin, R. E., On the flow–concentration relationship for traffic. Proceedings of the Second International Symposium on the Theory of Traffic Flow (London, 1963). OECD (1965), pp. 120–128.

Keefer, L. E., Speed–density study. Chicago Area Transportation Study. C.A.T.S. Res. News, 1 (13):6–10 (1957).

Leutzbach, W., Testing the applicability of the theory of continuity on traffic flow at bottlenecks. Proceedings of Third International Symposium on Theory of Traffic Flow (New York, June 1965). Published as *Vehicular Traffic Science,* American Elsevier, New York (1967), pp. 1–13.

May, A. D., Jr., and Wagner, F. A., Jr., Headway characteristics and interrelationships of fundamental characteristics of traffic flow. Proc. Highw. Res. Board, 39:542–547 (1960).

Walker, W. P., Speed and travel time measurement in urban areas. Highw. Res. Board Bul. 156 (1956), pp. 27–44.

Welding, P. I., Time series analysis applied to traffic flow. Proceedings of the Second International Symposium on the Theory of Traffic Flow (London, 1963). OECD (1965), pp. 60–72.

Underwood, R. T., Some aspects of the theory of traffic flow. Austr. Road Res., 1 (2): 35. (1962).

Underwood, R. T., Traffic flow models. Traffic Eng. Control, 5 (12):699–701 (1964).

4.9 PROBLEMS

1. For the speed–concentration model of Figure 4.12, compute the coordinates and plot curves of speed–flow and flow–concentration on rectilinear graph paper.

2. From the flow–concentration curve of Figure 4.19, use radius vectors to obtain speeds and plot a speed–flow curve.

Chapter 5

DRIVER INFORMATION PROCESSING CHARACTERISTICS

5.1 INTRODUCTION

Model representation of traffic behavior indirectly deals with human behavior. Several traffic flow models to be discussed in later chapters contain parameters used to account for various characteristics of the driver in the driver–vehicle system. Some models deal with traffic as a deterministic phenomenon even though the driver portion of the system, at least, is highly stochastic.

Although the whole field of human factors in traffic could, of course, be the subject of a very lengthy treatise, such as that by Forbes,[40] it is hoped that this chapter will provide some insight into the way drivers receive and use information and that this knowledge can then be applicable in various traffic flow models. It should be borne in mind that driver actions are highly stochastic, but that it is often possible to represent stochastic data either by expected values or by worst cases.

The discussion starts with a brief examination of the driving task and its information requirements. Then the ways in which a driver receives and processes information are examined. Finally, several miscellaneous items of driver behavior are considered.

5.2 NATURE OF THE DRIVING TASK

The principal goal of the driver is to guide his vehicle from origin to destination in a safe manner. He may have some additional goals such as arriving at his destination at the earliest possible time; he may also have certain goals concerning the environment through which he passes during his trip. At various points in the trip there may be special subtasks (e.g., parking).[1]

The task of accomplishing the driver's goals may be broken down into categories of action: perception, judgment, decision, and control.[2,41] The driver's control actions are limited to control of acceleration (braking and accelerating) and control of heading (steering or tracking).

The tracking or steering subtasks can be described in terms of a servo system (Figure 5.1) and several authors have proposed various models to describe the driver's actions as part of the servo loop.[2–6] To accomplish his steering function the driver attempts:

1. To select a reference (point, line, angle, curve, etc.) from which to determine the vehicle misalignment or deviation (i.e., error).
2. To detect such errors.
3. To establish an error criterion.
4. To respond to the detected error (via the steering wheel) in such a manner as to maintain the vehicle within the established criterion limits throughout the duration of the steering task.

The acceleration control subtask consists of detecting differences in velocity and/or spacing and taking actions that will prevent unsafe conditions and fulfill both the driver's goal of proceeding at a particular speed and such other goals as he may have.

5.2.1 Driver Information Needs

The information needs of the driver vary with the portion of the trip and the immediate maneuver to be accomplished. The trip may start with a planning phase in which macro-information provided by maps, weather reports, and road conditions is required. Once the driving has started, however, information comes from observations of the roadway, other traffic, signs and signals, and instruments. Generally, cues to the driver may be visual, auditory, tactile, kinesthetic, and even olfactory, but by far the most important sense is visual. This is followed by hearing, the ability to sense accelerations, and the ability to sense vibrations. Some characteristics of these senses are given in Table 5.1. (At one time auditory cues were an important source of information to the driver. However, with the present trend to make cars very quiet and to drive with the

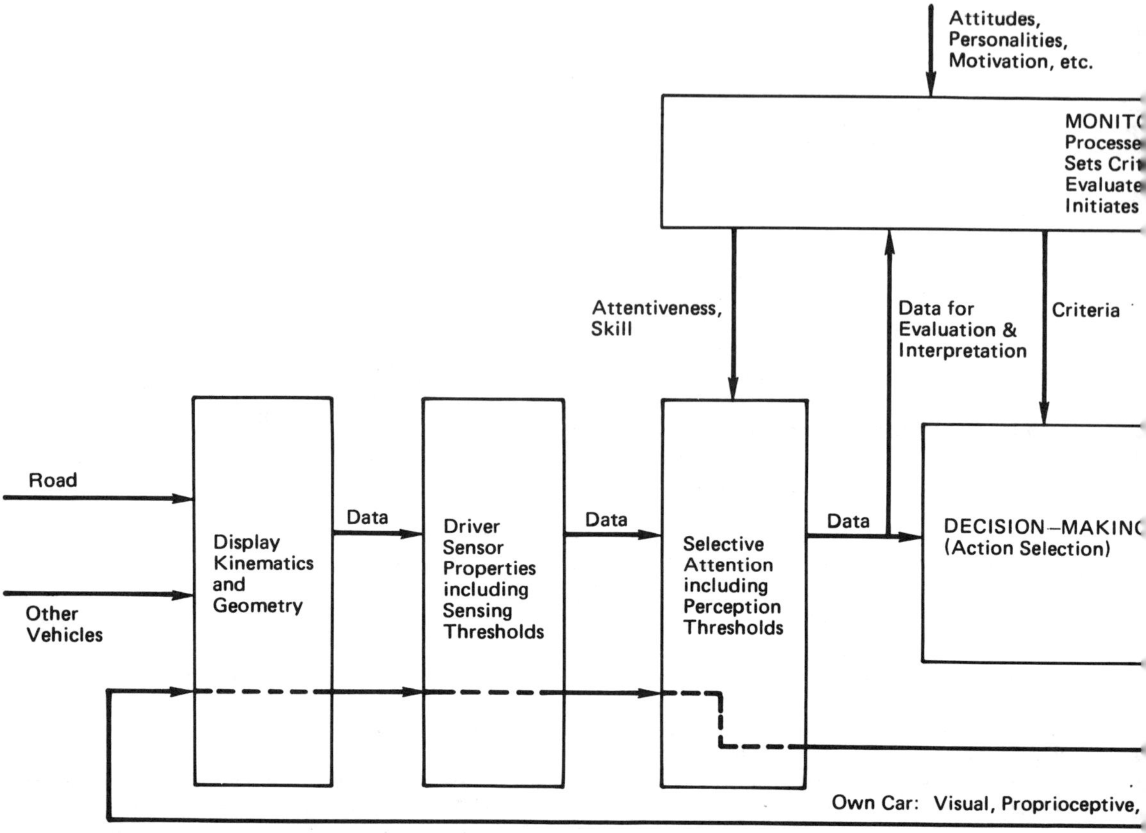

Figure 5.1 Generalized block diagram of the car-driver-roadway system (adapted from Rockwell[6]).

windows closed and air conditioning on (and perhaps a radio or tape player operating), many auditory cues cannot be heard.)

The behavior of the driver can be likened to the observer who must scan several displays. The driver, in addition to watching the roadway, must watch the car ahead and the cars in adjacent lanes; he must scan one or more rearview mirrors and from time to time must check the instrument panel. At times it is necessary to actively scan the roadside area for directional signs and other information such as landmarks. In many situations the driver must perform a filtering function to extract the information needed from the surrounding features (visual noise). Thus, it is important to study the ways the driver processes and responds to the information at hand.

5.3 HUMAN RESPONSE TO STIMULI

The human response to various stimuli varies with the type of stimulus and the level of the stimulus (intensity and/or frequency) with respect to the related threshold value. Furthermore, because each stimulus results in the transmission of certain information to the brain, response may be moderated by the simultaneous receipt of other information. As in any instrument system, the human system contains an inherent transmission delay between the stimulus and the response. Also, a finite time is required to process information. The processing time is added to the transmission time to create a total reaction time to the various stimuli.*

* Private communication from S. F. Hulbert.

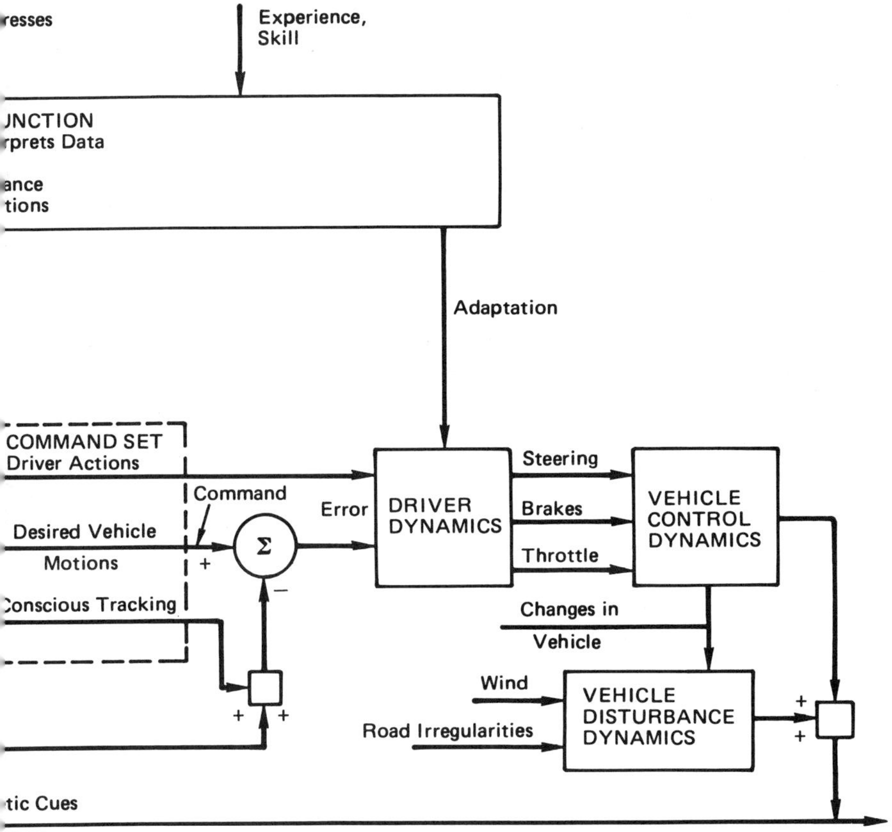

5.3.1 Simple Response Models

The ability of an observer to detect a stimulus is often stated by giving the threshold for the specified stimulus.[11] The (absolute) threshold may be defined as that amount of energy that will result in the correct identification of the signal 50 percent of the time. More recent research has used the "detection-effectiveness parameter." Another (relative) measure of stimulus is the "just noticeable difference" (JND). This has led to Weber's law, which may be stated †

$$\text{JND} = \Delta s / s = \text{constant}$$

where s is the previous stimulus energy value and Δs is the change in stimulus energy level. Experiments have shown that this relationship is a good approximation over about 99.9 percent of the usual range of sensory perception.

† Private communication from T. W. Forbes.

Based on Weber's law, Fechner suggested a generalization:

Magnitude of sensation
= k log (magnitude of stimulus)

A more accurate function was developed by Stevens.[43, 44]

$$\Psi = k(\phi - \phi_0)^n$$

where Ψ = sensation magnitude; ϕ = stimulus magnitude; k = a constant; and n = an exponent for a given sense area (e.g., vision, hearing).

5.3.2 Driver as Sampled-Data System

In observing any object to extract information, it is necessary for a person to fixate his view on one portion at a time. Thus, normal viewing is a process of sampling data. Further-

TABLE 5.1 Human Senses of Principal Importance to Drivers [a]

			Intensity Range		Intensity Discrimination	
Sense	Sensor	Medium	Smallest	Largest	Relative	Absolute
Vision	Eye	Some electro-magnetic waves Mechanical pressure Interrupted white light	2.2–5.7 $\times 10^{-10}$ ergs	Approx 10^9 times threshold	Approx 5% discernible levels in white light	3–5 discernible levels in white light of 0.1–50 ml
Hearing	Ear	Some amplitude and frequency variations of pressure of surrounding media (e.g., air) Interrupted white noise	1×10^{-9} ergs/cm^2	Approx. 10^{14} times threshold	At 2000 Hz Approx 325 discernible levels	With pure tones about 3–5 identifiable levels
Linear acceleration	Muscles; semi-circular canals		Vertical 4–12 cm/sec^2 Horizontal 12–20 cm/sec^2		No data	
Angular acceleration	Semi-circular canals		0.2 degrees/sec^2		No data	
Mechanical vibration	Skin		Varies with portion of body stimulated	Varies with portion of body stimulated		3–5 steps

more, because the driving task requires observing not only the roadway but also other targets and displays, the sampling nature is increased. Figure 5.2 shows histograms of fixations in normal traffic and while passing. Bekey [13] has investigated some of the phenomena associated with humans as data samplers. Some experiments to determine the characteristics of this sampling process have been conducted by Senders et al.[14–16]; these results are discussed in Section 5.3.4.

5.3.3 Driver Information Processing

Tasks, such as driving, can be thought of as responses to information-containing stimuli. It is convenient to quantify this information by the number of "bits" it contains; i.e., the number of mutually exclusive decisions that must be made to correctly execute the task.[17,46] Driving involves stimuli that are continuously changing; hence, it presents the driver with a flow of information in bits per second. If this rate is slow enough, drivers can respond correctly and process all information presented to them. Drivers, however, are limited in their capacity for processing information. If the rate of information presentation exceeds this capacity, the excess will not get processed and may even confuse the driver. Thus, the driving environment should not present information at a rate that exceeds the information processing capacity of drivers.

The driver is considered to have a single information channel of fixed capacity in which tasks are linearly additive.[19] (Several studies undertaken to measure the capacity of this channel are discussed in the following paragraphs.) Interestingly, the level to which a driver is stimulated can have a beneficial effect up to a certain point, but thereafter further stimulation can degrade his performance.[20]

5.3.4.1 Short-Viewing-Interval Experiments. Experiments reported by Szafran [21] using short viewing intervals have demonstrated the limits of the human information processing ability. As the amount of information in-

Wavelength or Frequency Range		Wavelength or Frequency Discrimination		
Lowest	Highest	Relative	Absolute	Maximum Information Rate
300 nm	1500 nm	Approx 128 discernible hues at moderate intensities	12–13 hues	4.6 × 10^6 bits/sec
Unnlimited	At intensities with duty cycle of 50% approx 50 interruptions per sec	At moderate intensities with 50% duty cycle, 375 rates between 1 and 45 interruptions per sec	5 or 6 interruption rates	
20 Hz	20,000 Hz	Between 20 and 20,000 Hz at 60 db approx 1,800 discernible steps	4 or 5 tones	8,000 bits/sec
Unlimited	At moderate intensities with 50% duty cycle approx 2000 Hz	At moderate intensities with 50% duty cycle 460 steps in range of 1–45 interrupts/sec	Unknown	
Unlimited	Unknown but reported as high as 10 kHz	Between 1 and 320 Hz 180 discernible steps	Unknown	
Unlimited	Unknown but reported as high as 10 kHz	Between 1 and 320 Hz 180 discernible steps	Unknown	

[a] Source of data: References 7, 8, 9, 10

creases, or when there is divided attention or unwanted or excessive information (visual noise), the additional information may be only marginally processed. The results are illustrated in Figure 5.3.

5.3.3.2 Measuring the Spare Mental Capacity of Drivers. Brown and Poulton [22] have described experiments by which they sought to measure the spare mental capacity of drivers under certain tasks. Although they demonstrated a method, they did not obtain a specific rate in bits per second.

5.3.3.3 Drivers' Channel Capacity Related to Accident Experience. Groups of drivers having high and low accident incidences and high and low violation experiences were tested to determine, by measuring decision time, their abilities to process information (Table 5.2). It will be noted that the subjects with high accident rates had a very low information processing capacity. One unexpected result of the experiments is that drivers with zero accidents but a high number of violations had the highest information processing capacity. Fergenson [23] offers the following interpretation:

> Possibly those individuals with many violations are involved in more critical situations due to their driving habits; but they still avoid accidents because their information processing capability is not overloaded (they have large channel capacity.

TABLE 5.2 Average Information Processing Ability of Subjects With and Without Violations and With and Without Accidents [a]

	Bits per Second	
	Zero Violations	High Violations
Zero accidents	26.09	38.67
High accidents	21.13	15.69

[a] From Fergenson.[23]

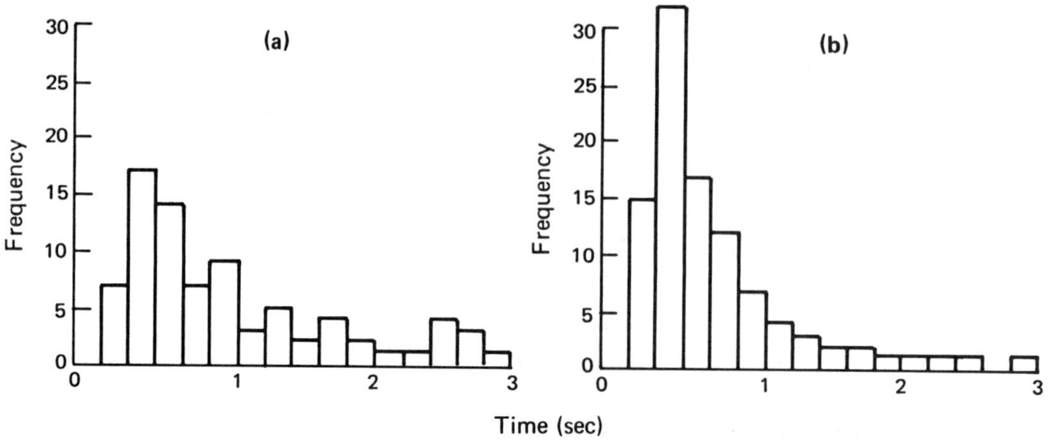

Figure 5.2 **Frequency distributions of eye fixation duration (a) during passing and (b) during normal driving (from Whalen et al. as given by Briggs [18]).**

5.3.4 Models of Driver Information Processing

An interesting experiment directly related to the information processing capacity of drivers has been reported by Senders and Ward.[15] The experimental program consisted of observing the relationship between a driver's maintained speed and the amount of time his vision was interrupted. Interruption was accomplished by dropping an opaque visor over the eyes of the driver for varying lengths of time.

In one variation of the experiment the frequency and duration of viewing time was fixed while the driver adjusted his speed to the amount of information he received. In a second variation the driver maintained a constant speed but varied the frequency of viewing time. Two types of roadway were used: an unopened segment of I-495 in Massachusetts was considered an "easy" road for driving; the other, a closed-circuit sports-car course, was considered difficult.

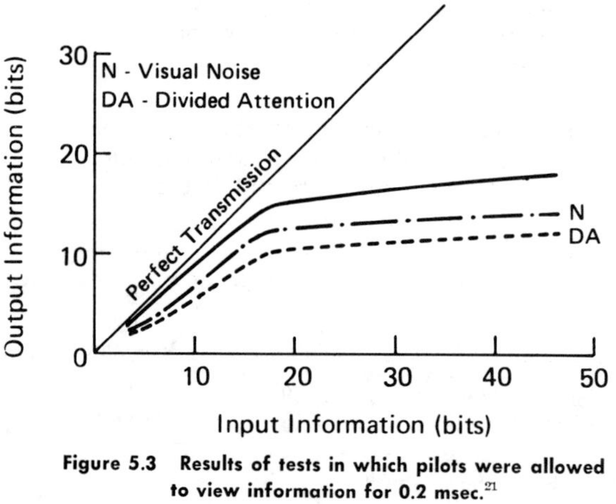

Figure 5.3 **Results of tests in which pilots were allowed to view information for 0.2 msec.[21]**

Senders and co-workers developed a model of information based on the assumption that the road had a certain number of information bits per mile. The faster one travels, the more bits he must process per unit of time. If the driver were to see the road only at fixed intervals, he would develop uncertainty about his position on the road and what events have appeared on the road since his last observation. For a given sampling rate of information it can be hypothesized that a driver will adjust his "maximum" speed to prevent the uncertainty between samples from becoming too great. (Under normal driving situations it is observed that a driver tends to other tasks while driving, such as tuning radios, lighting cigarettes, reading signs, and talking to passengers.)

Consider the situation in which a driver's vision is periodically occluded, as shown in Figure 5.4:

> At $t=-T_1$ vision is unobstructed and the driver absorbs information. The maximum amount of information is in store at $t=0$. Vision is obstructed for the following T_d seconds [Figure 5.4], during which time the store of information continually diminishes. The minimum amount of information is in store at

$t = T_d$, at which time vision is restored and the cycle repeats.[15]

The following assumptions and derivations of a driver–information processing model are made:

1. The driving situation is in the steady state. The vehicle proceeds at a constant velocity (V miles/sec); the timing of looks is periodic such that vision is allowed for T_l sec and occluded for T_d sec.

2. The road has a constant information density of H bits/mile.

3. The information density of the driver's stored image is $H\,e^{-x/D}$ bits/mile, where D is a weighting factor in miles and x is the distance from the start of occlusion.

4. The period of view is sufficient for the driver to absorb all the information available. The amount of information stored at $t=0$ is $H\,D$ bits.

5. Information is forgotten at the rate of $I_r(t)/F$ and becomes obsolete at the rate of $I_r(t)B/D$ bits/sec. The amount of information in storage t sec after the onset of occlusion is $H\,D\,e^{-(V/D+1/F)t}$ bits, where I_r is the bits of information stored, F is the time constant (sec), and V is speed (miles/sec).

The driver is assumed to adjust his velocity (or the occlusion time) so that his uncertainty $U(T_d)$ at the end of occlusion is less than some critical value U_c.

The uncertainty is made up of two elements: (1) uncertainty about the road and (2) uncertainty about the vehicle position and orientation of the vehicle. With the introduction of uncertainty, the model of driver behavior becomes

$$U(T_d) = H\cdot D[1 - e^{-(V/D+1/F)}T_d] + K_n\,V^2(T_d)^{3/2} \leq U_c \qquad (5.1)$$

where K_n includes the power density spectrum related to drivers' uncertainty about vehicle position, as well as other scaling factors; other variables are as previously defined.

In the first experiment, on an unopened section of I-495, drivers were given views of the roadway of 0.25-, 0.50-, or 1.0-sec duration. Between views the occlusion time was varied from 1.0 to 9.0 sec, and drivers were expected to adjust speed to reflect the information received.

When Eq. 5.1 was calibrated against observed driving data on the highway, the following ranges of results were obtained from five different drivers (viewing time=0.50 sec):

H= information bits per mile 12.0–34.0

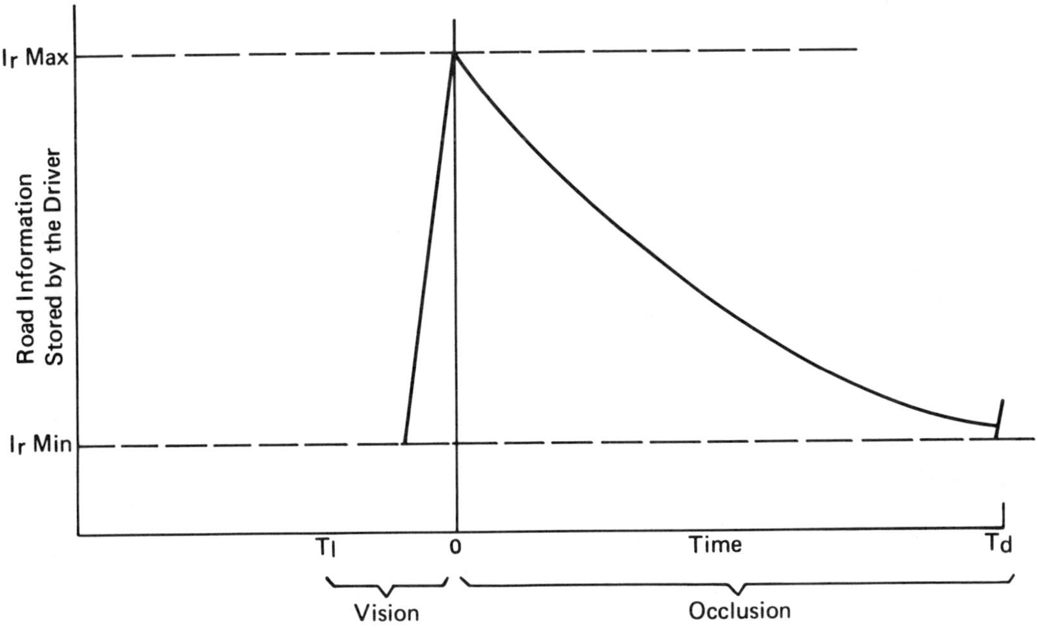

Figure 5.4 Timing of events.[15]

TABLE 5.3 Model Parameters for Various T_l [a,b]

	CWD [c]			JWS			DHK	DCM			CWD2			WVD
T_l	0.25	0.50	1.0	0.25	0.50	1.0	0.50	0.25	0.50	1.0	0.25	0.50	1.0	0.50
D	0.26	0.42	0.32	0.20	0.18	0.20	0.28	0.30	0.30	0.30	0.22	0.46	0.20	0.32
H	14.0	12.0	20.0	6.0	6.0	4.0	18.0	20.0	20.0	12.0	12.0	34.0	26.0	24.0
T^d	6.0	10.0	7.5	9.5	6.0	5.0	3.5	4.5	5.0	3.5	8.0	5.0	7.0	4.5
K	0.0002	0.0002	0.0002	0.0002	0.0002	0.0002	0.0002	0.0002	0.0002	0.0002	0.0002	0.0002	0.0002	0.0002
U	3.13	3.76	5.22	0.99	1.07	1.13	4.93	7.16	5.96	4.29	2.57	13.47	8.24	6.99

[a] Results of an experiment on an unopened section of I-495, with T_l at specified values and T_a controlled by the experimenter.
[b] Source: Senders et al.[14]
[c] Initials identify subjects.
[d] T=number of seconds after the start of occlusion at which the position of the vehicle is measured.

D=	distance ahead over which driver weighs information (miles)	0.18–0.50
F=	time (sec) to forget bits of information	3.5–10.0
U=	maximum amount of uncertainty accumulated in information bits	1.07–13.47

A listing of all of the model parameters for the six drivers is given in Table 5.3.

In general, the results of the experiments, using two different modes of operation and two different classes of roads, indicated that *the less frequent the observations, or the shorter the period of observation, the slower will be the speed that the driver can maintain.* Conversely, the greater the level at which the speed is fixed the more often a driver must look at the road. Differences between the two roads were modifiers in which the more complicated road required lower speeds for constant viewing and occlusion time or more frequent viewing for constant speed and viewing time.

Verification of adequacy of the information driving model suggests that trained drivers might be used to identify and quantify excessively demanding road configurations and that this technique might be used on "calibrated" roadways to make a preliminary classification of drivers in terms of skill level.

5.3.5 Driver Perception–Response Time

One of the important measures of driver response to information received is his reaction time. The best known of these responses is brake reaction time.

In early experiments brake reaction times were measured in the laboratory, and the results listed values that are now considered relatively short. Recent experiments in real traffic have produced brake reaction times that may be considered more reliable.[24] In these experiments the time measured represented the sum of the perception time (time to perceive the need for braking) and the time to move the foot from the accelerator to the brake pedal. Some drivers were tested under conditions of both surprise and limited anticipation. A large group was tested under limited anticipation alone (Figure 5.5). To correct for surprise, results in Figure 5.5 should be multiplied by 1.35. *For surprise situations, then, the median is 0.9 sec, with 10 percent of the reactions at 1.5 sec or longer.*

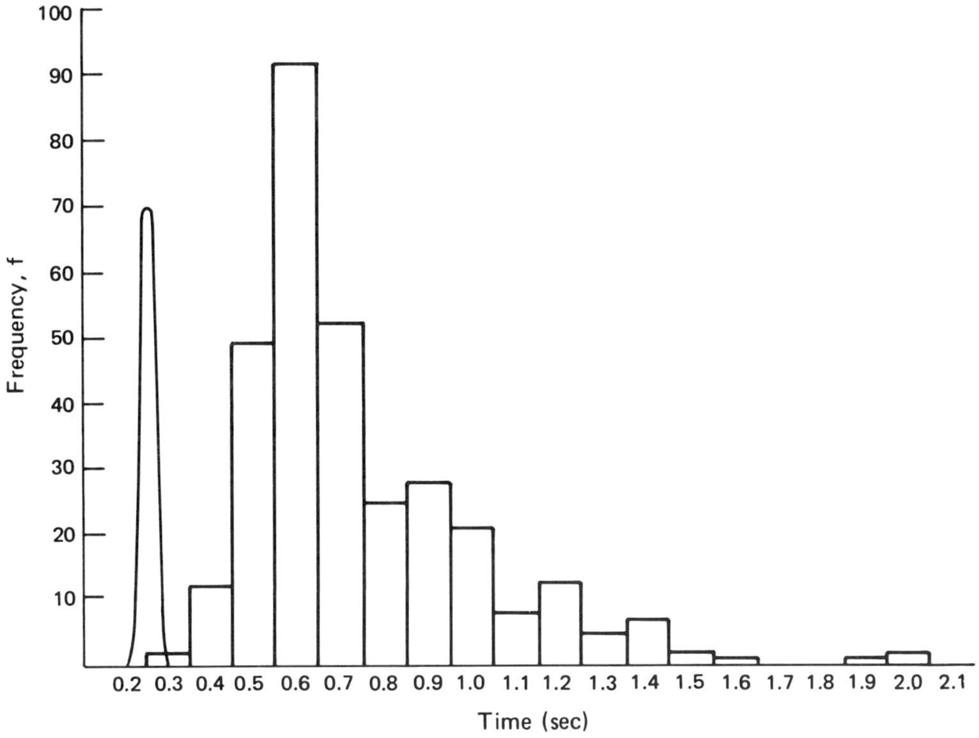

Figure 5.5 Histogram of brake reaction times of 321 drivers. The sharp peak at the left is the distribution of the reaction times of the measuring process.[24]

5.4 DETECTING VELOCITY DIFFERENCES IN CAR FOLLOWING

One of the most important driver tasks from the viewpoint of traffic flow theory is detection of differences of velocity between the lead car and that which the driver operates. Michaels [25] has given one approach to quantifying this task; his work is discussed later. Another approach to the problem is the ability of the driver to detect acceleration. Under general conditions, the human thresholds for sensing acceleration are [8] linear–horizontal (12–20 cm/sec^2), linear–vertical (4–12 cm/sec^2), and angular (0.2 degrees/sec^2).

For car following at night, Todosiev and Fenton [26] found the velocity threshold generally to be smaller than the corresponding day threshold.

An important aspect of velocity detection is the reaction or latency time. Braunstein and Laughery [27] have measured response latencies of drivers in detecting accelerations and decelerations of the lead car (Figure 5.6). Hoffmann [28] has proposed a model for latency as follows:

$$t = \left[\frac{\ell}{4(1 \pm \frac{1}{8})\,|a|} \right]^{1/2} \tag{5.2}$$

where

t = latency time;
$|a|$ = magnitude of acceleration in ft/sec^2; and
ℓ = vehicle separation in ft.

The minus sign represents acceleration; the plus sign, deceleration.

Field observations have led Lee and Jones [29] to make the following statement:

> The mean time lag for the queue-forming condition is less than that for the queue-releasing condition. In the queue-forming condition the driver of the following car must make quick decisions to avoid collision as the following car gets nearer to the vehicle ahead. In the queue-releasing condition there is less urgency for decisive action.

5.4.1 Estimated Driving Speed

Estimating the speed actually being driven is a problem related to velocity differences in car following. Experiments have shown that a driver's estimate of his speed depends on the speed at which he has been traveling previously. Denton [30] has stated, "Drivers underestimate their speed when decelerating and overestimate it when accelerating." Beers and Hulbert [42] have found that accuracy of speed estimation increases with age, but that older drivers tend to underestimate speed while younger drivers overestimate.

5.4.2 Behavior in Platoons *

Forbes [31] has derived the flow–concentration curve from driver characteristics. The curves on the left of Figure 5.7 would result if drivers maintained the same speed as concentration increases. At low concentration where cars do not interfere with each other, flow would increase linearly with concentration. If there is one car per mile moving at 25 mph, the flow is 25 cars per hour. If concentration is doubled to two cars per mile, flow is also doubled (in this case to 50 cars per hour). It may be seen from Figure 5.8 that Forbes' predictions are validated by actual traffic flow data from tunnels and expressways. However, as concentration increases, the driver is increasingly impeded by congestion, and flow departs progressively from the predicted lines.

Forbes' predictions of flow based on fixed reaction time headways are shown in the negatively sloped portions of the flow–concentration curves. As concentration increases, drivers move more slowly to preserve reaction time headway. (Haight, Bisbee, and Wojcik [49] have shown that the driver cannot reduce time headway below man–machine reaction time if the risk of collision with the car ahead is to be avoided. The derivation assumes that the driver's braking is as effective as that of the driver in front.) For example, if concentration increases from 120 cars to 240 cars per lane per mile, distance headway decreases from 44 to 22 ft, and clearance between 17-ft cars decreases from 27 to 5 ft. In this example, the driver must reduce speed to less than a fifth to equalize his time heading at the two concentration levels.

Flow data from highway sites are also shown in Figure 5.8. Curves a to d show data obtained at four freeway stations. The Edsel Ford Expressway (curve c) is fitted by a time headway between 1.0 and 1.5 sec; the Lincoln Tunnel (curve b) and the Penn–Lincoln Parkway (curve d) are predicted by headways between 1.5 and 2.0 sec. The traffic flow taken at another station on the Edsel Ford Expressway and shown in curve a did not reach a slowdown or stoppage, although flow volume reached as high as 2,000 vehicles per hour. It may be seen that the congested-flow data can be fairly well fitted by straight lines representing the assumption that drivers' average time heading is constant and independent of traffic density.

Forbes also showed that drivers reacted to sudden braking of the car ahead by dropping back to increase their time headways. If the leader of a three-car queue reduced speed by about 10 mph and after several seconds returned to cruising speed, the driver immediately behind dropped back to increase his time headway. The effect at various road sites differing in horizontal and vertical curvature is shown in Table 5.4. It may be seen that the rear driver increased time headway 0.8 sec or more after the deceleration maneuver. This change in time headway is observed after deceleration of actual traffic and affects flow rate.

Forbes' experimental work under dense traffic is described in Forbes and Simpson.[46]

5.4.2.1 Platoon Behavior as a Weber–Fechner Relationship. Daou [32] suggests that

* Material in section 5.4.2 is quoted from Gordon [48] with appropriate renumbering of reference citations and figures.

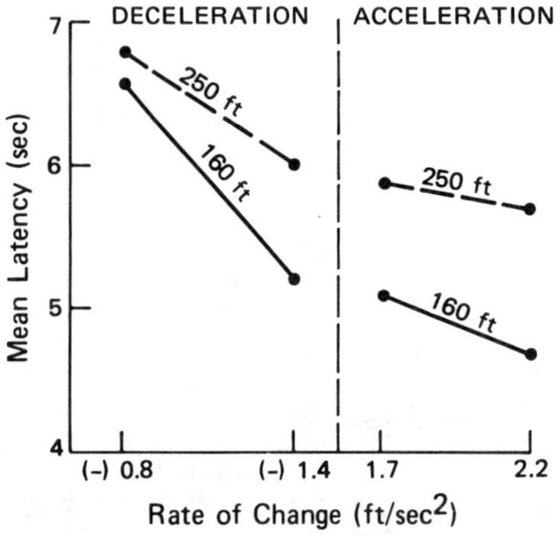

Figure 5.6 Effects of rate of change and initial distance on response latency.[27]

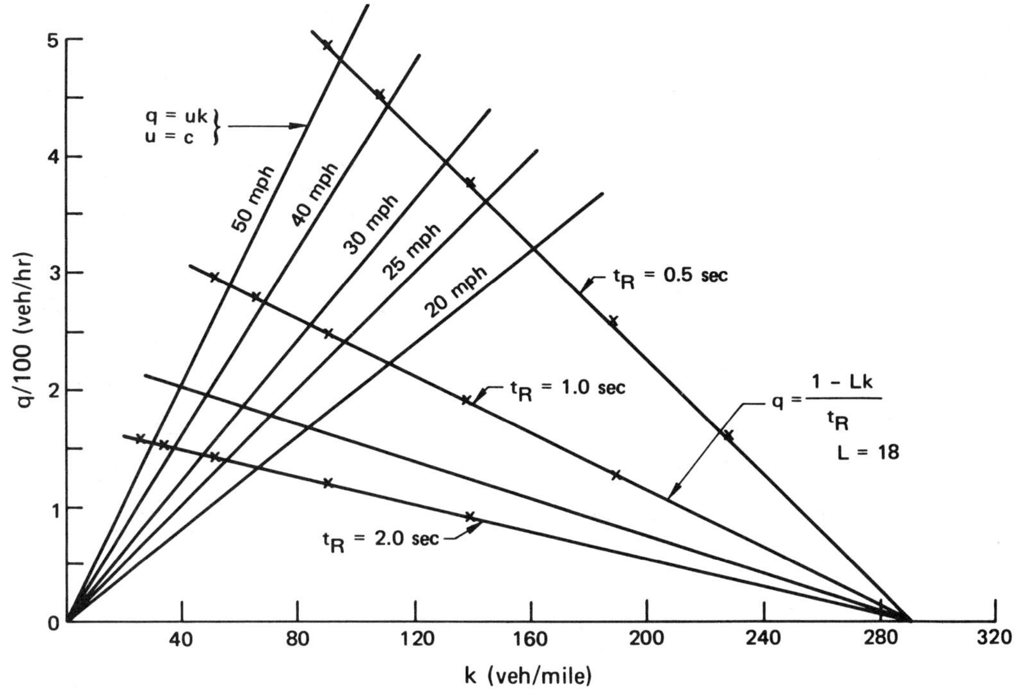

Figure 5.7 Relation of volume, speed, and concentration from driver response time as a limit (right) and scattered free flow (left).[31]

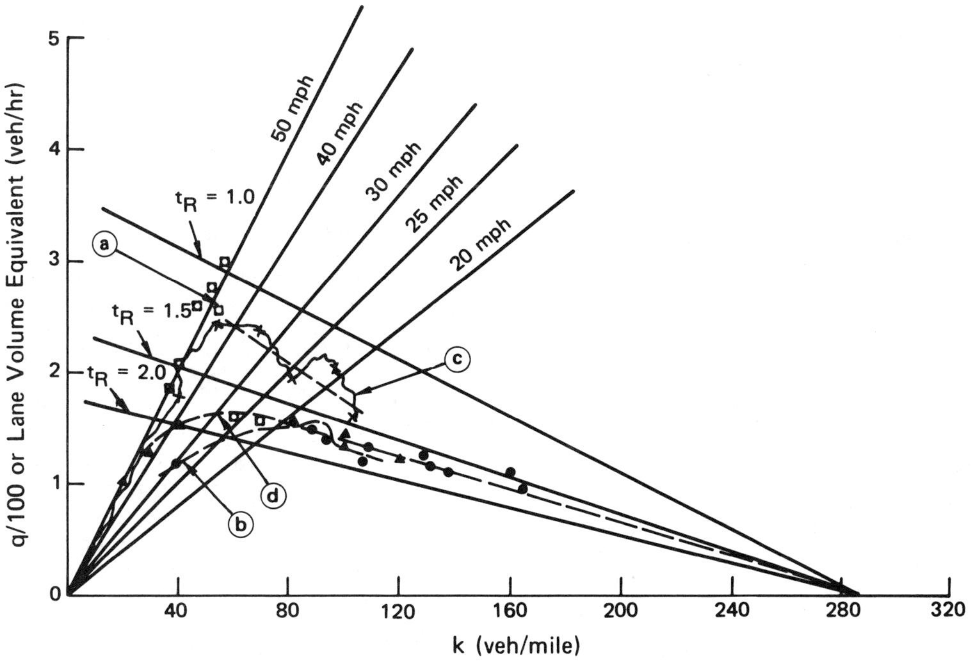

Figure 5.8 Four sets of empirical data plotted over theoretical trend lines of Figure 5.7.[31]

driver performance in platoons may be a Weber–Fechner relationship. Specifically, he suggests that headway and speed may be related by

$$h = t_r + \frac{(L+B)}{u} \quad (5.3)$$

where

h = headway (sec);
t_r = reaction time (taken as 1.488 sec);
L = length of lead vehicle (ft);
B = buffer distance between vehicles (ft);
$L+B$ = 35 ft; and
u = speed (ft/sec).

This relationship is illustrated in Figure 5.9. The low value of $L+B$ indicates a low speed. Such a speed would imply operation on the lower portion of the speed–flow curve. (See, for example, Figure 4.23.)

TABLE 5.4 Time Headways Before and After Deceleration of Lead Car [a]

Site	Lead Car Average Speed [b] (mph)	Average Time Headway (sec)	
		Before [b] Deceleration	After [c] Deceleration
1	30	1.8	2.5
2	30	1.5	2.6
3	30	1.5	2.4
4	33	1.5	2.3
5	28	1.6	2.6

[a] Compiled from Forbes et al.[47]
[b] Forbes' Table 2 (daylight).
[c] From Forbes' Table 3 (daylight).

5.5 MISCELLANEOUS DATA CONCERNING DRIVER PERFORMANCE

5.5.1 Overtaking and Passing

Figure 5.2 shows that driver eye fixations during passing tend to be of longer duration inasmuch as the driver has a greater level of concentration. Table 5.5 indicates some typical times for passing maneuvers.[33] Further information regarding passing may be found in references 34, 35, 36, 37, 38.

5.5.2 Information Content of Signs

The viewing time for a sign is often taken as about 1 sec. This limits the feasible content of a sign message to three or four short or easily recognized words.[39] The recent introduction (in the United States) of symbolic signs is expected to greatly facilitate information extraction from signs.

5.6 SUMMARY

The driving task consists of receiving information from the roadway, other cars, and the environment and of reacting to the various stimuli received by control of heading (steering) or control of acceleration (acceleration or braking). The driver uses a variety of senses to gather information required for the driving task, the most important of which is visual inputs. Whereas the information rate of the human eye can be up to 4.6×10^6 bits/sec, the human information processing channel has a maximum rate in the neighborhood of 25–35 bits/sec. Thus visual input could completely swamp the information processing channel if the driver did not do selective filtering.

TABLE 5.5 Typical Passing Times [a]

Type of Pass	Overtaken Car Speed (Medians)		Overtaken Car Speed (Means)					Overtaken Car Speed (Means)			
	30	50	0–19	20–29	30–39	40–49	50–59	25	30	40	50
Accelerative:											
Voluntary	10.0	11.5	8.7	8.8	9.8	10.8	10.5				
Forced	8.0	9.5	7.7	8.0	8.8	9.4	8.4	7.8	8.1	8.9	9.6
Flying:											
Voluntary	10.5	12.0	10.0	9.9	11.0	11.9	9.6				
Forced	8.0	10.5	8.1	8.9	9.8	11.8	9.3				

[a] Compiled by Rockwell[33] from four sources.

Indications are that the driver normally scans the roadway approximately 1.5 sec ahead.[18] If, however, the situation (highway and control task) is changing relatively slowly, the driver may look farther ahead and thereby increase his information input.

Indications are that the motor response time (e.g., move the foot from the accelerator to the brake) is relatively constant and that judgment and decision time (including perception and information processing) varies with the complexity of the situation and the decision to be made.

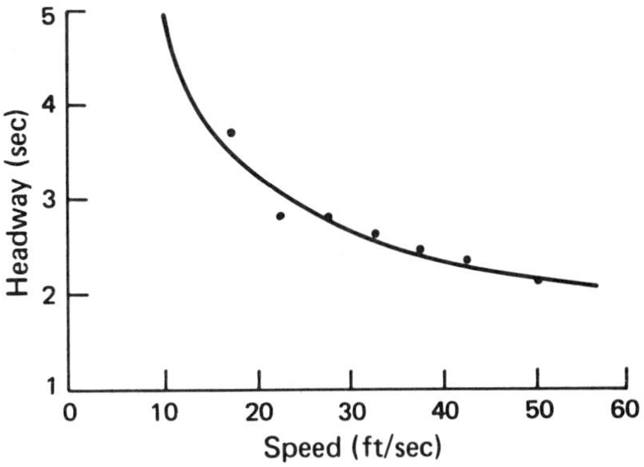

Figure 5.9 Plot of Daou's relationship for platoon behavior.[32]

5.7 REFERENCES

1. Allen, T. M., Lunenfeld, H., and Alexander, G. J., Drivers information needs. Highw. Res. Rec. 366 (1971), pp. 102–115.
2. Cumming, R. W., The analysis of skills in driving. Austr. Road Res., 1(9): 4–14 (1964).
3. Wohl, J. G., Man–machine steering dynamics. Human Fact., 3(6): 222–228 (1961).
4. McRuer, D., and Weir, D. H., Theory of manual vehicular control. Ergonomics, 12(4): 599–633 (1969).
5. Rashevsky, N., Mathematical biology of automobile driving. Bull. Math. Biophys., 29(1): 181–186 (1967).
6. Rockwell, T., Skills, judgment and information acquisition in driving, pp. 133–164. In T. W. Forbes (ed.) *Human Factors in Highway Traffic Research.* Wiley, New York (1971).
7. Mowbray, G. H., and Gebhard, J. W., Man's senses as informational channels, pp. 115–149. In H. W. Sinaiko (ed.) *Selected Papers on Human Factors in the Design and Use of Control Systems.* Dover, New York (1961).
8. Fogel, L. J., *Biotechnology: Concepts and Applications.* Prentice-Hall, Englewood Cliffs, N.J. (1963), p. 151.
9. Jacobson, H., The informational capacity of the human eye. Science, 113: 292–293 (1951).
10. Jacobson, H., The informational capacity of the human ear. Science, 112: 143–144 (1950).
11. Pew, R. W., Human information-processing concepts for system engineers. In R. S. Machol, W. P. Tanner, Jr., and S. N. Alexander (ed.) *System Engineering Handbook.* McGraw-Hill, New York (1965).
12. Fogel, L. J., *Biotechnology: Concepts and Applications.* Prentice-Hall, Englewood Cliffs, N.J. (1963), pp. 200–203.
13. Bekey, G. A., The human operator as a sampled-data system. IRE Trans. Human Fact. Electron., HFE-3(2): 43–51 (1962).
14. Senders, J. W., Kristofferson, A. B., Levison, W., Dietrich, C. W., and Ward, J. L., *An Investigation of Automobile Driver Information Processing.* Bolt, Beranek, and Newman Inc., Report No. 13335, April 26, 1966. Submitted to U.S. Bureau of Public Roads under contract CPR-11-0956. NTIS No. PB 170879.
15. Senders, J. W., and Ward, J. L., *Additional Studies on Driver Information Processing.* Bolt, Beranek, and Newman Inc., Final Report to Federal Highway Administration on contract CPR-11-5096, 15 April 1969. NTIS No. PB184499.
16. Senders, J. W., Kristofferson, A. B., Levison, W. H., Dietrich, C. W., and Ward, J. L., The attentional demand of automobile driving. Highw. Res. Board Rec. 195 (1967), pp. 15–33.
17. Shannon, C. E., and Weaver, W., *The Mathematical Theory of Communication.* University of Illinois Press, Chicago (1949).
18. Briggs, G. E., Driving as a skilled performance. *Proceedings of the Second Annual Automobile Insurance Industry*

Traffic Safety Research Symposium. (1968), pp. 123–143.

19. Senders, J. W., The estimation of operator workload in complex systems, pp. 207–216. In K. B. DeGreene (ed.) *Systems Psychology.* McGraw-Hill, New York (1970).
20. Welford, A. T., Arousal, channel-capacity and decision. Nature, 194 (4826): 365–366 (1962).
21. Szafran, J., Limitations and reliability of the human operator of control systems to process information. Aerosp. Med., 239–242 (Mar. 1966).
22. Brown, I. D., and Poulton, E. C., Measuring the spare "mental capacity" of car drivers by a subsidiary task. Ergonomics, 4 (1):35–40 (1961).
23. Fergenson, P. E., *A Study of the Relationship Between Driver Decision Time and Driving Accident and Violation Record.* Davidson Laboratory, Stevens Institute of Technology, Report SIT–DL–69–1329 (1968). NTIS No. PB188159.
24. Johansson, G., and Rumar, H., Drivers' brake reaction times. Human Fact., 13 (1):23–27 (1971).
25. Michaels, R. M., Perceptual factors in car following. *Proceedings of the Second International Symposium on the Theory of Traffic Flow* (London, June 1963). Published by OECD (1965), pp. 44–59.
26. Todosiev, E. P., and Fenton, R. E., Velocity thresholds in car-following at night. Highw. Res. Rec. 122 (1966), pp. 94–104.
27. Braunstein, M. L., and Laughery, K. R., Detection of vehicle velocity changes during expressway driving. Human Fact., 6 (4):327–331 (1964).
28. Hoffmann, E. R., Note on detection of vehicle velocity changes. Human fact., 8 (1):139–141 (1966).
29. Lee, J., and Jones, J. G., Traffic dynamics: visual angle car following models. Traffic Eng. Control, 9(7):348–350 (1967).
30. Denton, G. G., A subjective scale of speed when driving a motor vehicle. Ergonomics, 9 (3):203–210 (1966).
31. Forbes, T. W., Human factor considerations in traffic flow theory. Highw. Res. Rec. 15 (1963), pp. 60–66.
32. Daou, A., On flow within platoons. Austr. Road Res., 2 (7):4–13 (1966).
33. Rockwell, T., Skills, judgment, and information acquisition in driving, pp. 133–164. In T. W. Forbes (ed.) *Human Factors in Traffic Safety Research.* Wiley, New York (1972).
34. Farber, E., Silver, C. A., and Landis, D., Knowledge of closing rate versus oncoming-car speed as determiners of driver passing behavior. Highw. Res. Rec. 247 (1968), pp. 1–6.
35. Gordon, D. A., and Mast, T. M., Drivers' decisions in overtaking and passing. Highw. Res. Rec. 247 (1968), pp. 42–50.
36. Farber, E., and Silver, C. A., Behavior of drivers performing a flying pass. Highw. Res. Rec. 247 (1968), pp. 51–56.
37. Silver, C. A., and Farber, E., Driver judgment in overtaking situations. Highw. Res. Rec. 247 (1968), pp. 57–62.
38. Jones, H. R., and Heimstra, N. W., Ability of drivers to make critical passing judgments. Highw. Res. Rec. 122 (1966), pp. 89–92.
39. Forbes, T. W., Visibility and legibility of highway signs, pp. 95–109. In T. W. Forbes (ed.) *Human Factors in Traffic Safety Research.* Wiley, New York (1972).
40. Forbes, T. W. (ed.), *Human Factors in Traffic Safety Research.* Wiley, New York (1972).
41. Forbes, T. W., and Katz, M. S., *Summary of Human Engineering Research Data and Principles Related to Highway Design and Traffic Engineering Problems.* American Institute for Research, (1957), mimeo.
42. Beers, J., and Hulbert, S. F., *Judgment of Vehicle Speeds and Traffic Patterns.* UCLA Engineering Report 7281, June 1972, PB218925.
43. Stevens, S. S., The surprising simplicity of sensory metrics. Am. Psychol., 27: (1962).
44. Von Fieadt, G., *The World of Perception,* Dorsey Press (1966), pp. 169–170.
45. Forbes, T. W., and Simpson, M. E., Driver and vehicle response in freeway deceleration waves. Transp. Sci. 2 (1):77–104 (1968).
46. Attneave, F., *Applications of Information Theory to Psychology.* Holt, Rinehart and Winston, New York (1959), pp. 1–12.

47. Forbes, T. W., Zagorski, H. J., Holshouser, E. L., and Deterline, W. A., Measurement of driver reactions to tunnel conditions. Proc. Highw. Res. Board, 37:345–357 (1958).

48. Gordon, D. A., *Psychological Contributions to Traffic Flow Theory*. Draft report to FHWA staff (June 1973).

49. Haight, F. A., Bisbee, E. F., and Wojcik, C., Some mathematical aspects of the problem of merging. Highw. Res. Board Bull. 356 (1962), pp. 1–14.

5.8 RELATED LITERATURE

Baker, C. A., and Steedman, W. C., Perceived movement in depth as a function of luminance and velocity. Human Fact., 3:166–173 (1961).

Brown, I. D., Human factors in the control of road vehicles. Electron. Power, 14:275–279 (1968).

Bryant, J. F. M., Progress of research in human factors. Austr. Road Res., 3 (1):38–47 (1967).

Duckstein, L. A., Unwin, E. A., and Boyd, E. T., Variable perception time in car following and its effect on model stability. IEEE Trans. Man–Mach. Sys., MMS-11 (3):149–156 (1970).

Forbes, T. W., Factors in highway sign visibility. Traffic Eng., 39 (12):20–27 (1969).

Good, M. C., Rolls, K. J., and Joubert, P. N., Driver–vehicle behaviour in free-path turns. Transp. Res., 3 (1):43–68 (1969).

Gordon, D. A., Experimental isolation of the driver's visual input. Highw. Res. Rec. 122 (1966), pp. 19–34.

Gordon, D. A., The driver in single lane traffic. Highw. Res. Rec. 349 (1971), pp. 31–40.

Gordon, D. A., Traffic interactions and delays in freeway traffic. Highw. Res. Rec. 336 (1970), pp. 76–91.

Gordon, D. A., Experimental isolation of the driver's visual input, Human Fact., 8 (1): 129–137 (1966).

Haight, F. A., A mathematical model of driver alertness. Ergonomics, 15 (4):367–378 (1972).

Hoffman, E. R., The interaction between the driver, vehicle and road. Austr. Road Res. 3 (2):4–26 (1967).

Hulbert, S. F., and Burg, A., Human factors in transportation systems, pp. 471–509. In K. B. DeGreene (ed.) *Systems Psychology*. McGraw-Hill, New York, (1970).

King, G. F., and Lunenfeld, H., *Development of Information Requirements and Transmission Techniques for Highway Users*. National Cooperative Highway Research Program Report 123. Highway Research Board, Washington, D.C. (1971), p. 239.

Kolers, P. A., *Aspects of Motion Perception*. Pergamon Press, New York (1972).

Lazarus, R. S., Deese, J., and Osler, S. F., The effects of psychological stress upon performance. Psychol. Bull., 49 (4):293–317 (1952).

Llewellyn, K. R., Visual guidance of locomotion. J. Exp. Psychol., 91 (2):245–261 (1971).

Rashevsky, N., A note on the mathematical biology of automobile driving. Bull. Math. Biophys., 29 (1):187–188 (1967).

Rockwell, T. H., and Snider, J. S., *An Investigation of Variability in Driving Performance on the Highway*, Report RF–1450 17–31. Systems Research Group, Ohio State University, Columbus (1965).

Senders, J. W., Elkind, J. I., Grignetti, M. C., and Smallwood, R., *An Investigation of the Visual Sampling Behaviour of Human Observers*. Bolt, Beranek and Newman Inc., Report NASA Cr-434. Submitted to National Aviation and Space Administration under contract NAS 1–3860, April 1966. NITS No. N66–22910.

Senders, J. W., Human performance. Int. Sci. Technol., No. 55:58–60, 62, 64, 66–68, 89 (1966).

Snider, J. N., The capability of automobile drivers to sense vehicle velocity. Highw. Res. Rec. 159 (1967), pp. 25–35.

Steedman, W. C., and Baker, C. A., Perceived movement in depth as a stimulus size. Human Fact., 4:349–354 (1962).

Weir, D. H., and McRuer, D. T., A theory for driver steering control of motor vehicles. Highw. Res. Rec. 247 (1968), pp. 7–28.

Olson, P. L. The driver's reference point as a function of vehicle type, direction and radius of turn. Human fact., 6(4):319–325 (1964).

Wierwille, W. W., Gagne, G. A., and Knight, J. R., An experimental study of human operator models and closed-loop analysis methods for high-speed automobile driving. IEEE Trans. Human Fact. Electron., HFE-8 (3): 187–201 (1967).

Wooler, J., The measurement of driver performance. Ergonomics, 15 (1):81–87 (1972).

5.9 PROBLEMS

1. Assume that a lead car decelerates for a "stop" sign at a rate of 0.25 *g*. What is the latency in detecting the deceleration?

2. What would be the modal brake reaction time under surprise?

3. According to Daou, in a platoon traveling at 40 mph, what headway would you expect?

4. (a) Consider the 1-sec observations shown in Table 5.3 and determine the rate (in bits/sec) at which each driver acquires roadway information. (b) If a driver looks at the roadway for 1 sec, how many bits of roadway information does he have stored when he changes his fixation to, say, his rearview mirror? (Use median values of H and D.)

Chapter 6

CAR FOLLOWING AND ACCELERATION NOISE

6.1 INTRODUCTION

In Chapter 5 it was pointed out that the individual driver's actions result from his interpretation of the information received and his decision to make some response to that information. The driver's actions are limited to the control of acceleration (braking and accelerating) and the control of heading (steering). This chapter is concerned with the dynamics of a stream of traffic resulting from a series of drivers attempting to regulate their accelerations to accomplish a smooth safe trip.

The earliest attempts at estimating the capacity of a single lane on a roadway were based on assumptions of car-following behavior of individual drivers. The first edition of the *Highway Capacity Manual* [1] contains a synopsis of 23 early studies of highway capacity as reported between 1924 and 1941.

Nearly all of the calculations were based on the formula:

$$C = 5{,}280\, V/S \qquad (6.1)$$

where

C = capacity of a single lane (vehicles/hr);
V = velocity (miles/hr); and
S = average spacing (ft) from front bumper to front bumper of moving vehicles.

For the greater number of calculations of lane capacity the spacings were arrived at by assuming a car-following law in which the following driver adjusted his position relative to the lead vehicle in anticipation of a "brick-wall' stop by the lead vehicle. If the lead vehicle were to come to an instantaneous stop the following driver is assumed to have allowed himself a distance for stopping that included the distance traveled during braking, the distance traveled during reaction time, and the length of the lead vehicle. The average spacing is then given by:

$$S = aV^2 + bV + c \qquad (6.2)$$

where

S = average spacing, as previously defined;
V = velocity (mph);
a = a constant that is a function both of the assumed deceleration rate and conversion from mph units to fps units;
b = a constant that is a function both of the assumed reaction time and conversion from mph units to fps units; and
c = assumed distance (ft) front bumper to front bumper of stopped vehicles.

The models noted, however, were applicable only to the case of uniform velocity for each vehicle in the traffic stream and provided no insight into the behavior of a line of traffic when one of the vehicles in a line accelerates or decelerates and the following vehicles attempt to maintain some desired spacing. With the work of Reuschel [2,3] in 1950 and Pipes [4] in 1953, the analysis of car-following models was formalized, and operations research techniques were used to develop models of car following. This work was further extended by Kometani and Sasaki [5–8] in Japan and by Herman and his associates [9–13] at the General Motors Research Laboratories.

A principal effort of car-following studies has been that of trying to understand the behavior of a single-lane traffic stream by examining the manner in which individual vehicles followed one another. Studies of this nature have been used to examine control and communication techniques that will minimize the occurrence of rearend chain collisions in dense traffic.

6.2 DEVELOPMENT OF A CAR-FOLLOWING MODEL

Car-following models are a form of stimulus–response equation, where the response is the reaction of a driver to the motion of the vehicle immediately preceding him in the traffic stream. The response of successive drivers in the traffic stream is to accelerate or decelerate in proportion to the magnitude of the stimulus at time t and is begun after a time lag T.

The basis equation of these models is of the form

$$\text{Response } (t+T) = \text{Sensitivity} \times \text{Stimulus } (t)$$

What is the nature of the driver's response? To what stimulus does he react and how do we measure his sensitivity?

Consider the case of drivers in "dense" traffic who, because of traffic restrictions (no lane-changing permitted) or high volumes in adjacent lanes, are forced to follow the car immediately in front without passing.

It is assumed that a driver will space himself at a distance $s(t)$ from the lead vehicle such that in the event of an emergency stop by the first vehicle, the second vehicle will come to rest without striking the lead vehicle. The second driver allows himself a reaction time T, measured from the time t at which the lead driver initiates his stop, until the second driver initiates his own stopping maneuver.

The relative positions of the two vehicles at the time t, measured from the front bumper of each vehicle, are shown in Figure 6.1(a), where vehicle n is the lead car and $n+1$ the following car.

The relative positions of the two vehicles after the stopping maneuver are given in Figure 6.1(b), where

$x_n(t)$ = position of vehicle n at time t;
$s(t)$ = spacing between vehicles at time $t = x_n(t) - x_{n+1}(t)$;
d_1 = distance traveled by vehicle $(n+1)$ during reaction time $T = T\,u_{n+1}(t)$;
d_2 = distance traveled by vehicle $(n+1)$ during deceleration maneuver $= [u_{n+1}(t+T)]^2/2a_{n+1}(t+T)$;
d_3 = distance traveled by vehicle (n) during deceleration maneuver $= [u_n(t)]^2/2a_n(t)$;
L = distance from front bumper to front bumper at rest;
$u_i(t)$ = velocity of vehicle i at time t; and
$a_i(t)$ = acceleration of vehicle i at time t.

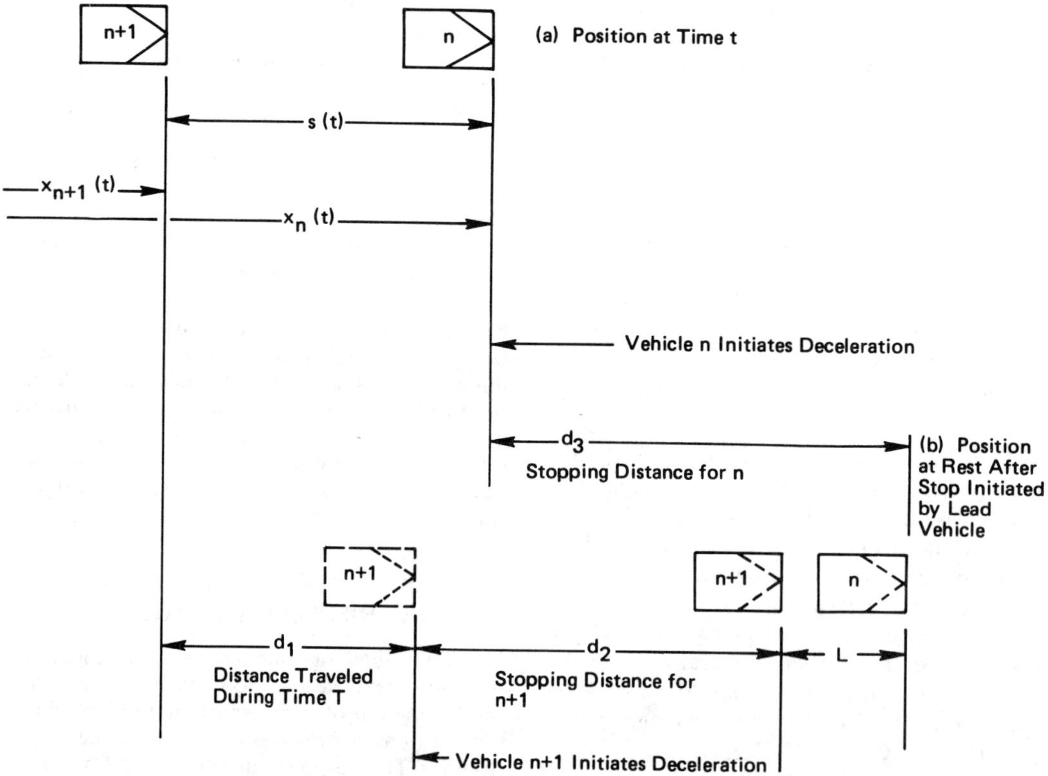

Figure 6.1 Positions of lead and following vehicles for emergency stop condition.

The desired spacing at time t, such that no collision will occur in event of a sudden stop, is then

$$s(t) = x_n(t) - x_{n+1}(t) = d_1 + d_2 + L - d_3 \quad (6.3)$$

Defining the velocity of a vehicle as

$$u(t) = \frac{dx(t)}{dt} = \dot{x}(t) \quad (6.4a)$$

and the acceleration as

$$a(t) = \frac{d^2x}{dt^2} = \ddot{x}(t) \quad (6.4b)$$

and substituting in Eq. 6.3 the appropriate relations for d_1, d_2, and d_3, gives

$$\begin{aligned} s(t) &= x_n(t) - x_{n+1}(t) \\ &= T\dot{x}_{n+1}(t) \\ &\quad + [\dot{x}^2_{n+1}(t+T)]/[2\ddot{x}_{n+1}(t+T)] \\ &\quad + L - [\dot{x}^2_n(t)]/[2\ddot{x}_n(t)] \quad (6.5) \end{aligned}$$

If the stopping distances and velocities of the two vehicles are assumed equal, so that $d_2 = d_3$, the spacing becomes

$$x_n(t) - x_{n+1}(t) = T\dot{x}_{n+1}(t+T) + L \quad (6.6)$$

which is the distance traveled by the following vehicle during reaction time (T), d_1, plus the separation between the front bumpers at rest, L. It will be observed that this spacing is less than that assumed in Eq. 6.2.

Differentiating with respect to (t),

$$\dot{x}_n(t) - \dot{x}_{n+1}(t) = T[\ddot{x}_{n+1}(t+T)] \quad (6.7)$$

so that the acceleration of the $n+1$st vehicle at time $(t+T)$ becomes

$$\ddot{x}_{n+1}(t+T) = T^{-1}[\dot{x}_n(t) - \dot{x}_{n+1}(t)] \quad (6.8)$$

which is of the form given previously,

Response $(t+T)$ = Sensitivity × Stimulus (t).

The response of the $n+1$st driver, which takes place at time $(t+T)$, is to accelerate (decelerate) by an amount proportional to the positive (negative) difference in the relative velocity of the nth and $n+1$st driver and the measure of sensitivity is given by T^{-1} (sec).

How well does a simple car-following model of the type given in Eq. 6.8 describe driver behavior? Assume two vehicles waiting at a traffic signal, with the front bumper of the second vehicle positioned 25 ft from the front bumper of the lead vehicle. The reaction time (T) of the drivers is taken to be 1.0 sec and the sensitivity (T^{-1}) is 1.0. At time 0.0, shortly after the signal changes to green, the first vehicle immediately (a physically impossible situation that is assumed because further calculations are simplified) moves away at 30.0 ft/sec. The second vehicle follows the first according to the rule given in Eq. 6.8.

Substituting appropriate values for T and T^{-1}, Eq. 6.8 may be written as:

$$\ddot{x}_{n+1}(t+1) = 1.0[\dot{x}_n(t) - \dot{x}_{n+1}(t)] \quad (6.9)$$

A direct analytical solution of Eq. 6.9 is cumbersome; it is easier to find a numerical solution. As an example, consider the case where the second vehicle attempts to follow a first vehicle that has instantly attained a velocity of 30.0 ft/sec from a stopped position as developed below. For this solution the position of the first vehicle is advanced 30 ft during each 1-sec time slice. At each time t separated by an increment of time Δt, the acceleration of the second vehicle is calculated (Eq. 6.9). All distances measured are from the stop line at location 0 ft.

Assuming that during each time slice Δt (1.0 sec in this example) the acceleration is uniform and equal to the average of the accelerations calculated at the start and end of each time slice, the equations of velocity $u_2(t)$ and position $x_2(t)$ are given by

$$\begin{aligned} \dot{x}_2(t) &= \dot{x}_2(t-\Delta t) \\ &\quad + \tfrac{1}{2}[\ddot{x}_2(t-\Delta t) + \ddot{x}_2(t)]\Delta t \quad (6.10) \end{aligned}$$

and

$$\begin{aligned} x_2(t) &= x_2(t-\Delta t) + \dot{x}_2(t-\Delta t)\Delta t \\ &\quad + \tfrac{1}{2}[\ddot{x}_2(t-\Delta t) + \ddot{x}_2(t)]\Delta t^2/2 \\ &= x_2(t-\Delta t) \\ &\quad + \tfrac{1}{2}[\dot{x}_2(t-\Delta t) + \dot{x}_2(t)]\Delta t \quad (6.11) \end{aligned}$$

where Δt is the time increment between successive calculations (sec).

A numerical solution of these three formulas is given in Table 6.1, where the relationship is solved at 1.0-sec increments. It is evident that vehicle 2 quickly reaches the speed of the lead vehicle and settles down to follow at a distance of about 55.0 ft from it, with only minor corrections in velocity and spacing after 7 or 8 sec have elapsed. The spacing of 55.0 ft is the same as that given in Eq. 6.6 when the reaction time is assumed to be 1.0 sec and the stopped spacing L between vehicles is 25 ft.

So far concern has been with the behavior of only the first and second cars waiting in a queue at a stop signal. What of a vehicle that is not immediately behind the lead vehicle but is the fourth or fifth vehicle in line at the moment the lead vehicle moved at 30 ft/sec? The behavior of a series of vehicles could be

TABLE 6.1 Car-Following Calculations

Time (sec)	$\dot{x}_1$ (ft/sec)	$\ddot{x}_2$ (ft/sec^2)	$\dot{x}_2$ (ft/sec)	$\dot{x}_1-\dot{x}_2$ (ft/sec)	x_1 (ft)	x_2 (ft)	x_1-x_2 (ft)
0.0	30.0	0.0	0.0	30.0	0	−25.0	25.0
1.0	30.0	30.0	0.0	30.0	30	−25.0	55.0
2.0	30.0	30.0	30.0	0	60	−10.0	70.0
3.0	30.0	0.0	45.0	−15.0	90	27.5	62.5
4.0	30.0	−15.0	37.50	−7.5	120	68.8	51.2
5.0	30.0	−7.50	26.25	3.75	150	100.6	49.4
6.0	30.0	3.75	24.375	5.625	180	125.9	54.1
7.0	30.0	5.625	29.062	0.938	210	152.7	57.3
8.0	30.0	0.938	32.345	−2.345	240	183.4	56.6
9.0	30.0	−2.345	31.641	−1.641	270	215.4	54.6
10.0	30.0	−1.641	29.648	0.352	300	246.0	54.0
11.0	30.0	0.352	29.004	0.996	330	275.3	54.7
12.0	30.0	0.996	29.678	0.332	360	304.7	55.3
13.0	30.0	0.322	30.342	−0.342	390	334.7	55.3
14.0	30.0	−0.342	30.332	−0.332	420	365.0	55.0
15.0	30.0	−0.332	29.995	0.005	450	395.2	54.8
16.0	30.0	0.005	29.831	0.169	480	425.1	54.9
17.0	30.0	0.169	29.918	0.082	510	455.0	55.0
18.0	30.0	0.082	30.046	−0.046	540	484.9	55.1
19.0	30.0	−0.046	30.064	−0.064	570	515.0	55.0
20.0	30.0	−0.064	30.009	−0.009	600	545.0	55.0

$$\ddot{x}_2(t+1) = 1.0[\dot{x}_1(t) - \dot{x}_2(t)]$$

$$\dot{x}_2(t) = x_2(t-1) + \frac{1}{2}[\ddot{x}_2(t-1) + \ddot{x}_2(t)]$$

$$x_2(t) = x_2(t-1) + \frac{1}{2}[\dot{x}_2(t-1) + \dot{x}_2(t)]$$

extended by extending the type of arithmetic calculations given in Table 6.1, but the calculations are lengthy and numerical accuracy dictates that the calculations be done at 0.1-sec intervals (see Fox and Lehman [14] for a discussion of solving the car-following problem with an electronic computer).

An alternate method for calculating the behavior of a platoon of vehicles for the car-following situation is to apply the Laplace transform to the solution of the problem. The particular problem postulated here, the behavior of a platoon of vehicles starting from rest when the velocity of the lead vehicle instantaneously becomes u_0 at time $t \geq 0$, has been solved by Kometani and Saski.[7] Further discussion of the application of the Laplace transform will be found in references 9, 15, 16.

A numerical solution of the behavior of the first five vehicles in a platoon, based on the equations of Kometani and Sasaki [7] is given in Table 6.2. The assumptions are the same as for the two-car situation, with $T=1.0$ sec and vehicles starting from a queue at rest with 25-ft spacing between vehicles, except these calculations are done at 0.1-sec intervals. The results are presented at 1.0-sec intervals, and the reader will observe that the location of the second vehicle x_2 is not exactly that calculated by using 1.0-sec intervals as in Table 6.1.

Although the simple rule for car following developed in Eq. 6.9 gave a reasonable pattern for the motion of the second vehicle relative to the first, it will be observed that the third and fourth vehicles will be involved in a rear-end collision about 90 ft and 7.0 sec from the starting point when the spacing (front bumper to front bumper) is reduced to less than 18.0 ft. It is evident that if all drivers were to follow the postulated behavior there would be a rash of rearend collisions at signalized intersections. The amplitude of the response to the instant change in velocity of vehicle 1 becomes greater for each successive vehicle and the system is said to be unstable.

6.3 TRAFFIC STABILITY

The question of stability in a platoon of vehicles is important in reviewing different patterns of car-following behavior. If driver behavior is to be modified, or the mechanical

TABLE 6.2 Solution of Car-Following Model

Time (sec)	$\dot{x}_1$ (ft/sec)	x_1 (ft)	$\dot{x}_2$ (ft/sec)	x_2 (ft)	$\dot{x}_3$ (ft/sec)	x_3 (ft)	$\dot{x}_4$ (ft/sec)	x_4 (ft)	$\dot{x}_5$ (ft/sec)	x_5 (ft)
0.0	30	0.0	0.00	−25	0.00	−50.0	0.00	−75.0	0.00	−100.0
1.0	30	30.0	0.00	−25	0.00	−50.0	0.00	−75.0	0.00	−100.0
2.0	30	60.0	30.00	−10	0.00	−50.0	0.00	−75.0	0.00	−100.0
3.0	30	90.0	45.00	27.5	15.00	−42.5	0.00	−75.0	0.00	−100.0
4.0	30	120.0	35.00	67.5	50.00	−10.0	5.00	−72.5	0.00	−100.0
5.0	30	150.0	23.75	96.9	58.75	44.4	36.25	−51.9	1.25	−99.4
6.0	30	180.0	25.25	121.4	29.00	88.2	78.00	−4.8	19.00	−89.2
7.0	30	210.0	31.71	149.8	6.96	106.23 [a]	109.75	89.1 [a]	70.08	−44.7

[a] Front ends of vehicles are separated by less than 18.0 ft (collision imminent).

devices in the vehicle or the signal system changed, it is important to determine that the system is stable; that is, that a change in velocity by the lead vehicle of a platoon will not be amplified by successive vehicles in the platoon.

The stimulus–response equation exemplified in Eq. 6.8 can be generalized as

$$\ddot{x}_{n+1}(t+T)=\alpha[\dot{x}_n(t)-\dot{x}_{n+1}(t)] \quad (6.12)$$

This is a linear car-following model because the response, acceleration (deceleration), is directly proportional to the stimulus, a positive (negative) difference in relative velocity of the nth and $n+1$st vehicles. Although more complex car-following models give better descriptions of observed traffic flows, the linear model given in Eq. 6.12 is most amenable to a theoretical analysis of stability.

Herman and co-workers [10,11] have discussed two conditions of stability—local and asymptotic. Local stability is concerned with the response of a vehicle to the change in motion of the vehicle immediately in front of it. It can be demonstrated by the pattern of spacing between vehicle 1 and vehicle 2 as in Table 6.1. In tracking the lead vehicle, the second vehicle first lost ground, then "overshot" at 5.0 sec, and continued to oscillate in damped amplitude about an ultimate "steady-state" spacing of 55 ft.

The manner in which a fluctuation in the motion of the lead vehicle is propagated through a line of vehicles is a function of asymptotic stability. From the data of Table 6.2 it is evident that the motion introduced to the first vehicle is propagated through the line of vehicles in a pattern of increasing amplitude, leading to the rearend collision between vehicles 3 and 4.

6.3.1 Local Stability

Herman and his associates [10,12] have identified the following situations for local stability in which:

$0 \leq (C=\alpha T) < 1/e$ (0.368)	spacing is nonoscillatory
$1/e \leq (C=\alpha T) < \pi/2$ (1.571)	damped oscillation of spacing
$(C=\alpha T)=\pi/2$	spacing is oscillatory with undamped oscillation
$(C=\alpha T)>\pi/2$	increasing amplitude in oscillation of spacing

As the value of C increases, the spacing between the two vehicles becomes increasingly unstable. If one reacts too strongly (large α, reflecting excessive throttle or brake-pedal response) to an event that occurred too far in the distant past (large response lag T), the situation at the moment of response may have changed to the point where the response is actually in the wrong direction.

The influence of the parameter $C(=\alpha T)$ on local stability is demonstrated in Figure 6.2. For values of 0.50 and 0.80, the spacing shows damped oscillation; at $\alpha T=\pi/2(=1.57)$, the spacing is oscillatory and undamped; at $\alpha T=1.60$, the spacing is oscillatory with increasing amplitude. The lead vehicle first decelerated and then accelerated back to its initial velocity, with an initial spacing of 70 ft between vehicles. Positions of the two vehicles were then calculated by Eq. 6.12 and the results plotted as shown.

For the example given in Table 6.1, $\alpha T=1.0(=T^{-1}\times T)$, indicating that the spacing

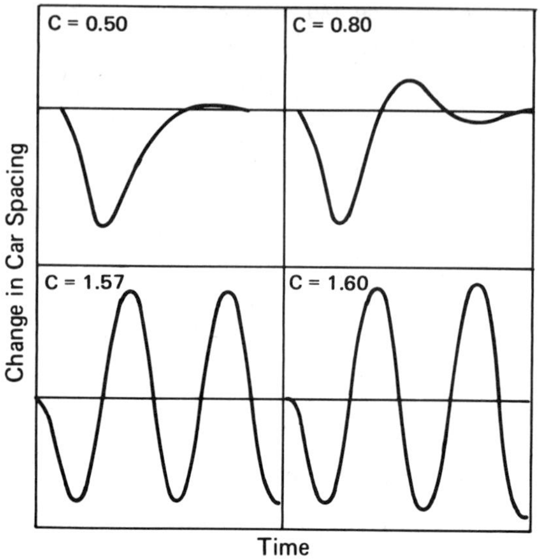

Figure 6.2 Change in car spacing of two cars.[10]

is oscillatory but damped, which can be readily verified by examining the final column of Table 6.1.

6.3.2 Asymptotic Stability

The limit for asymptotic stability has been investigated and reported by Chandler, Herman, and Montroll.[9] A line of traffic is asymptotically stable only when $C=\alpha T<\frac{1}{2}$, which may be compared with the limit for local stability, where $C=\frac{1}{2}$ indicated that the spacing is oscillatory but damps very quickly (Figure 6.2). The fluctuation in the motion of the lead car is propagated down the line of vehicles at a rate of α^{-1} sec/car.

A value of $\alpha T>\frac{1}{2}$ will propagate a disturbance with increasing amplitude, ultimately resulting in a rearend collision as demonstrated between vehicles 3 and 4 at about 7 sec elapsed time in Table 6.2. This effect is further demonstrated in Figures 6.3 and 6.4.

Figure 6.3 shows the spacing of successive pairs of vehicles for different values of αT and for a condition where the first car decelerated and then accelerated back to its initial velocity, with an initial spacing of 70 ft between vehicles.

Figure 6.4 shows the variation in spacing of a group of nine vehicles relative to a phantom lead car moving with constant speed at an initial spacing of 40 ft. Vehicle 1 decelerates and then accelerates back to its original velocity. In the process his spacing from the phantom vehicle is increased but is stable. Successive vehicles make corrections at a later time and with increasing amplitude of oscillation until a collision occurs between vehicles 7 and 8. For this example $\alpha=0.8$ and $T=2.0$ sec.

6.3.3 Propagation of a Disturbance

Herman and Potts[12] have reported a series of three experiments designed to investigate the way in which a disturbance is propagated down a line of cars. Eleven cars were driven in a line down a test track at about 40 mph. The lead car was suddenly braked, and the elapsed time between the appearance of the brake lights on the lead and sixth car t_6 and the time between the lead and eleventh car t_{11} were recorded. In experiment A the drivers were instructed to react only to the brake light of the car immediately in front of them; in B the drivers were to react to any braking stimulus; in C the brake lights were disconnected on all but the first and last vehicles. The results are summarized in Table 6.3.

The shortest time of propagation occurred in experiment B; the longest (about 1.0 sec/car), in experiment C. The difference in time between experiments A and C demonstrates the value of the brake light as a means of communicating the act of deceleration. It was pointed out earlier that the rate of propagation is α^{-1} sec/car. The propagation rate for experiment C ($\alpha^{-1}=1.01$) agrees closely with the assumptions made in developing the car-following example discussed in Section 6.1 ($\alpha^{-1}=1.0$).

6.4 NONLINEAR CAR-FOLLOWING MODEL

In the development of a linear car-following model and a brief review of stability, it has been assumed that for the relationship, Response = Sensitivity × Stimulus, the sensitivity α is a constant value. This would imply that for a given difference in velocity between a driver and the lead vehicle, his response would be independent of the spacing between the two vehicles. Gazis, Herman, and Potts[11] developed a more realistic model, in which they

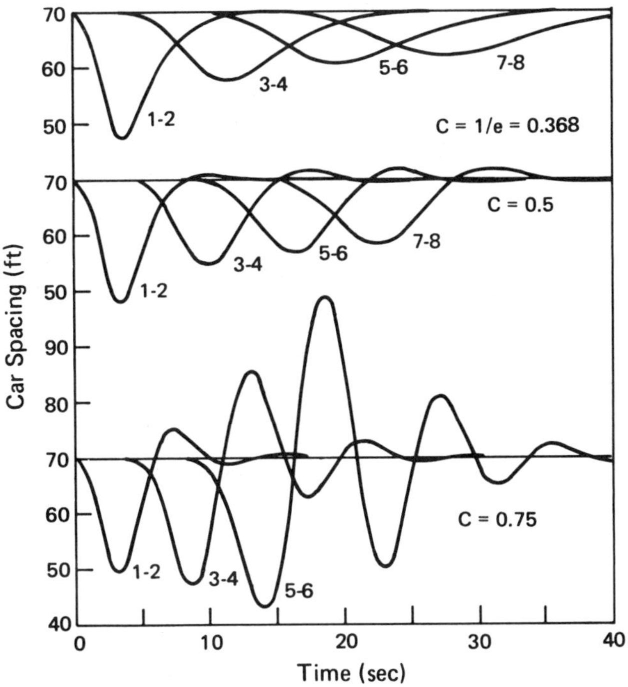

Figure 6.3 Car spacings of a line of cars with constant control for varying values of C ($=\alpha T$) and for a condition where the first car decelerated and then accelerated back to its initial velocity, with an initial spacing of 70 ft between vehicles.[10]

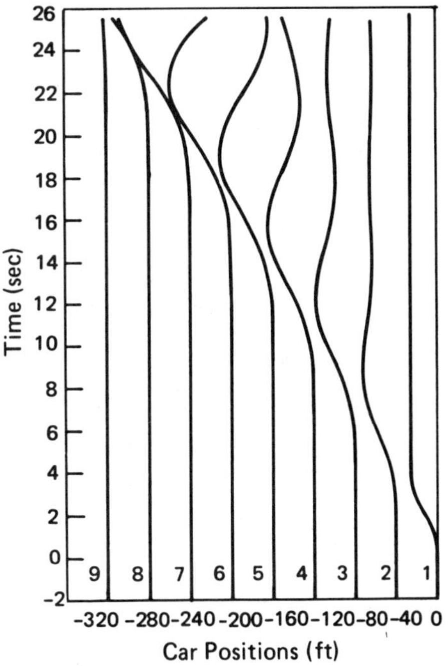

Figure 6.4 Asymptotic instability of a line of nine cars under the influence of velocity control ($\alpha = 0.8$, $T = 2.0$) relative to a phantom lead car moving with constant speed at an initial spacing of 40 ft. Vehicle 1 decelerates and then accelerates back to its original velocity. In the process its spacing from the phantom vehicle is increased but is stable. Successive vehicles make corrections at a later time and with increasing amplitude of oscillation until a collision occurs between vehicles 7 and 8.[10]

proposed that the sensitivity be inversely proportional to the spacing so that

$$\ddot{x}_{n+1}(t+T) = \{\alpha_0/[x_n(t) - x_{n+1}(t)]\} [\dot{x}_n(t) - \dot{x}_{n+1}(t)] \qquad (6.13)$$

where $\{\alpha_0/[x_n(t) - x_{n+1}(t)]\}$ is a measure of

TABLE 6.3 Time for Propagation (sec) of Fluctuation Down a Line of Cars

	Experiment A		Experiment B		Experiment C
Run Number	t_6	t_{11}	t_6	t_{11}	t_{11}
1	3.00	5.95	2.33	5.70	10.90
2	3.00	6.05	1.49	6.85	9.95
3	3.05	5.75	2.68	6.50	12.00
4	3.44	6.75	1.68	6.10	10.20
5	2.73	7.80	2.26	3.72	9.35
6	—	—	—	—	8.30
Avg. per car	0.61	0.65	0.42	0.58	1.01

the sensitivity and the units of α_0 are distance/time.

A number of car-following experiments were performed in the Holland and Lincoln Tunnels in New York and at the General Motors test track. Two vehicles were connected by a length of wire on a reel, permitting direct measurement of spacing and relative velocity simultaneously with measurements of the acceleration and velocity of the following vehicle. Different drivers were given instructions to follow the lead vehicle as they would under normal driving conditions. A summary of the values of α_0 and T found in the car-following experiments is given in Table 6.4.[12] As will be seen in section 6.5, α_0 is representative of the speed associated with maximum traffic volumes.

The values given are the averages for all of the drivers tested. In the General Motors experiment, for example, the value of T ranged from 1.0 to 2.2 sec.

6.5 FROM CAR-FOLLOWING TO TRAFFIC STREAM MODELS

The relationship between car-following models and traffic stream models of the type discussed in Chapter 4 was first examined by Gazis et al.[11] Assume that the lead vehicle in a stream of cars is proceeding at a constant velocity u and that each following vehicle proceeds at the same velocity and is separated from the preceding vehicle by a distance dictated by the drivers' perception and interpretation of a "safe" following distance. The platoon of cars will move along the roadway in a "steady-state" condition for which one can observe flow rate q, density k, and velocity u. Gazis et al. demonstrated that it is possible to derive equations of traffic stream flow directly from the laws of motion that are suggested by car-following theory. Basically, the procedure is to integrate an expression for the acceleration of the (n+1st) vehicle, giving an expression for the velocity of that vehicle, which in turn is the steady-state velocity of the traffic stream. The resulting equation of the velocity can then be solved for known boundary conditions, determining the constant of integration and then substituting in terms of the appropriate quantities, q and k, used in the definition of flow ($q=uk$) (Chapter 2).

TABLE 6.4 Summary of Car-Following Experiments

Location	Number of Drivers	α_0 (mph)	T (sec)
General Motors test track	8	27.4	1.5
Holland Tunnel	10	18.2	1.4
Lincoln Tunnel	16	20.3	1.2

Consider the application of the procedure to the simple linear car-following model given in Eq. 6.12:

1. Express the acceleration for the ($n+1$st) vehicle,

$$\ddot{x}_{n+1}(t+T)=\alpha[\dot{x}_n(t)-\dot{x}_{n+1}(t)]$$

2. Integrate the expression to obtain the velocity of the $n+1$st vehicle (the velocity of the traffic stream)

$$\dot{x}_{n+1}(t+T)=\alpha[x_n(t)-x_{n+1}(t)]+C_0$$

Because under steady state the velocity at time $(t+T)$ is the same as the velocity at time (t), the lag time T can be disposed of such that

$$\begin{aligned}\dot{x}_{n+1}=u&=\alpha[x_n-x_{n+1}]+C_0\\&=\alpha s+C_0\end{aligned}\qquad(6.14)$$

where $[x_n-x_{n+1}]$ is the average spacing between vehicles, $s(=1/k)$. The expression for velocity and spacing are for the average vehicle in the traffic stream.

3. Determine the constant of integration (C_0) by solving Eq. 6.14 for a known condition—in this case, when velocity $u=0$, spacing$=s_j=$jam spacing$=1/k_j=$(jam density)$^{-1}$. Therefore, $0=\alpha[1/k_j]+C_0$ and $C_0=-\alpha/k_j$.

4. Express u in terms of k by substituting for the value of C_0 and recalling that $s=1/k$ in Eq. 6.14—in this instance $u=\alpha[1/k-1/k_j]$.

5. Recalling the relationship ($q=uk$), we arrive at the steady-state equation

$$\begin{aligned}q&=k\alpha[1/k-1/k_j]\\&=\alpha[1-k/k_j]\end{aligned}\qquad(6.15)$$

6. Determine the proportionality constant α by solving for a known condition; in this case, when $k=0$ the value of q in Eq. 6.15 will be a maximum q_m so that α will equal q_m.

Note that the unit of α is time^{-1}=flow rate. The traffic stream model given in Eq. 6.15 indicates that the flow rate q is a maximum at density $k=0$. If $\alpha=(1\text{ sec})^{-1}$ then the maximum flow rate at $k=0$ will be 3,600 vehicles/hr. This is inconsistent with observed traffic stream data (see Figures 4.15 and 4.17), indicating that the car-following model of Eq. 6.12 is not a realistic model, at least at low densities.

The same procedure may now be applied to Eq. 6.13.

1. $\ddot{x}_{n+1}(t+T)=\alpha_0 \dfrac{\dot{x}_n(t)-\dot{x}_{n+1}(t)}{x_n(t)-x_{n+1}(t)}$

2. Integration $\left(\text{this is of the form } C\int \frac{du}{u}\right)$ results in $u=\alpha_0 \ln[x_n-x_{n+1}]+C_0$ $=\alpha_0 \ln[s]+C_0=\alpha_0 \ln[1/k]+C_0$.

3. Since for $u=0$, $k=k_j$
$$0=\alpha_0 \ln[1/k_j]+C_0,$$
$$C_0=-\alpha_0 \ln[1/k_j].$$

4.
$$u=\alpha_0 \ln[1/k]-\alpha_0 \ln[1/k_j]$$
$$=\alpha_0 \ln[1/k-\ln 1/k_j]$$
$$=\alpha_0 \ln[k_j/k] \qquad (6.16)$$

5.
$$q=u\,k$$
$$=\alpha_0 k \ln(k_j/k) \qquad (6.17)$$

6. To determine the proportionality constant α_0, refer to Figure 4.15 and Section 4.3.2 to determine the known physical conditions. Observe that slope $dq/dk=0$ at maximum volume q_m. Differentiation of Eq. 6.17 yields:

$$dq/dk=\alpha_0[k(k/k_j)(-k_j/k^2)+\ln(k_j/k)]=0$$
$$=\alpha_0[-1+\ln(k_j/k)]=0$$
$$=\alpha_0 \ln[k_j/ke]=0$$

where e = base of natural logarithms.

Assuming that $\alpha_0 \neq 0$ and defining k_m as the density at maximum flow q_m yields $\ln(k_j/k_m e)=0$ so that $k_j/k_m e=1$ and $k_m=k_j/e$.

Defining u_m as the velocity at q_m, $q_m=u_m k_m=u_m k_j/e$.

Substituting u_m and k_m in Eq. 6.16, gives $u_m=\alpha_0 \ln[k_j/k_m]=\alpha_0 \ln[k_j e/k_j]=\alpha_0 \ln e$, so that

$$\alpha_0=u_m.$$

Eq. 6.16 was first proposed by Greenberg,[17] who developed the relationship from the equation describing unidimensional flow in fluids and that denoting conservation of matter. The relationship was confirmed from experimental observations of flow, density, and velocity. The Greenberg equation of state (section 7.4.1), using the application of a fluid flow analogy, is based on a macroscopic approach and is mathematically equivalent to Eq. 6.13, which is based on the principles of car following, a microscopic approach. This model has been verified by experimental data (Figure 4.15).

6.6 GENERAL EXPRESSION FOR CAR-FOLLOWING MODELS

Continuing examination of the relationship between microscopic and macroscopic models led to a generalized form of the car-following equation:

$$\ddot{x}_{n+1}(t+T)=\alpha_0 \frac{\dot{x}^m_{n+1}(t+T)}{[x_n(t)-x_{n+1}(t)]^l}[\dot{x}_n(t)-\dot{x}_{n+1}(t)] \qquad (6.18)$$

where l and m are constants.

First proposed by Gazis, Herman, and Rothery,[13] the general expression has been further examined by May and Keller.[18]

Integration of the generalized equation (Eq. 6.18) by Gazis et al.[13] has given

$$f_m(u)=\alpha_0 f_l(s)+C_0 \qquad (6.19)$$

where u is the steady-state speed of a stream of traffic, s is the constant average spacing, and α_0 and C_0 are appropriate constants consistent with physical restrictions. Further,

$$f_m(u)=u^{l-m} \quad \text{for } (m\neq 1)$$

and

$$f_m(u)=\ln u \quad \text{for } (m=1) \qquad (6.20)$$

$$f_l(s)=s^{1-l} \quad \text{for } (l\neq 1)$$

and

$$f_l(s)=\ln s \quad \text{for } (l=1) \qquad (6.21)$$

The value of C_0 is related either to free speed u_f (the velocity of a single vehicle whose speed is not influenced by interaction with other vehicles) or velocity at jam spacing, $s_j=1/k_j$.

The values of C_0 are shown to be

$$C_0=f_m(u_f) \quad \text{for } m>1,\ l\neq 1 \text{ or } m=1,\ l>1 \qquad (6.22)$$

and

$$C_0=\alpha_0 f_l(s_j) \quad \text{for all other combinations of } m \text{ and } l \text{ (except } m=1,\ l>1 \text{ not bounded)} \qquad (6.23)$$

The traffic stream model equations resulting from the application of Eqs. 6.19–6.23 are shown in Table 6.5 and Figure 6.5. These steady-state solutions were originally derived independently of car-following assumptions. However, as shown previously, they have a

TABLE 6.5 Steady-State Flow Equations (q–k)[a]

l	Equation of State	Reference
	$m=0$	
0	$q=\alpha[1-k/k_j]$ $\alpha=q_m=1/\text{reaction time}$	Chandler et al.[9] Pipes [4]
1	$q=\alpha k \ln[k_j/k]$ $\alpha=$ velocity at optimum flow (u_m)	Greenberg [17] Gazis et al.[13]
3/2	$q=\alpha k[1-(k/k_j)^{1/2}]$ $\alpha=$ velocity at free flow (u_f)	Drew [19]
2	$q=\alpha k[1-k/k_j]$ $\alpha=u_f$	Greenshields [20]
	$m=1$	
2	$q=\alpha k e^{(k/k_0)}$ $\alpha=u_f$; $k_0=$density at optimum flow	Edie [21]
3	$q=\alpha k e-\frac{1}{2}(k/k_0)^2$ $\alpha=u_f$	Drake et al.[22]

[a] Based on May and Keller.[18]

direct correspondence with car-following models. A generalized discussion of the steady-state flow for various values of m and l is given in Gazis et al.,[13] but only those combinations presented in Table 6.5 have been verified by observations of vehicle flow on roadways.

May and Keller [18] have also examined the case for noninteger values of m and l, proposing a model with $m=0.8$, $l=2.8$ when fitted to data on the Eisenhower Expressway in Chicago. The following steady-state flow equation results:

$$q=ku_f[1-(k/k_j)^{1.8}]^5 \quad (6.24)$$

Values for m and l have evolved as various investigators attempted to fit observed data to proposed models of driver and/or stream flow behavior. The case $m=0$, $l=0$ evolved from the "simple" linear car-following situation developed in Eq. 6.8. The same model had been developed by Pipes [4] and analyzed for stability by Herman et al.[10,12]

Further examination of experimental data based on car-following experiments led to the hypothesis that a driver does not have a constant reaction to a stimulus by the lead vehicle but that the reaction will vary inversely as the distance between the subject vehicles.[11] This is the case for $m=0$, $l=1$, which led directly to a comparison with the Greenberg fluid flow analogy data.

Greenshields' [20] analysis of traffic flow, which corresponds to the case $m=0$, $l=2$, was first developed from photographic observations of traffic flow made in 1934. Values for flow rate and mean velocity were calculated for 100-vehicle groupings; density was calculated from the flow rate and velocity information. A simple straight-line fit between velocity and density was deduced from a plot of the data.

Although the steady-state flow equation resulting from this analysis does not fit observed data as well as some other models listed here, it does have the advantage of being amenable to calculation and manipulation.

An independent verification of the Greenshields model is reported by Pipes and Wojcik [23] in which perceptual factors are related to the car-following model. (A similar derivation is shown in Fox and Lehman.[14] In Figure 6.6 let θ be the visual angle subtended by the lead vehicle. Taking w as the width of vehicle (and neglecting the distance L, the length of the lead vehicle),

$$\theta=\frac{w}{s} \quad (6.25)$$

Differentiating with respect to t,

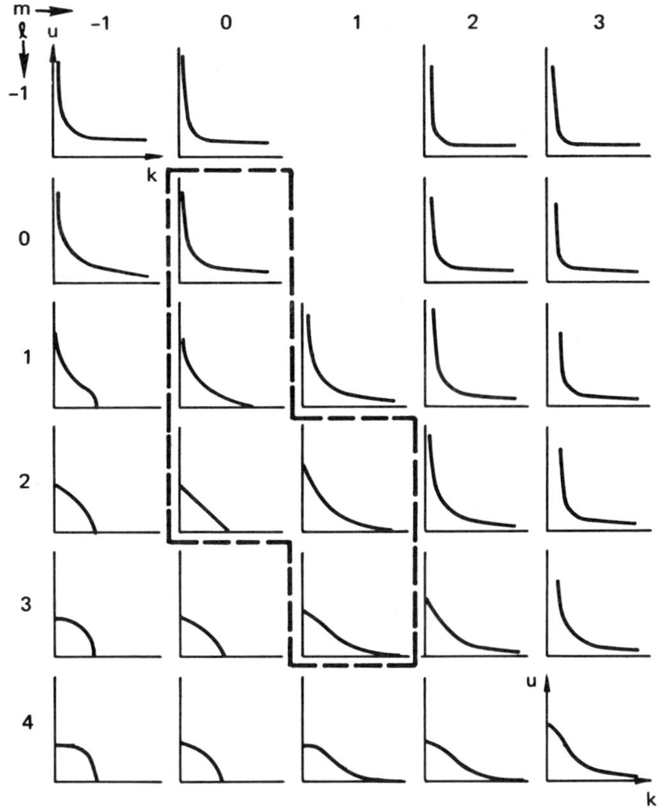

Figure 6.5 Matrix of speed-density relationships for various *m*, *l* combinations of the general car-following equation.[18] **(Dashed lines enclose limiting values of *l* and *m* used in Table 6.5.)**

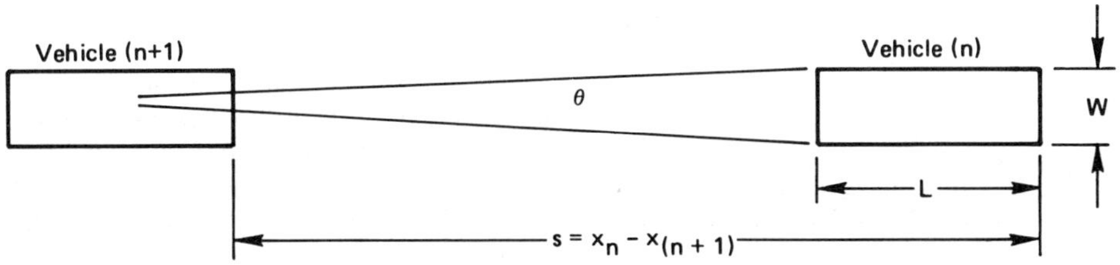

Figure 6.6 Conditions of the Pipes and Wojcik[23] **verification of the Greenshields traffic flow model.**

$$\frac{d\theta}{dt}=\dot{\theta}=-\frac{w}{s^2}\frac{ds}{dt} \qquad (6.26)$$

If it is assumed that the acceleration for the following vehicle is proportional to the driver's perception of the rate of change of the visual angle, $\dot{\theta}$,

$$\ddot{x}_{n+1}=c(-\dot{\theta}) \qquad (6.27)$$

which implies acceleration if θ is negative and deceleration if $\dot{\theta}$ is positive. Because $s=x_n-x_{n+1}$ and $ds/dt=\dot{x}_n-\dot{x}_{n+1}$, substituting Eq. 6.26 in Eq. 6.27 yields

$$\dot{x}_{n+1}=c\frac{(\dot{x}_n-\dot{x}_{n+1})}{(x_n-x_{n+1})^2} \qquad (6.28)$$

where $c=CW$, a constant.

Further discussion concerning the rate of change in θ for the acceleration and deceleration case is given by Michaels.[24] Drew [19] developed a generalized equation of steady-state flow (for which the Greenberg and Greenshields models are special cases) including a "parabolic" model as well as the Greenberg and Greenshields models. This equation was tested on the Gulf Freeway in Houston, Texas. Time-lapse aerial photography was used to measure speed and concentration.

Statistical analysis of the regression equations for the relationship between u and k indicated that all three models yielded satisfactory fits to the data. The car-following model ($m=0$, $l=3/2$) is the microscopic equivalent of the parabolic model.

Edie [21] noted that the car-following model of Gazis et al.[11] and Greenberg's equivalent model [17] were not applicable at low density, with the free-flow velocity approaching infinity as the density approached zero. His proposal would increase driver sensitivity as the driver velocity increases and decreases sensitivity as the square of the separation.

Edie demonstrated the validity of the relationship by relating his modification of the car-following model to data reported by Gazis et al.[11]

The model of Drake et al.[22] ($m=1$, $l=3$) is based on a regression analysis of speed–density–flow data obtained on the Eisenhower Expressway in Chicago. A further discussion of this model by May and Keller [18] suggests that this model may have particular value when flow rates are greater than 1,800 vehicles/hr/lane.

6.7 EXTENSIONS AND MODIFICATIONS OF CAR-FOLLOWING MODELS

6.7.1 Three-Car Experiments

In the previous development the assumption was made that the following driver receives clues from the immediately preceding vehicle. It can also be hypothesized that a driver also receives clues from the second car ahead. Eq. 6.12, the linear car-following model, can be modified to include the influence of a second lead vehicle, as follows:

$$\ddot{x}_{n+2}(t+T)=c_1[\dot{x}_{n+1}(t)-\dot{x}_{n+2}(t)]+c_2[\dot{x}_n(t)-\dot{x}_{n+2}(t)] \qquad (6.29)$$

where c_1 is the sensitivity to the velocity difference between the third and second vehicles and c_2 is the sensitivity to the velocity difference between the third and first vehicles.

Herman and Rothery [25] report on experiments in which three cars were physically connected by steel wires permitting direct measurement or calculation of relative speed and distance between the vehicles as they were driven on a test track. Experiments involving only two vehicles had indicated that the value of T for drivers on the test track was about 1.6 sec. Although it might be anticipated that knowledge of a second car would permit drivers to decrease their reaction time, such was not the case; the reaction time derived from the three-vehicle experiments was still 1.6 sec. Best correlations between the model and observed data for response between the first and third vehicles only occurred when T equaled 2.3 sec. The authors conclude that although it is not possible to show that a driver follows only the immediately preceding vehicle, the stimulus provided by that lead vehicle is probably the most significant input in the car-following model.

Fox and Lehman [14] incorporated the effect of a second lead car in their computer simulation model, as follows:

$$\ddot{x}_{n+2}(t+T)=\alpha_1\dot{x}_{n+2}(t+T)\left\{\frac{W_1[\dot{x}_{n+1}(t)-\dot{x}_{n+2}(t)]}{[x_{n+1}(t)-x_{n+2}(t)]^2}+\frac{W_2[\dot{x}_n(t)-\dot{x}_{n+2}(t)]}{[x_n(t)-x_{n+2}(t)]^2}\right\} \qquad (6.30)$$

where α_1 is a sensitivity factor and $W_1+W_2=1$. When $W_1=1$, there is no effect for a second lead car; and in several conditions, which were simulated for later comparison with field data, this was the case. The influence of following a second car ahead is discussed in considerable detail by the authors.

6.7.2 Asymmetry for Acceleration and Decleration

In all previous discussions the implicit assumption has been made that a driver adjusting to a change in velocity will accelerate and decelerate at the same rate for a given stimulus. In extrapolating from personal experience it is seen that the deceleration capabilities of most passenger cars are greater than the acceleration capabilities and that in congested traffic greater action is taken when the vehicle(s) in front is (are) decelerating rather than accelerating relative to the subject vehicle. Herman and Rothery [25] examined this hypothesis by weighting the negative relative speeds by values in the range 1 to 1.25. The linear model (Eq. 6.12) was modified so that

$$\ddot{x}_{n+1}(t+T)=\alpha_{+}[\dot{x}_n(t)-\dot{x}_{n+1}(t)] \quad \text{for relative velocity positive} \tag{6.31a}$$

and

$$\ddot{x}_{n+1}(t+T)=\alpha_{-}[\dot{x}_n(t)-\dot{x}_{n+1}(t)] \quad \text{for relative velocity negative} \tag{6.31b}$$

Herman and Rothery tested this model for the ratio range (α_{-}/α_{+}) between 0.5 and 1.5. Data from 40 car-following test runs were fitted to the model, and an improvement in fit was found to occur in the range $1 \leq (\alpha_{-}/\alpha_{+}) \leq 1.25$. The average weighting factor was ≈ 1.1 for runs made in the Holland and Lincoln Tunnels of New York City and on a test track.

Forbes and his associates [26,27] have reported on simulated studies of platoon behavior and air photo records of urban freeway sections.[28] As a result of the simulation runs (two vehicles following a lead vehicle "closely but safely as if anxious to get home in heavy traffic"), the authors found that time headways after an experimental (and unexpected) deceleration by the lead vehicle were about twice as large as the time headways before the deceleration. As the lead car accelerated after the slowdown the following drivers allowed a greater time and distance gap to develop, from which it may be inferred that reaction time to acceleration is greater than for deceleration.

Newell,[29,30] building on the findings of Forbes et al., has hypothesized and discussed a car-following model that considers different forms for acceleration and deceleration. Newell considers that delay in responding to a stimulus may be a "consequence of laziness or intentional failure of drivers to respond to every stimulus, rather than some inherent limitation on reaction times."

Newell's approach to the car-following model is illustrated in Figure 6.7. Consider a vehicle that has been following a lead vehicle at a constant velocity and at a desired spacing. The lead vehicle accelerates but the following vehicle is in no hurry to close the spacing, being satisfied with the longer spacing until some further change takes place in the velocity (and therefore spacing) of the lead vehicle. When the lead car decelerates, the following car allows some of the excess gap to dissipate before reacting to the deceleration.

In Figure 6.7 it is assumed that there are two velocity–headway relationships of the form

$$\dot{x}_{n+1}(t+T)=G_{n+1}[x_n(t)-x_{n+1}(t)] \tag{6.32}$$

where G_{n+1} is a function selected to represent the empirical relationship between velocity and headway (one for acceleration and a second for deceleration, as shown by the solid lines). The solid lines are connected by a family of curves, one through each point in the region of the velocity–headway space between the two curves. These connecting curves (dashed line) have small (or zero) slope, indicating a small (or zero) change in velocity as a function of the change in spacing.

The following example illustrates the use of the curves. Assume two cars have an initial velocity V_1 and a spacing given by the deceleration curve (point 1). The lead car accelerates to a velocity V_2. The following car accelerates only slightly, letting the spacing increase as shown at point 2. When that occurs the following driver accelerates according to the acceleration curve shown until he has reached a velocity V_2 and the associated spacing (point 3). When the lead car decreases velocity to V_1, the following driver decelerates only slightly, allowing the gap to decrease to point 4. At point 4 the driver initiates a more affirmative deceleration movement (along the deceleration curve) until the velocity and spacing have returned to the initial condition (point 1).

It is important to note that in this model the stimulus to the driver is spacing (not relative velocity as previously discussed) and that there are different relationships for the acceleration and decleration cases.

Newell further relates his hypothesis to the formation of shock waves in dense traffic, showing his model to be consistent with observations of shock waves in tunnel flow. Newell has indicated only the structure of the theory; he has not attempted to obtain quantitative results.[29, p. 53]

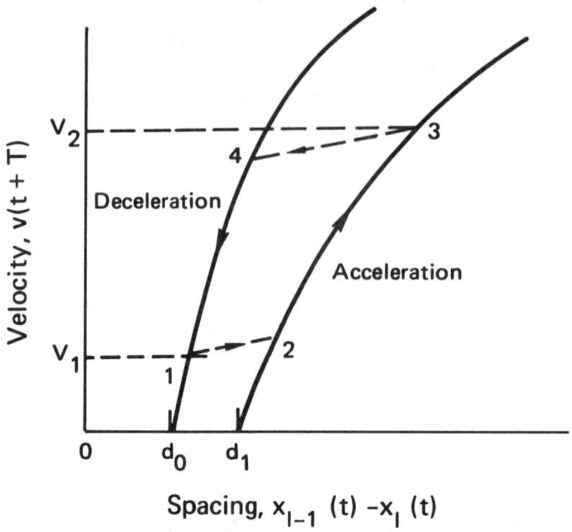

Figure 6.7 Velocity-headway relationships during acceleration and deceleration.[30]

6.8 FURTHER APPLICATION OF CAR-FOLLOWING MODELS

In previous sections of this chapter applications have related to the explanation of observed traffic phenomena, particularly in conditions of dense traffic such as observed in tunnels or on crowded highways. Car-following models have also been used to evaluate aids to driver car following, to examine the behavior of platoons of buses as may be operated on an exclusive freeway lane, to anticipate the effect of short cars on the flow and speed of downtown traffic, and to examine safety in car following.

6.8.1 Applications to Driver-Aided Car Following

One method of improving lane capacity would be to give a driver more information, permitting him to decrease his reaction time and therefore follow the lead vehicle at a lesser headway. Fenton and Montano [31] report the results of an experiment in which additional information about the lead car was given to a driver by means of a tactile device built into the single control stick used for steering, acceleration, and deceleration (as opposed to a steering wheel, accelerator pedal, and brake pedal as normally used). The tactile device was a finger flush with the control stick at the correct spacing, recessed when headways were too small and protruding when headways were greater than desired. Drivers were able to reduce tracking error, very nearly duplicating the motion of the lead driver, but at the same time it was observed that asymptotic stability was not obtained under the conditions for optimum tracking.

An earlier experiment in improving driver information is reported by Bierley.[32] Drivers were given a visual display that showed spacing in one instance and a combination of spacing and relative velocity in a second instance. The display was used to supplement any clues the driver might normally obtain from observing a lead vehicle.

In the first instance (spacing information only added to normal clues) there was improvement (reduction) in absolute spacing error and in variability in spacing error. However, there was no significant reduction in reaction times or maximum spacing change, indicating that the simple display was no real improvement over no display at all.

In the second instance (spacing + relative velocity information added to normal clues) following performance was significantly improved. Absolute spacing error, absolute spacing changes, variability in spacing error, and reaction time were all reduced. Bierley does not make an analysis of asymptotic stability for this case, but the results of Fenton and Montano would suggest that the improved car following is not coincident with asymptotic stability. Bierley suggests a car-following model of the form:

$$\ddot{x}_{n+1}(t+T)=\alpha[\dot{x}_n(t)-\dot{x}_{n+1}(t)] + k[x_n(t)-x_{n+1}(t)] \qquad (6.33)$$

where α and k are sensitivity constants for relative speed and relative spacing, respectively.

In a comprehensive study and review of the "car-following" problem as applied to improving flow (safety and volume) Rockwell and Treiterer [33] suggest a control system in which the acceleration of the following vehicles is given by:

$$\ddot{x}_{n+1}(t+T)=\alpha[\dot{x}_n(t)-\dot{x}_{n+1}(t)]+K\ddot{x}_n(t) \qquad (6.34)$$

The relationship suggests that the acceleration, after some lag (T), is a function of difference in velocity plus a term indicating that the following car exactly duplicates the acceleration of the lead car after a lag of T sec. (This latter component is termed "acceleration control" by Rockwell and Treiterer.) The parameter K varies from 0 to 1.0, representing the proportion of the control that is acceleration control. For $K=1.0$ (total acceleration control) each vehicle would exactly duplicate the acceleration of the preceding vehicle after a lag of T sec, a system that cannot be controlled by drivers without the aid of supplementary devices.

6.8.2 Single-Lane Bus Flow

An application of car-following theory to an analysis of single-lane flow of heavy vehicles (large buses were used in the experiment) is reported by Rothery et al.[34] Car-following experiments, similar to those involving pairs of automobiles, were conducted with pairs of buses directly connected by the apparatus used to measure the required parameters of car-following models.

Three versions of the general expression for car following (Eq. 6.18) were analyzed. The three models evaluated were for the following cases:

Model	*l*	*m*
linear	0	0
reciprocal spacing	1	0
reciprocal spacing–speed	2	1

Of the three models tested, the reciprocal spacing model provided the best fit to the data. For the 22 drivers involved, the value of lag time (T) ranged from 0.4 to 1.6 sec (the higher value occurred during a test conducted in the rain). These values can be compared with those found in Table 6.4, where the average value of T is about 1.4 sec. The value of α_0 for the reciprocal spacing model was approximately 36 mph for expressway facilities, substantially greater than the values of about 20 mph found for passenger vehicles in the vehicular tunnels as reported in Table 6.4.

An analysis of stability for the bus-following experiment showed that all of the data points were in the region of asymptotic stability, whereas in previous automobile experiments only about one-half of the data points fell in this region.

Applicability of the car-following model to steady-state flow was tested by observations of platoons ranging from two to ten buses. The platoon experiments verified the predictions of the bus-following model; i.e., maximum flows at a rate of 1,450 buses/hr at a constant speed of 33 mph were measured.

6.8.3 Effect of Small Cars on Downtown Traffic

Two major advantages are usually claimed for small cars: they occupy less space when parked and they will reduce congestion. McClenahan and Simkowitz[35] have estimated the anticipated reduction in congestion by computer simulation of a single lane of traffic consisting of a number of cars with identical acceleration and performance characteristics, but with varying lengths.

The simulation modeled the behavior of a file of cars down one lane of a street with arbitrarily fixed-cycle traffic signals. Driver behavior was based on a car-following model plus a model of a lead-car driver's behavior. The model was designed to reflect driver reaction to the traffic signal indication if it took precedence over the stimulus received from the lead car. When the lead car provided the stimulus, the driver behavior was based on the model given in Eq. 6.13 (the reciprocal spacing model). The results of the simulation are synopsized in Table 6.6. They indicate that the flow increased by 70 percent and the velocity by 57 percent if all small cars (10 ft long) are substituted for all long cars (20 ft long), and congestion is such that there is a queue of 15 cars at each light.

The validity of the simulation model was verified by data collected during a Friday rush-hour period on a three-block section of Walnut Street in downtown Philadelphia. Results of the field experiment corresponded closely with the simulation results.

6.8.4 Safety in Car Following

A further example of simulating car-following behavior on a digital computer is given by Fox and Lehman.[14] The purpose of the study was to investigate those driver and vehicle characteristics that are most important in eliminating rearend collisions. The car-following model used is that given by Eq. 6.18 with $m=1$, $l=2$. Refinements included the option of including the next-ahead vehicle as a stimulus, a driver sensitivity factor that reflected whether the driver was accelerating or decelerating, and a distance "threshold" beyond which the driver did not adjust his velocity as dictated by the car-following equation.

The implications drawn from the simulation studies were concerned with the reduction in accidents that might follow by sharpening

TABLE 6.6 Flow and Velocity as a Function of Traffic Concentration and Vehicle Size [a]

Number of Cars at Each	Percentage of 10-ft Cars (remaining cars are 20 ft)											
	0		10		20		33		50		100	
Light	Flow [b]	Vel [c]	Flow [b]	Vel [c]	Flow [b]	Vel [c]	Flow [b]	Vel [c]	Flow [b]	Vel [c]	Flow [b]	Vel [c]
5	1154	13.2	—	—	—	—	—	—	—	—	1400	14.4
10	955	6.3	—	—	—	—	—	—	—	—	1390	8.4
15	732	3.5	745	3.4	855	4.0	855	4.0	1005	4.5	1240	5.5
20	745	3.6	—	—	—	—	—	—	—	—	1154	4.0

[a] From McClenahan and Simkowitz.[35]
[b] Flow converted to vehicles/hr of green.
[c] Velocity in ft/sec.

drivers' perceptions and minimizing response lag. The simulation demonstrated that the quantities most efficient for system behavior are (1) driver reaction time, (2) desired spacing, and (3) threshold boundary for relative velocity perception. Specific suggestions for improving or minimizing the effect of these three quantities are given in Fox and Lehman, Chapter 4.[14]

6.9 ACCELERATION NOISE

It is reasonable to assume that a driver will attempt to maintain a uniform velocity when he is traveling along a roadway. Even at low volumes on a limited-access roadway, however, he will fluctuate from his desired velocity. In the presence of high volumes, where his velocity and acceleration are a function of the car-following laws, or in the urban situation, where traffic controls dictate his velocity, there will be greater fluctuations about his desired velocity.

A measure of the fluctuations of a driver is given by the standard deviation σ of the acceleration about the mean acceleration and is defined as the acceleration noise. The mathematical definition of this quantity, assuming mean acceleration to be zero, is

$$\sigma = \left\{ (1/T) \int_0^T [a(t)]^2 dt \right\}^{1/2} \qquad (6.35)$$

where $a(t)$ is the acceleration (positive or negative) at time t and T is the total time in motion. An alternative form, in which acceleration is sampled at successive time intervals (Δt) becomes

$$\sigma = [(1/T)\Sigma[a(t)]^2 \, \Delta t]^{1/2} \qquad (6.36)$$

A "smooth" trip will have minor deviations, a "rough" trip greater deviations from the mean acceleration.

The concept of acceleration noise developed as a result of car-following studies [9, 10] and was further analyzed by Jones and Potts [36] who suggested that the parameter might be used to give partial answers to such questions as, "How much safer and more economical is a four-lane dual highway than a twisty two-lane road?"; or "Are teenage drivers more reckless than other drivers?"; or "How much congestion is produced by increasing traffic volume and the general side activity generated by a shopping area?"

6.9.1 Calculation of Acceleration Noise

Eqs. 6.35 and 6.36 do not lend themselves to ease of calculation of data collected in field studies. An equation for acceleration noise, which is adaptable to reduction and analysis of data, is

$$\sigma = \left[\frac{(\Delta u)^2}{T} \sum_{i=1}^{K} \frac{n_i^2}{\Delta t_i} - \left(\frac{V_T - V_0}{T} \right)^2 \right]^{1/2} \qquad (6.37)$$

where T is the time in motion for the trip segment, Δu is taken as a constant increment of velocity change (mph), Δt_i is the time interval (sec) for a change in velocity of magnitude $n_i \Delta u$ (n is integer), and V_0 and V_T represent the velocity (mph) at the start and end of the trip segment. K represents the number of segments of uniform acceleration,

$$\left(\sum_{i=1}^{K} \Delta t_i = T \right)$$

where Δt_i and n_i are measured along a trace of velocity versus time, with measurements being recorded for every change in acceleration. For a long trip, or a trip where V_T and V_0 are nearly identical, the second term of the equation can be neglected. The derivation of this equation is given by Drew, Dudek, and Keese.[37]

Data may be collected by connecting a recording pen directly to a vehicle speedometer cable or to a fifth wheel and tracing the velocity directly on a moving strip of paper, where the distance the paper moves is proportional to time. A hypothetical trace of this type is shown in Figure 6.8. Data may also be collected by recording the total distance traveled (ft) at fixed intervals (about 1 sec) and then calculating velocity and acceleration from these distance–time measurements. The distance may be recorded on film, advancing at one frame per second, or on tape, where the distance is printed or punched at the proper time interval. Except for punch tape that may be read directly into a computer, the data reduction is lengthy and tedious.

A trace of a velocity–time graph for a hypothetical vehicle proceeding on a rural highway and then entering an urban area at about 5.0 min from the start of the record is shown in Figure 6.8. The initial velocity V_0 is 54 mph and at 30 sec the driver begins to decelerate for a curve with an advisory speed limit of 45 mph, gradually accelerating again to 60 mph at 96 sec. He continues at this velocity until 132 sec, at which time he is forced to adjust his speed to a slower-moving vehicle. He continues with minor velocity fluctuations until 246 sec and then decelerates to adjust to urban traffic and a speed-zone restriction. For the second 5-min interval the driver is subject to a rapid series of accelerations and decelerations, including a complete stop for a traffic signal 450 sec after the start of the record. The acceleration noise for the first 5 min of the record is determined as in Table 6.7.

In the example it is assumed that $\Delta u = 2.0$ mph and the acceleration noise is calculated for the first 5-min interval as shown in Figure 6.8. It is convenient to use a table of values of $n^2/\Delta t$ (such as Table 6.8) in order to calculate the value of $n^2/\Delta t$ progressively on a desk calculator. Substituting the appropriate values in

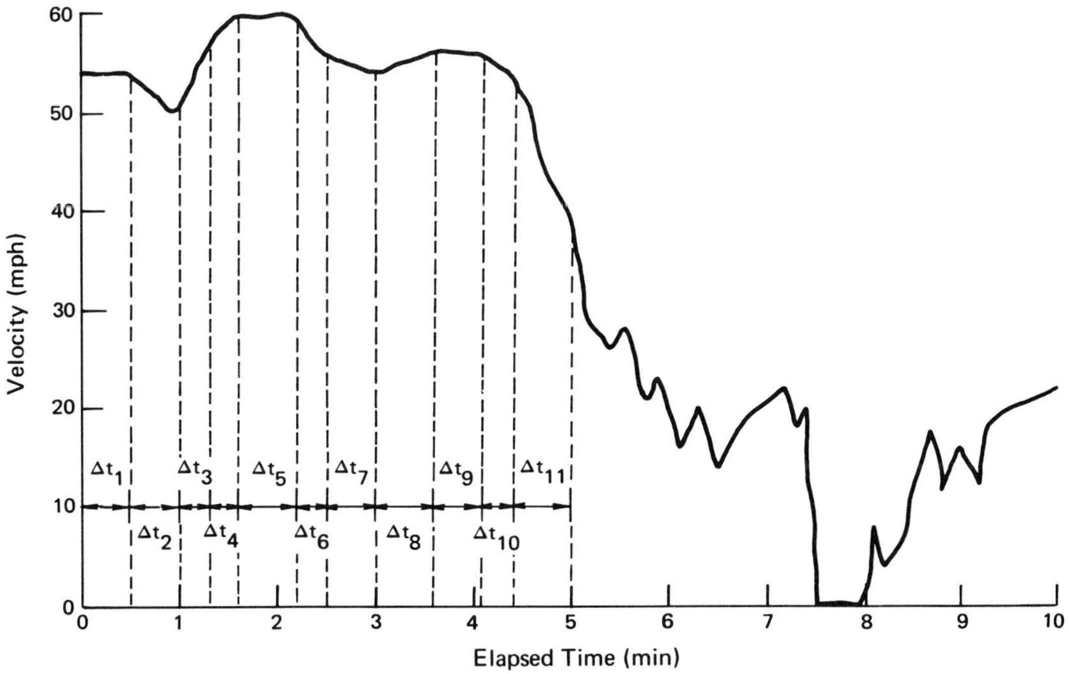

Figure 6.8 Velocity trace over 10-min time interval.

Eq. 6.37, recalling that V_0 and V_T must be converted from mph to ft/sec, $\sigma = \{(8.60/300) \times 2.55 - [(40\text{-}54)/300]^2 \times (88/60)^2\}^{1/2} = (0.068)^{1/2} = 0.26$ ft/sec².

By contrast, the acceleration noise for the second 5-min period is ≈ 0.88 ft/sec.² For the second 5-min period, $T = 270$ sec; because T is defined as the time in motion, the interval from 450 to 480 sec is not included in the calculation of acceleration noise.

TABLE 6.7 Example Showing Calculation of Acceleration Noise

Interval	Elapsed Time at End of Interval (sec)	Velocity u at End of Interval (mph)	n	Δt_i (sec)	$n_i^2/\Delta t_i$
0	0	54	—	—	—
1	30.0	54	0	30.0	0.00
2	60.0	50	2	30.0	0.13
3	78.0	56	3	18.0	0.50
4	96.0	60	2	18.0	0.22
5	132.0	60	0	36.0	0.00
6	150.0	56	2	18.0	0.22
7	180.0	54	1	30.0	0.03
8	216.0	56	1	36.0	0.03
9	246.0	56	0	30.0	0.00
10	264.0	54	1	18.0	0.06
11	300.0	40	7	36.0	1.36
TOTAL				300.0	2.55

If Δt is in seconds, the running time T in seconds, and $\Delta u = 2.0$ mph, $(\Delta u)^2 = (2.08 \times 88/66)^2 \approx 8.60$ ft²/sec.²

TABLE 6.8 Values of $n^2/\Delta t$

	Value of $n^2/\Delta t$							
Δt	$n=1$	$n=2$	$n=3$	$n=4$	$n=5$	$n=6$	$n=7$	$n=8$
1	1.00	4.00	9.00	16.00	25.00	36.00	49.00	64.00
2	0.50	2.00	4.50	8.00	12.50	18.00	24.50	32.00
3	0.33	1.33	3.00	5.33	8.33	12.00	16.33	21.33
4	0.25	1.00	2.25	4.00	6.25	9.00	12.25	16.00
5	0.20	0.80	1.80	3.20	5.00	7.20	9.80	12.80
6	0.17	0.67	1.50	2.67	4.17	6.00	8.17	10.67
7	0.14	0.57	1.29	2.29	3.57	5.14	7.00	9.15
8	0.13	0.50	1.13	2.00	3.13	4.50	6.13	8.00
9	0.11	0.44	1.00	1.78	2.78	4.00	5.44	7.11
10	0.10	0.40	0.90	1.50	2.50	3.60	4.90	6.40
11	0.09	0.36	0.82	1.45	2.27	3.27	4.45	5.82
12	0.08	0.33	0.75	1.33	2.08	3.00	4.08	5.33
13	0.08	0.31	0.69	1.23	1.92	2.77	3.77	4.92
14	0.07	0.29	0.64	1.14	1.79	2.57	3.50	4.57
15	0.07	0.27	0.60	1.07	1.67	2.40	3.27	4.27
16	0.06	0.25	0.56	1.00	1.56	2.25	3.06	4.00
17	0.06	0.24	0.53	0.94	1.47	2.12	2.88	3.76
18	0.06	0.22	0.50	0.89	1.39	2.00	2.72	3.56
19	0.05	0.21	0.47	0.84	1.32	1.89	2.58	3.37
20	0.05	0.20	0.45	0.80	1.25	1.80	2.45	3.20

6.9.2 Acceleration Noise Related to Roadway and Traffic

Acceleration noise is mainly influenced by three factors—the driver, the road, and traffic conditions. An aggressive driver, with frequent and relatively large speed changes, will have greater "noise" than a passive driver. A narrow, winding road or a signalized urban street will show greater and more frequent changes in velocity than will occur on a multilane freeway. Finally, a driver in congested traffic will generate more acceleration noise than that obtained at low traffic volumes, as demonstrated by the example shown in Figure 6.8.

Jones and Potts [36] measured acceleration noise over different roads, varying traffic conditions, and different drivers. They reported the following conclusions:

1. For two roads through hilly country, σ is much greater for a narrow 2-lane road than for a 4-lane dual highway.
2. For a road in hilly country, σ is greater for a downhill journey than for an uphill one.
3. For two drivers driving different speeds below the design speed of a highway, σ is much the same.
4. If one or both drivers exceed the design speed, σ is greater for the faster driver.
5. Increasing traffic volume increases σ.
6. Increasing traffic congestion produced by parking cars, stopping busses, cross traffic, crossing pedestrians, etc., increases σ.
7. The value of σ may be a better measure of traffic congestion than travel times and stopped times.
8. High values of σ indicate a potentially dangerous situation.

Although Jones and Potts caution against the use of arbitrary interpretation of values of σ, they make the observation that $\sigma = 0.7$ ft/sec^2 is a low value and $\sigma = 1.5$ ft/sec^2 is a high value.

Helly and Baker [38] modified the calculation of acceleration noise by proposing a parameter

$$G = \frac{\sigma}{\nu} \tag{6.38}$$

where G is the velocity gradient, σ is the acceleration noise, and ν is the mean velocity on the trip. It was reasoned that the acceleration noise alone does not describe the quality of the trip, that a fast trip and a slow trip could both have the same value of σ but the fast trip would be more desirable. The time T used in the calculations of σ was the total trip time, including stop time.

Underwood [39] reported results of observations made in Australia that suggest that G seems to be a better measure of traffic congestion than σ, but contradicts item 7 of the conclusion of Jones and Potts [36] by questioning whether the velocity gradient is any better than total travel time as a measure of congestion.

6.9.3 Acceleration Noise of a Vehicle in Traffic

The acceleration noise of an isolated vehicle was discussed in Section 6.9.1. In Sections 6.1–6.7, several simple car-following laws for traffic were exhibited. Clearly, the total acceleration noise of a vehicle in traffic is a superposition of its natural noise (the acceleration noise at very low traffic volume) and its response to that of its predecessors through the law of following. The total acceleration noise of vehicles at different locations in a platoon has been measured by Herman and Rothery [40] (Figure 6.9). It is noted that traffic has broadened the acceleration distribution function so that the acceleration noise far down the platoon is about three times that of the lead car, which is effectively moving freely on the road. Figure 6.9 also shows that if the traffic stream is proceeding in a stable manner, where stability means the damping out of small

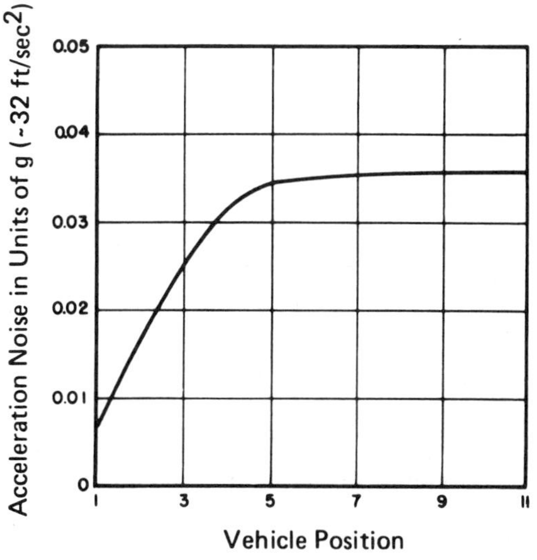

Figure 6.9 Acceleration noise of vehicles at different locations in a platoon.[40]

disturbances, as discussed in Section 6.3.2, the increase of the noise of a given lead vehicle is damped out by the time the signal of its motion has propagated down to the fifth or sixth car behind it. Traffic broadens the acceleration distribution increasing the value of σ, the broadening being smaller for the conservative driver who is satisfied to follow the stream than for the "cowboy" who by weaving attempts to drive 5–10 mph faster than the stream. This is shown in Figure 6.10 for traffic on Woodward Avenue in Detroit.[10]

The traffic broadening is not large for smoothly flowing traffic, but the dispersion increases rapidly at the onset of congestion. For stop-and-go traffic the dispersion is small because cars are unable to accelerate to appreciable speeds.

The broadening of the acceleration distribution by traffic depends on the parameters of the law of following, because, as previously noted, the acceleration of the nth car at time t is a superposition of its natural acceleration noise and its response to the motion of its predecessor. Montroll [41] has shown that in smoothly moving traffic separation distance varies only slightly from the equilibrium distance s; hence, Eq. 6.13 can be linearized so that addition of the natural acceleration $\beta(t)$ gives

$$\ddot{x}_{n+1}(t+T)=C[\dot{x}_n(t)-\dot{x}_{n+1}t]+\beta(t) \quad (6.39)$$

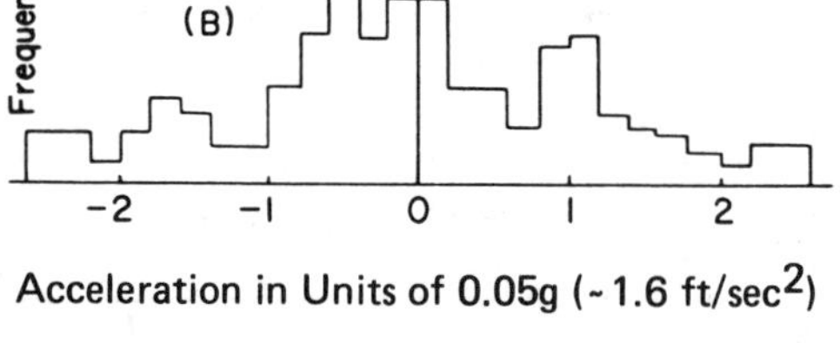

Figure 6.10 Acceleration distribution functions for a driver (A) moving with a traffic stream at approximately 35 mph and (B) attempting to drive 5 to 10 mph faster than the stream average.[10]

in which

$$C=\alpha_0/[x_n(t)-x_{n+1}(t)]=\alpha_0/s \quad (6.40)$$

The $\beta(t)$ is a random value of the acceleration a, whose value at time t is not specified. It is determined by its distribution function $f(a)$ so that $f(a)da$ is the probability that $\beta(t)$ has a value between a and $a+da$ at time t. For simplicity, assume that $\beta(t)$ has the same distribution for all drivers on the road of interest. One can use the standard methods of the theory of Brownian motion to determine the statistical differences of properties of $a_n(t)=\ddot{x}_n(t)$ from those of $\beta(t)$ in terms of C and T (lag time). If the acceleration noise is peaked in the low-frequency range, one finds that the dispersion σ of the distribution function of $a_n(t)$ (as $n\to\infty$; i.e., for cars far from the beginning of a platoon) is related to the natural noise dispersion σ_0 of $\beta(t)$ by

$$\sigma=\sigma_0/(1-2CT)^{1/2} \qquad \text{if } 2CT<1 \quad (6.41)$$

The stability condition ($CT<1/2$) again makes its appearance. (See section 6.3.2.) The closer the traffic reaches the limit of stability ($2CT\to 1$), the larger is the traffic broadening of the acceleration noise.

If Eq. 6.40 is substituted in Eq. 6.41, the average spacing may be expressed as

$$s=2CT/[1-(\sigma_0/\sigma)^2] \quad (6.42)$$

This equation was checked with the Holland Tunnel observations of Herman, Potts, and Rothery. The traffic broadening of the acceleration noise dispersions σ/σ_0 in the tunnel varied from about 1.50 to 1.75, depending on the density during the experiment. The average time lag of 1.5 sec, which was observed in car-following experiments, was substituted in Eq. 6.42, as was the observed ratio σ/σ_0. The computed values of s were then converted into appropriate densities ($s=1/k$), which were compared with the observed densities made at the same time as σ/σ_0 was determined. These calculated values generally did not deviate from the measured ones by more than 10 or 15 percent.

6.10 REFERENCES

1. *Highway Capacity Manual.* U.S. Government Printing Office, Washington, D.C. (1950), pp. 2–3.
2. Reuschel, A., Vehicle movements in a platoon . Oesterreichishes Ing.-Arch., 4: 193–215 (1950) (In German).

3. Reuschel, A., Vehicle movements in a platoon with uniform acceleration or deceleration of the lead vehicle. Z. Oesterreichischen Ing.-Archit.-Ver., 95: 59–62, 73–77 (1950) (In German).
4. Pipes, L. A., An operational analysis of traffic dynamics. J. Appl. Phys., 24 (3): 274–281 (1953).
5. Kometani, E., and Saski, T., On the stability of traffic flow (report-I). Oper. Res. Soc. Jap., 2 (1): 11–26 (1958).
6. Kometani, E., and Sasaki, T., Dynamic behavior of traffic with a nonlinear spacing–speed relationship. Proceedings, Symposium on the Theory of Traffic Flow. Elsevier, Amsterdam (1961), pp. 105–119.
7. Kometani, E., and Sasaki, T., A safety index for traffic with linear spacing. Oper. Res., 7 (6): 704–720 (1959).
8. Kometani, E., and Sasaki, T., Car following theory and stability limit of traffic volume. Oper. Res. Soc. Jap., 3 (4): 176–190 (1961).
9. Chandler, R. E., Herman, R., and Montroll, E. W., Traffic dynamics: studies in car following. Oper. Res., 6 (2): 165–184 (1958).
10. Herman, R., Montroll, E. W., Potts, R. B., and Rothery, R. W., Traffic dynamics: analysis of stability in car-following. Oper. Res., 7 (1):86–106 (1959).
11. Gazis, D. C., Herman, R., and Potts, R. B., Car-following theory of steady state flow. Oper. Res., 7 (4):499–505 (1959).
12. Herman, R., and Potts, R. B., Single-lane traffic theory and experiment. Proceedings, Symposium on the Theory of Traffic Flow. Elsevier, Amsterdam (1961), pp. 120–146.
13. Gazis, D. C., Herman, R., and Rothery, R. W., Nonlinear follow-the-leader models of traffic flow. Oper. Res., 9 (4):545–567 (1961).
14. Fox, P., and Lehman, F. G., *Safety in Car-Following—A Computer Simulation.* Newark College of Engineering, New Jersey (1967), 173 pp.
15. Chow, T., Operational analysis of a traffic dynamics problem. Oper. Res. 6 (6): 827–834 (1958).
16. Oliver, R. M., Catoire, B., and Skeates, R. S., On a mathematical function of traffic flow theory. Proceedings, Second International Symposium on the Theory of Road Traffic Flow (London, 1963). OCED (1965), pp. 12–26.
17. Greenberg, H., An analysis of traffic flow. Oper. Res., 7 (1):79–85 (1959).
18. May, A. D., Jr., and Keller, H. E. M., Non-integer car-following models. Highw. Res. Rec. 199 (1967), pp. 19–32.
19. Drew, D. R., Deterministic aspects of freeway operations and control. Highw. Res. Rec. 99 (1965), pp. 48–58.
20. Greenshields, B. D., A study in highway capacity. Proc. Highw. Res. Board, 14: 448–477 (1934).
21. Edie, L. C., Car-following and steady-state theory for noncongested traffic. Oper. Res., 9 (1):66–76 (1961).
22. Drake, J. S., Schoefer, J. L., and May, A. D., Jr., A statistical analysis of speed density hypotheses. Highw. Res. Rec. 154 (1967), pp. 53–87.
23. Pipes, L. A., and Wojcik, C. K., A contribution to theory of traffic flow. SAE Conference Proceedings—Analysis and Control of Traffic Flow Symposium (Detroit, 1968), pp. 53–60.
24. Michaels, R. M., Perceptual factors in car following. Proceedings, Second International Symposium on the Theory of Road Traffic Flow (London, 1963). OECD (1965), pp. 44–59.
25. Herman, R., and Rothery, R. W., Car following and steady state flow. Proceedings, Second International Symposium on the Theory of Road Traffic Flow (London, 1963). OCED (1965), pp. 1–11.
26. Forbes, T. W., Zagorski, H. J., Holshouser, E. L., and Deterline, W. A., Measurement of driver reactions to tunnel conditions. Proc. Highw. Res. Board, 37: 345–357 (1958).
27. Forbes, T. W., Human factor considerations in traffic flow theory. Highw. Res. Rec. 15 (1963), pp. 60–66.
28. Forbes, T. W., and Simpson, M. E., Driver and vehicle response in freeway deceleration waves. Transp. Sci., 2 (1): 77–104 (1968).
29. Newell, G. F., Theories of instability in dense highway traffic. J. Oper. Res. Soc. Jap., 5:9–54 (1963).
30. Newell, G. F., Instability in dense highway traffic, a review. Proceedings, Second International Symposium on the Theory

of Road Traffic Flow (London, 1963), Organization for Economic Co-operation and Development (1965), pp. 73–83.

31. Fenton, R. E., and Montano, W. B., A study of driver-aided car-following. Proceedings of Fourth International Symposium on Theory of Traffic Flow (Karlsruhe, Germany, June 1968). Published as *Beitrage zur Theorie des Verkehrsflusses, Strassenbau und Strassenverkehrstechnik* (1969), Vol. 86, pp. 1–7.
32. Bierley, R. L., Investigation of an intervehicle spacing display. Highw. Res. Rec. 25 (1963), pp. 58–75.
33. Rockwell, T. H., and Treiterer, J., Sensing and communication between vehicles. National Cooperative Highway Research Program Report 51. Highway Research Board, Washington, D.C. (1968).
34. Rothery, R., Silver, R., Herman, R., and Torner, C., Analysis of experiments on single-lane bus flow. Oper. Res., 12 (6):913–933 (1964).
35. McClenahan, J. W., and Simkowitz, H. J., The effect of short cars on flow and speed in downtown traffic: a simulation model and some results. Transp. Sc., 3 (2):126–139 (1969).
36. Jones, T. R., and Potts, R. B., The measurement of acceleration noise—a traffic parameter. Oper. Res., 10 (6): 745–763 (1962).
37. Drew, D. R., Dudek, C. L., and Keese, C. J., Freeway level of service as described by an energy-acceleration noise model. Highw. Res. Rec. 162 (1967), pp. 30–85.
38. Helly, W., and Baker, P. G., Acceleration noise in a congested signalized environment. *Vehicular Traffic Science.* Proceedings of the Third International Symposium on the Theory of Traffic Flow (1965). American Elsevier, New York (1967), pp. 56–61.
39. Underwood, R. T., Acceleration noise and traffic congestion. Traffic Eng. Control, 10 (3):120–123, 130 (1968).
40. Herman, R., and Rothery, R. W., Microscopic and macroscopic aspects of single lane traffic flow. Jour. Oper. Res. Soc. Jap., 5:74–93 (1962).
41. Montroll, E. W., Acceleration noise and clustering tendency of vehicular traffic. Theory of Traffic Flow. Proceedings of the Symposium on the Theory of Traffic Flow (1959). Elsevier, New York (1961), pp. 147–157.

6.11 RELATED LITERATURE

Bender, J. G., and Fenton, R. E., On the flow capacity of automated highways. Transp. Sci., 4 (1):52–63 (1970).

Bexelius, S., An extended model for car following. Transp. Res., 2 (1):13–21 (1968).

Brill, E. A., A car-following model relating reaction times and temporal headways to accident frequency. Transp. Sci., 6 (4): 343–353 (1972).

Cosgriff, R. L., English, J. J., and Roeca, W. B., An automatic system for longitudinal control of individual vehicles. Highw. Res. Rec. 122 (1966), pp. 7–18.

Herman, R., Lam, T., and Rothery, R., Further studies on single-lane bus flow: transient characteristics. Transp. Sci., 4 (2):187–216 (1970).

Heyes, M. P., and Ashworth, R., Further research on car-following models. Transp. Res., 6 (3): 287–291 (1972).

Kapur, K. C., Vehicle-following behavior by calculus of variations. Transp. Res., 5(2): 69–73 (1971).

Leipnik, R. B., A solution of Pipes's vehicle following equation without lag. Transp. Res., 2(3): 270–281 (1968).

Lenard, M., Safety considerations for a high density automated vehicle system. Transp. Sci., 4(2): 138–158 (1970).

Pipes, L. A., Car following models and the fundamental diagram of road traffic. Transp. Res., 1(1): 21–29 (1967).

Rockwell, T. R., Ernst, R. L., and Hanken, A., A sensitivity analysis of empirically derived car following models. Transp. Res., 2(4): 363–373 (1968).

Todosiev, E. P., and Fenton, R. E., Velocity thresholds in car-following at night. Highw. Res. Rec. 122 (1966), pp. 94–104.

Treiterer, J., Improvement of traffic flow and safety by longitudinal control. Transp. Res., 1(3): 231–251 (1967).

Unwin, E. A., and Duckstein, L., Stability of reciprocal-spacing type car following models. Transp. Sci., 1(2): 95–108 (1967).

Weiss, G., On the statistics of the linear car following model (Letters to the Editor). Transp. Sci., 3(1): 88–89 (1969).

6.12 PROBLEMS

1. Repeat the example of the car-following calculations given in Table 6.1 using 0.5-sec increments. If a computer is available, you may wish to repeat the process at 0.1-sec increments. (Fox and Lehman [14] discuss problems that may be encountered with computer simulation of car following.)

2. Show that the steady-state flow equation corresponding to the car-following equation

$$\ddot{x}_{n+1}(t+T)=\alpha_0\frac{\dot{x}_n(t)-\dot{x}_{n+1}(t)}{[x_n(t)-x_{n+1}(t)]^2}$$

is equal to $q=u_f k[1-k/k_j]$. Use the technique outlined in Section 6.5.

3. Calculate the acceleration noise for the time interval 5–10 min shown in the velocity trace given in Figure 6.8. Recall that intervals of zero velocity (7:30 to 8:00) are not included in the calculations.

Chapter 7

HYDRODYNAMIC AND KINEMATIC MODELS OF TRAFFIC

7.1 INTRODUCTION

Because traffic involves flows, concentrations, and speeds, there is a natural tendency to attempt to describe traffic in terms of fluid behavior. In car-following models traffic is recognized as being made up of discrete particles and it is the interactions between these particles that have been examined. Applying to traffic those models which have been developed for fluids (i.e., continuum models) implies greater concern in the over-all statistical behavior of the traffic stream than in the interactions between particles. Because the sample size for traffic includes only a few particles, fluid models have certain shortcomings. Nevertheless, fluid models (i.e., models that treat traffic as a continuous medium) have certain uses, expressly the behavior of the stream rather than individual cars.

Although some writers have postulated an analogy between traffic and a real fluid, it is preferable to begin with fundamental observations and postulates concerning traffic and then to identify analogies with fluids as these analogies appear. To begin, the continuity equation for traffic is developed and its analogy to the continuity equation for fluids is delineated. Next, the concept of waves in traffic is developed and example applications to practical problems are provided. Thereafter, selected models of platoon diffusion are treated. Finally, the Boltzmann-like theory of traffic is briefly presented.

7.2 CONTINUITY EQUATION

Consider two traffic counting stations on a one-way link as shown in Figure 7.1. The stations are so situated that there are no traffic sources or sinks between the stations, and station 2 is downstream from station 1.

Let N_i be the number of cars passing station i during time Δt, q_i the flow (volume) passing station i during time Δt, Δx the distance between stations, and Δt the duration of simultaneous counting at stations 1 and 2. Suppose that $N_1 > N_2$. This implies that there must be a buildup of cars between station 1 and station 2, inasmuch as there is no traffic sink between the stations.

Let $(N_2 - N_1) = \Delta N$. (Thus, ΔN will be negative for a buildup.

$$N_1/\Delta t = q_1$$
$$N_2/\Delta t = q_2$$
$$\Delta N/\Delta t = \Delta q$$

Then, the buildup of cars between stations during the period Δt will be $(-\Delta q)(\Delta t)$. If Δx is of such a length that it is appropriate to use a uniform density (concentration) over this distance, let Δk = increase in concentration of cars between stations 1 and 2 during period Δt, or

$$\Delta k = \frac{-(N_2 - N_1)}{\Delta x}$$

Then, the buildup of cars may be expressed by

$$(\Delta k)(\Delta x) = -\Delta N$$

Under the assumption of conservation of cars,

$$-(\Delta q)(\Delta t) = (\Delta k)(\Delta x)$$

and

$$\frac{\Delta q}{\Delta x} + \frac{\Delta k}{\Delta t} = 0$$

If the medium is now considered continuous and the finite increments are allowed to become infinitesimal, in the limit

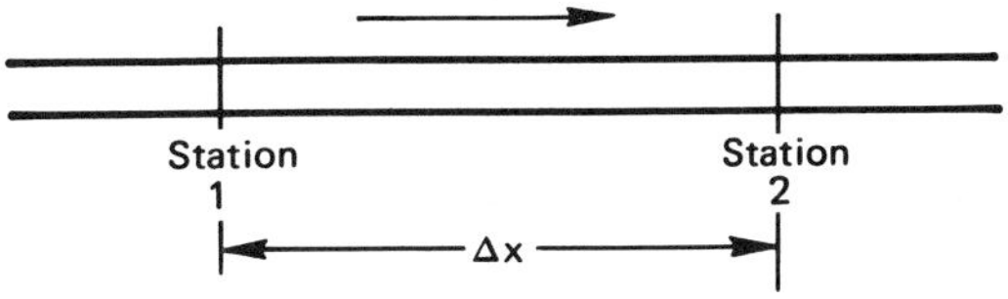

Figure 7.1 Sketch of two closely spaced measuring stations.

$$\frac{\partial q}{\partial x}+\frac{\partial k}{\partial t}=0 \qquad (7.1)$$

Eq. 7.1 will be recognized as the continuity expression for a fluid.

7.3 WAVES IN TRAFFIC

To anyone slightly familiar with the behavior of fluids, especially with shock waves in fluids, the behavior of traffic at a bottleneck appears to be acting in a shock wave-like manner. (In the late 1950s wave phenomena in traffic even became the subject of advertisements.[1]) The existence and behavior of traffic shock waves has been demonstrated by Lighthill and Whitham.[2] But the use of traffic wave analysis is not limited to shock waves. Lighthill and Whitham have also demonstrated that several traffic problems can be analyzed by assuming a system of traffic waves. In the following sections a variety of analytic techniques is used to demonstrate and analyze traffic waves.

7.3.1 Fundamental of Traffic Shock Wave Motion *

For the purposes of this discussion a shock wave is defined as the motion or propagation of a change in concentration and flow. Consider the movement of two distinct concentrations of traffic k_1 and k_2 along a straight highway (Figure 7.2). These concentrations are separated by the vertical line S, which has a velocity of u_w. This velocity is considered positive if the line moves in the direction of positive x as depicted by the arrow. With the following notations,

u_1 = space mean speed of vehicles in region A;

u_2 = space mean speed of vehicles in region B;

$U_{r1} = (u_1 - u_w)$ = speed of vehicles in region A relative to the moving line S; and

$U_{r2} = (u_2 - u_w)$ = speed of vehicles in region B relative to the moving line S,

it is evident that in time t the number of vehicles N crossing the dividing line S is

$$N = U_{r1} k_1 t = U_{r2} k_2 t$$

or

$$(u_1 - u_w) k_1 = (u_2 - u_w) k_2 \qquad (7.2)$$

Eq. 7.2 is a restatement of the conservation of matter applied to the vehicles that cross the line and may be written in the form

$$u_2 k_2 - u_1 k_1 = u_w (k_2 - k_1) \qquad (7.3)$$

If the rate of traffic flow in region A is q_1, and the rate of traffic flow in region B is q_2,

$$q_1 = k_1 u_1$$

and

$$q_2 = k_2 u_2.$$

On insertion of these values, Eq. 7.3 may be manipulated to

$$u_w = (q_2 - q_1)/(k_2 - k_1) \qquad (7.4)$$

If the rates of flow and the concentrations are nearly equal,

$$(q_2 - q_1) = \Delta q, \qquad (k_2 - k_1) = \Delta k$$

and Eq. 7.4 becomes

$$u_w = \Delta q/\Delta k = \mathrm{d}q/\mathrm{d}k \qquad (7.5)$$

which is the equation for the velocity u_w with which small *disturbances* in the traffic stream are propagated.

In the general case in which the differences in the concentrations on the two sides of the moving line S are not infinitesimally small, Eq. 7.4 may be written in the form

$$u_w = (u_2 k_2 - u_1 k_1)/(k_2 - k_1) \qquad (7.6)$$

Eq. 7.6 demonstrates that the speed u_w is the slope of the chord between points 1 and 2 on the flow–concentration diagram. This fact is used in several analyses that follow.

7.3.2 Accelerations in Traffic Stream Observations

Having developed the speed of propagation of small disturbances by two semiqualitative approaches, it is now possible to examine the various accelerations related to the traffic stream. This approach will be purely analytic.*

* This section and Figures 7.2–7.5 have been adapted from the manuscript prepared by L.A. Pipes for *Highway Research Board Special Report 79*.[3]

* This section is based on Pipes[4] but has been changed somewhat in notation and in order of presentation.

Consider the speed of the traffic stream, u. From the discussion of partial derivatives in any calculus text,

$$du = \frac{\partial u}{\partial t} dt + \frac{\partial u}{\partial x} dx$$

$$\frac{du}{dt} = \frac{\partial u}{\partial t} + \frac{\partial u}{\partial x} u \qquad (7.7)$$

Here $\frac{du}{dt}$ is the acceleration of an observer moving with the traffic stream, and $\frac{\partial u}{\partial t}$ is the acceleration of the traffic stream as viewed by an observer at a fixed point at the side of the road.

If it is assumed that u is a function of k,

$$u = u(k) \qquad (7.8)$$

$$\text{then } \left.\begin{aligned} \frac{\partial u}{\partial t} &= \frac{du}{dk}\frac{\partial k}{\partial t} \\ \frac{\partial u}{\partial x} &= \frac{du}{dk}\frac{\partial k}{\partial x} \end{aligned}\right\} \qquad (7.9)$$

Substituting Eq. 7.9 in Eq. 7.7 gives

$$\frac{du}{dt} = \frac{du}{dk}\frac{\partial k}{\partial t} + u \frac{du}{dk}\frac{\partial k}{\partial x} \qquad (7.10)$$

As a consequence of Eq. 7.8,

$$q = ku = ku(k) = q(k) \qquad (7.11)$$

$$\frac{\partial q}{\partial x} = \frac{dq}{dk}\frac{\partial k}{\partial x} = u_w \frac{\partial k}{\partial x} \qquad (7.12)$$

where

$$u_w \equiv \frac{dq}{dk} \qquad (7.13)$$

Eq. 7.1 may now be rewritten using Eq. 7.13 as

$$\frac{\partial k}{\partial t} = -\frac{\partial q}{\partial x} = -u_w \frac{\partial k}{\partial x} \qquad (7.14)$$

Noting that $q = ku$, u_w may be restated *

$$u_w = \frac{dq}{dk} = \frac{d}{dk}(ku) = u + k\frac{du}{dk} \qquad (7.15)$$

Substituting Eq. 7.14 in Eq. 7.10 gives

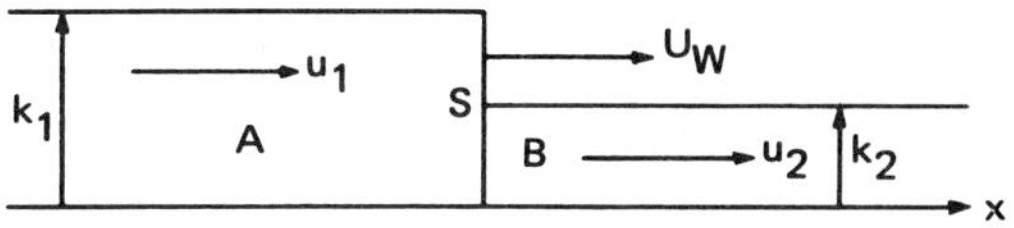

Figure 7.2 Movement of two concentrations.[3]

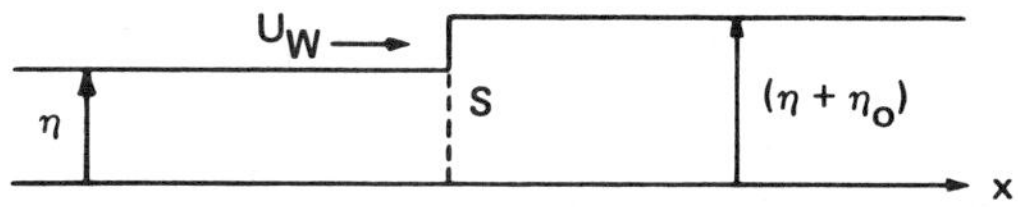

Figure 7.3 Small discontinuity in concentration.[3]

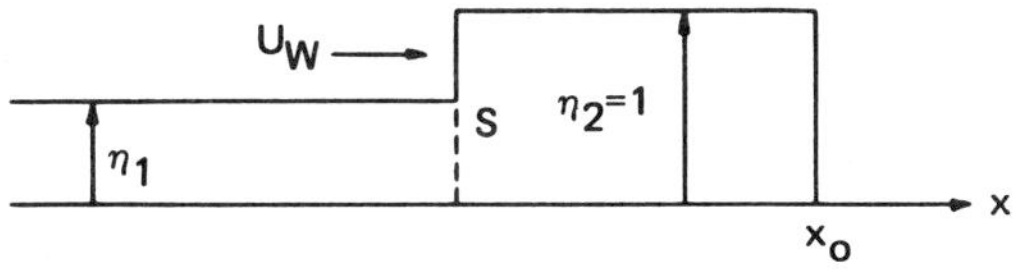

Figure 7.4 Shock wave caused by stopping.[3]

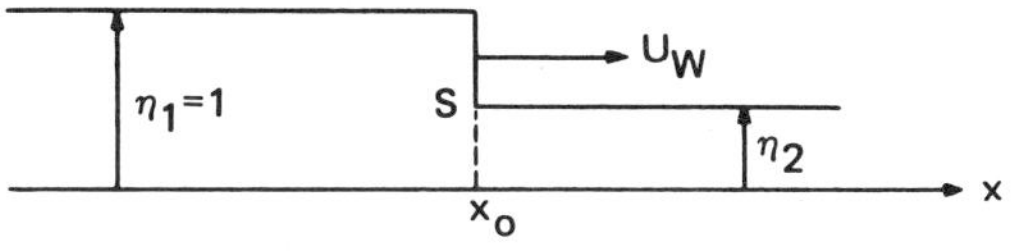

Figure 7.5 Shock wave caused by starting.[3]

$$\begin{aligned} \frac{du}{dt} &= \frac{du}{dk}\left(-u_w \frac{\partial k}{\partial x}\right) + u\frac{du}{dk}\frac{\partial k}{\partial x} \\ &= \frac{du}{dk}\frac{\partial k}{\partial x}[-u_w + u] \end{aligned} \qquad (7.16)$$

Substituting Eq. 7.15 in Eq. 7.16 gives

$$\frac{du}{dt} = -k\left(\frac{du}{dk}\right)^2 \frac{\partial k}{\partial x} \qquad (7.17)$$

Eq. 7.17 expresses the acceleration of an observer moving with traffic as a function of the density gradient $\frac{\partial k}{\partial x}$.

Regardless of the sign of $\frac{du}{dk}$, the square makes the value of the parentheses positive.†

* It should be noted that the wave velocity is 0 when headway equals reaction time, is positive when headway is greater than reaction time, and is negative when headway is less than reaction time.

† A review of Figures 4.3–4.8 and 4.10–4.14 reveals that $\frac{du}{dk}$ is zero or negative for all of the models discussed.

Thus, from Eq. 7.17, when an observer moving with the traffic fluid moves into a less dense region his acceleration is positive; when he moves into a more dense region his acceleration is negative.

The acceleration of traffic as seen by a fixed observer can be restated, using Eq. 7.14:

$$\frac{\partial u}{\partial t}=\frac{du}{dk}\frac{\partial k}{\partial t}=\left[-u_w\frac{du}{dk}\right]\frac{\partial k}{\partial x} \qquad (7.18)$$

where the quantity in brackets can take on positive, negative, or zero values.

7.3.3 An Application

Before considering an application, let us summarize:

u = speed of traffic stream (i.e., speed of an observer moving with the traffic stream);

du/dt = acceleration of traffic stream (i.e., acceleration of an observer moving with the traffic stream);

$\partial u/\partial t$ = acceleration of the traffic stream as seen by a fixed observer;

u_w = speed of propagation of small disturbance in concentration and flow $= u + k\dfrac{du}{dk}$; and

$\partial u/\partial x$ = speed gradient along the roadway.

$$du/dt=\left[-u_w\frac{du}{dk}\right]\frac{\partial k}{\partial x}+u\frac{\partial u}{\partial x} \qquad (7.19)$$

7.3.3.1 Numerical Example of Shock Wave Analysis. The following numerical example of shock wave analysis was suggested by L. C. Edie.* Consider traffic flowing at 1,000 vehicles/hr with a concentration of 20 vehicles/mile and a speed of 50 mph, as represented by point 1 in Figure 7.6. A truck with a speed of 12 mph (as represented by the slope of the radius vector 0–2 in Figure 7.6) enters the traffic stream and travels for 2 miles. Because it is impossible to pass, cars immediately behind the truck are forced to match his speed, so that a platoon forms with a platoon concentration of 100 vehicles/mile, a space mean speed of 12 mph, and a platoon flow of 1,200 vehicles/hr, as represented by point 2 of Figure 7.6. The rear of the platoon (i.e., the point at which the free-flowing traffic behind the platoon catches up with the cars in the platoon) moves with a speed represented by the slope of chord 1–2 in Figure 7.6, so

$$u_w=\frac{q_2-q_1}{k_2-k_1}=\frac{1{,}200-1{,}000}{100-20}=2.5 \text{ mph}.$$

Thus, the rear of the platoon (shock wave of increased concentration) is moving *forward* with a speed of 2.5 mph with respect to the roadway. But the front of the platoon (the truck) is moving forward at a speed of 12 mph. Therefore, the length of the platoon is growing at a rate of (12.0 − 2.5) = 9.5 mph. The truck requires ⅙ hr for a 2-mile trip so by the time the truck turns off, the length of the platoon is (⅙)(9.5) = 1.58 miles. At 100 cars/mile, the

* Private communication.

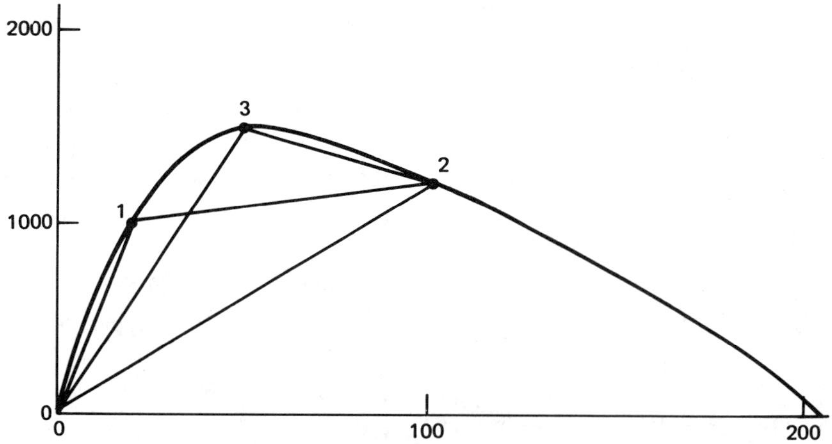

Figure 7.6 Sketch for numerical example of shock wave analysis.

platoon will contain 158 cars. Note that while the platoon is growing, although the rear moves forward at a speed of 2.5 mph, the rear is moving backward at a speed of 9.5 mph with respect to the cars in the platoon.

After the truck turns off, flow increases to optimum flow (capacity) of the facility. This condition is represented by point 3 of Figure 7.6, with a flow of 1,500 vehicles/hr, a concentration of 50 vehicles/mile, and a space mean speed of 30 mph. The front of the platoon moves with a speed represented by the slope of the chord 2–3 in Figure 7.6, or

$$u=\frac{q_3-q_2}{k_3-k_2}=\frac{1{,}500-1{,}200}{50-100}=-6.0 \text{ mph.}$$

With the rear of the platoon moving forward at 2.5 mph and the front of the platoon moving to the rear at 6.0 mph, the platoon (originally 1.58 miles long) will dissipate in 1.58/(2.5+6.0)=0.174 hr (=10.44 min). Thus, for approximately 10 min after the truck has turned off there is still a platoon of queued vehicles.

7.3.4 Shock Wave Behavior for Specific u–k Relationship

So far, the analysis has not considered any specific relation between the mean velocities u_1 and u_2 and the concentrations k_1 and k_2. If we now assume Greenshields' model,

$$u_i=u_f(1-k_i/k_j) \qquad (4.1)$$

If we further let

$$\eta_i=k_i/k_j \qquad (7.20)$$

we can now write

$$u_1=u_f(1-\eta_1) \text{ and } u_2=u_f(1-\eta_2) \qquad (7.21)$$

where u_f is the free-flow speed of the traffic stream and η_1 and η_2 are the normalized concentrations on both sides of the boundary line S. Substituting these values in Eq. 7.4 gives a wave speed of

$$u_w=\frac{[k_1u_f(1-\eta_1)]-[k_2u_f(1-\eta_2)]}{(k_1-k_2)} \qquad (7.22)$$

The relationships for η_1 and η_2 from Eq. 7.20 may be used to simplify Eq. 7.22 with the result

$$u_w=u_f[1-(\eta_1+\eta_2)] \qquad (7.23)$$

which gives the velocity of the line S in terms of the normalized concentrations on either side of the moving discontinuity.

7.3.4.1 The Case of Nearly Equal Concentrations. If the normalized concentrations η_1 and η_2 on both sides of the boundary line S are nearly equal, the situation shown in Figure 7.3 exists. The normalized concentration to the left of S is η, whereas the normalized concentration to the right of S is $(\eta+\eta_0)$, where $\eta+\eta_0\leq 1$. In this case, let

$$\eta_1=\eta, \qquad \eta_2=[\eta+\eta_0] \qquad (7.24)$$

and

$$[1-(\eta_1+\eta_2)]=[1-(2\eta+\eta_0)]=[1-2\eta] \qquad (7.25)$$

in which η_0 is neglected. If Eq. 7.25 is substituted in Eq. 7.23, the wave of discontinuity is propagated with the following velocity:

$$u_w=u_f[1-2\eta]$$

This is the equation for the propagation of shock waves obtained by Lighthill and Whitham [1] by a more elaborate analysis.

7.3.4.2 Stopping Waves. Consider a line of traffic moving with a normalized concentration η_1 and a mean vehicle velocity of

$$u_1=u_f[1-\eta_1] \qquad (7.26)$$

At a position $x=x_0$ on the highway, a traffic signal causes the traffic to halt, and the stream immediately assumes a saturated normalized concentration of $\eta_2=1$, as shown in Figure 7.4. To the left of the line S, the traffic is still moving with the mean velocity given by Eq. 7.26 at the original concentration of η_1. Under these conditions the shock wave velocity is obtained by substituting $\eta_1=\eta_1$ and $\eta_2=1$ in Eq. 7.23 to give

$$u_w=u_f[1-(\eta_1+1)]=-u_f\eta_1 \qquad (7.27)$$

which indicates that the shock wave of stopping travels backward with a velocity of $u_f\eta_1$. If the signal at $x=x_0$ turns red at $t=0$, then in time t later, a line of cars of length $u_f\eta_1t$ will be stopped behind x_0.

7.3.4.3 Starting Waves. To discuss the nature of the shock wave produced by the starting of a line of vehicles, assume that at $t=0$ a line of vehicles has accumulated behind a signal located at $x=x_0$. Because this line of vehicles is standing still, it has a saturated concentration of $\eta_1=1$ (Figure 7.5). Assume that at $t=0$ the signal at $x=x_0$ turns green and permits vehicles to move forward with a velocity of u_2. Because $u_2=u_f(1-\eta_2)$ there exists a concentration of

$$\eta_2=[1-(u_2/u_f)] \tag{7.28}$$

Therefore, a starting shock wave forms as soon as the line of vehicles begins to move. The velocity of this shock wave is obtained by substituting $\eta_1=1$ and $\eta_2=\eta_2$ in Eq. 7.23; thus,

$$u_w=u_f[1-(1+\eta_2)]=-u_f\eta_2=-(u_f-u_2) \tag{7.29}$$

Therefore, the shock wave of starting travels backward from x_0 with a velocity of (u_f-u_2). Because the starting velocity is small, it is seen that the shock wave of starting travels backward with a velocity essentially equal to $-u_f$.

Another way of deducing a wave motion is as follows: If $q=q(k,x)$, then

$$\frac{\partial q}{\partial x}=\frac{\partial q}{\partial k}\frac{\partial k}{\partial x}$$

If k is relatively constant with small variations about a mean value, then

$$\frac{\partial q}{\partial k}=u_w$$

when substituted in Eq. 7.1 gives

$$\frac{\partial k}{\partial t}+u_w\frac{\partial k}{\partial x}=0$$

This differential equation has the solution $k=\theta(x-u_wt)$, where θ is an arbitrary function. Thus small changes in concentration will be propagated in the direction of traffic flow with velocity u_w. (See, for instance, Pipes [5].)

7.3.5 Wave Flow Traffic Analysis

Thus far the discussion has centered around *shock* waves, which are observable in the field. However, analysis of several traffic situations can be performed by use of a postulated system of traffic waves that are not observable in the field, as first pointed out by Lighthill and Whitham.[1] (In a sense these waves are analogous to radio waves that cannot be seen and are thus difficult to comprehend.)

The flow–concentration curve is an essential part of any traffic wave analysis. In Figure 4.15 it was pointed out that for any point on the flow–concentration curve, the radius vector represents the traffic speed $\bar{u}_s$ and the tangent represents the wave velocity u_w. Figure 7.7 demonstrates the use of traffic waves to predict the occurrence of a shock wave. The left-hand side of the figure is a flow–concentration (q–k) curve; the right-hand side, a time–space diagram. On the q–k curve, point A represents a situation where traffic is flowing near capacity and the speed is reduced to a value well below free-flow speed. Point B represents a situation where traffic flows at a somewhat higher speed because of the lower density. Tangents at points A and B represent the wave velocities for these two situations. Now, if the faster flow of point B occurs later in time than that of point A, the waves of point B will eventually catch up with those of point A. This is shown in the time–space diagram of Figure 7.7. The intersection of these two sets of waves has a slope equal to the chord connecting the two points on the q–k curve, and this intersection represents the path of the shock wave.* Note that the velocity of the shock wave is often negative with respect to the roadway and is always negative with respect to the traffic.

At this point it is necessary to clarify that the waves on the time–space diagram in this analysis are not trajectories of vehicles but lines of constant flow and thus lines of constant speed. The vehicles have a greater velocity than the waves, because the speed of the vehicle stream is represented by the radius vector, whereas the velocity of the waves is represented by the tangent. Thus the reader may wish to regard these waves as imaginary but useful as an analysis tool.

7.3.5.1 Progress of a Traffic Hump.

An example of the application of such analysis techniques is the progress of a traffic "hump" as discussed by Lighthill and Whitham.[1] A hump is a parcel of increased density, such as might occur on a freeway, flowing at a constant level along the main stream, when there is a short-term influx of substantial proportions at one on-ramp.

Figure 7.8 portrays the traffic waves associated with the formation of a hump. The speed of the front of the hump can be stated immediately from Figure 7.8 as the velocity of

---

* The slope of the chord equals the mean velocity of the two waves for parabolic q–k curves. This is also approximately true for non-parabolic curves, except those containing a vertical tangent, because the following series expression applies:

$$\frac{q(k_2)-q(k_1)}{k_2-k_1}=\frac{1}{2}(q'(k_1)+q'(k_2)) - \frac{(k_2-k_1)^2}{24}[q'''(k_1)+q'''(k_2)]+\ldots$$

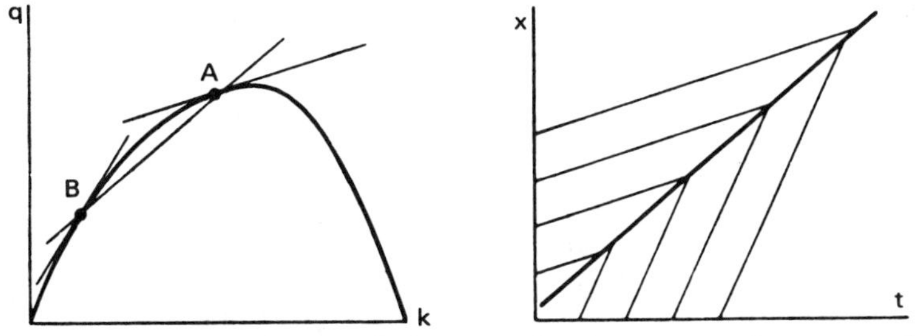

Figure 7.7 Analysis of traffic shock wave using flow-concentration curve and time-space diagram. The wave speeds in the space diagram are drawn parallel to the respective tangents of the flow-concentration diagram.[1]

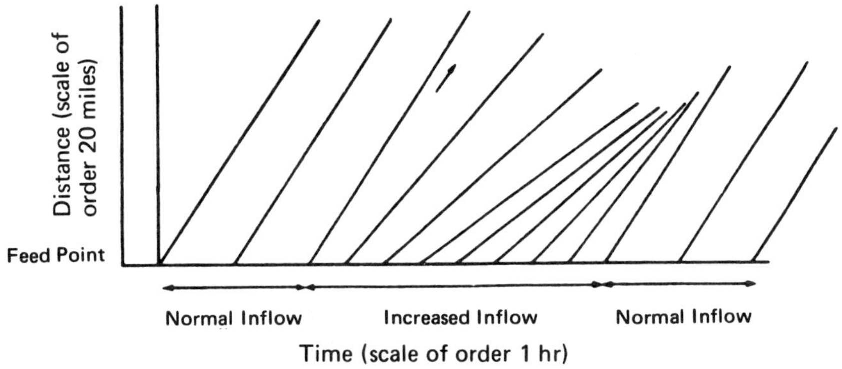

Figure 7.8 Wave forms in the traffic hump.[1]

the earliest wave representing the increased flow. The construction of the path of the shock wave is carried out as discussed in Figure 7.7 and is shown in Figure 7.9.

7.3.5.2 Behavior of Traffic at Bottlenecks. The study of traffic behavior at bottlenecks represents an important application of traffic wave analysis. A bottleneck is here defined as a stretch of roadway where the capacity is less than that of the roadway sections upstream and downstream from it. Figure 7.10 depicts the flow–concentration curves at various points within the bottleneck. Note the horizontal line at the left of the diagram. This line indicates how the speed of the stream suddenly drops as the bottleneck is reached. Figure 7.11 indicates the passage of the traffic waves through the bottleneck area.

When the approach flow attains the point where it exceeds the capacity of the bottleneck, the duration of such a condition would be finite. This situation might be associated with the passage of a hump as previously discussed or attributable to the buildup of traffic during the peak period, which is followed by dissipation.

Figures 7.12 and 7.13 illustrate this analysis. First, no wave carrying a flow greater than the capacity of the bottleneck can pass the bottleneck zone. (Here, it is necesary to reemphasize that waves are being discussed and not the actual traffic stream. Given time, the actual traffic stream will eventually pass through the bottleneck.) Consider the arrival of waves as shown in Figure 7.13. At the left-hand side waves of low flow (volume) are seen; hence, high-velocity waves are arriving at the bottleneck area. As they arrive, there is a jump from one flow–concentration q–k curve to another of lesser capacity (maximum flow) (Figure 7.12), resulting in a decrease in velocity. As the arriving flow increases, arriving wave velocities are less. Eventually, the condition is reached where the jump from one q–k curve to another results in a zero velocity of the wave within the bottleneck. This is indicated by the horizontal wave in Figure 7.13. As the flow continues to increase the situation typified by the wave ABC is reached. The wave enters the bottleneck with the velocity indicated by point A; this velocity is changed to zero by the bottleneck as indicated by point B. Then increased concentration forces operation to the right-hand side of the q–k curve resulting in a negative wave velocity, the wave leaving the bottleneck in a rearward direction with the velocity indicated by point C. As waves start toward the rear they meet arriving waves; the intersection of these two sets of waves forms a shock wave as indicated by the heavy line at the right of Figure 7.13. As long as arriving flow exceeds the capacity of the (permanent) bottleneck, the shock wave will continue to move upstream, carrying with it decreased flow.

7.4 TRAFFIC FLUID STATE CONSIDERATIONS

In Eq. 7.17 now let

$$F = -k\left(\frac{du}{dk}\right)^2 \tag{7.30}$$

where F may be regarded as a transfer function. This allows Eq. 7.17 to be rewritten

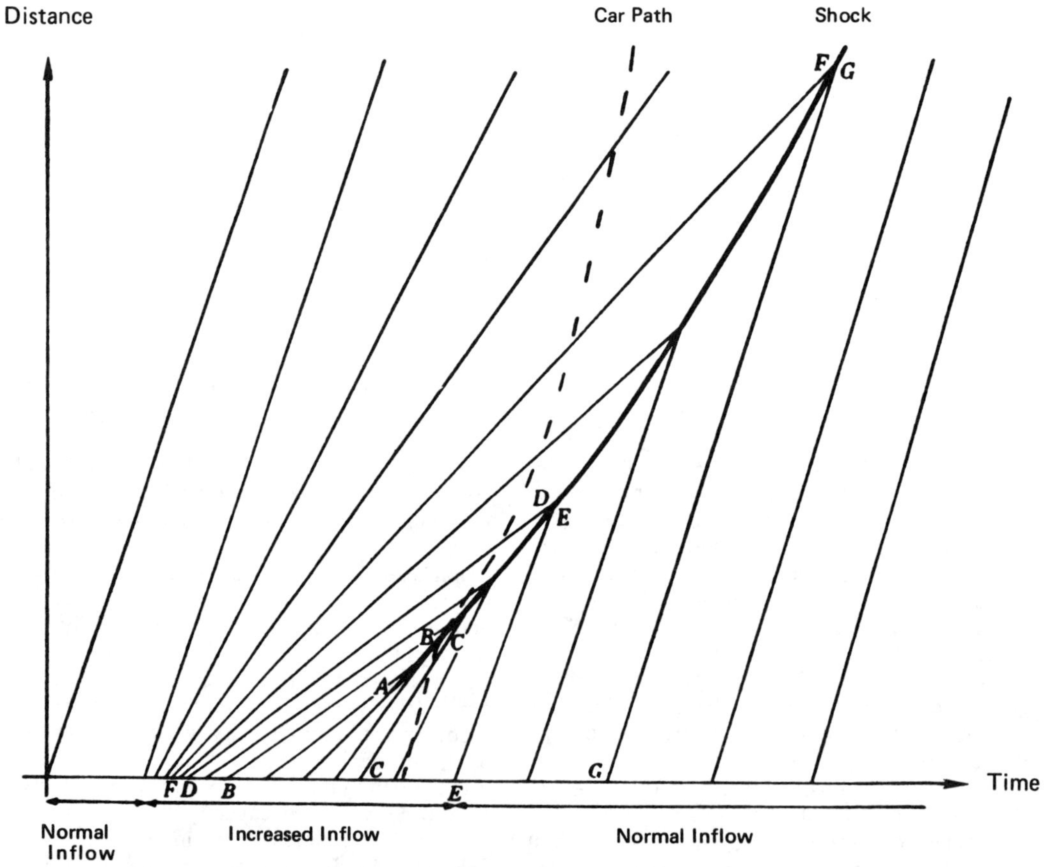

Figure 7.9 Progress of the traffic hump with time.[1]

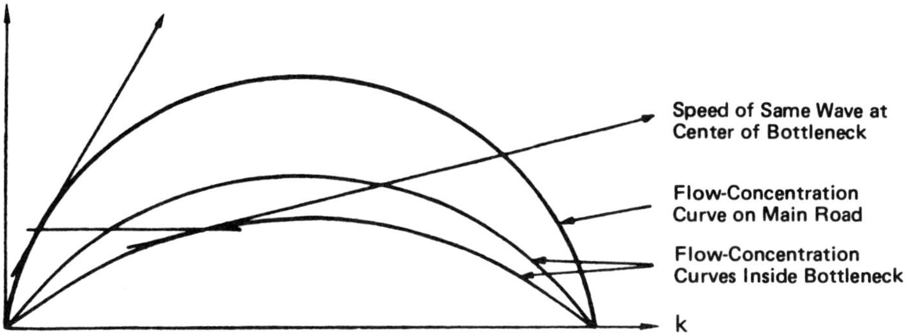

Figure 7.10 Changes in flow-concentration curve at a bottleneck.[1]

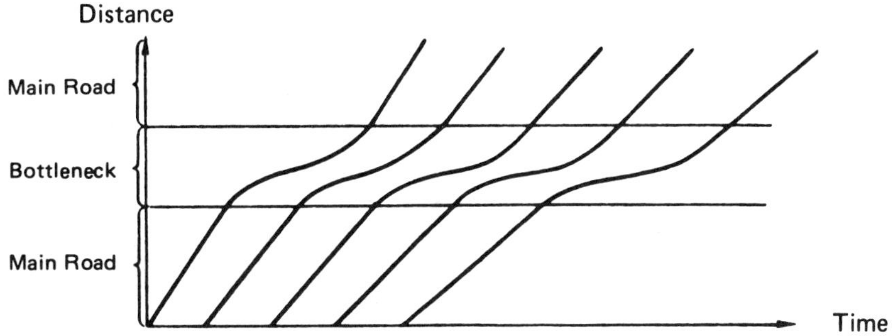

Figure 7.11 Passage of waves through a bottleneck whose capacity exceeds the arrival flow rate.[1]

$$\frac{du}{dt}=F\frac{\partial k}{\partial x} \quad (7.31)$$

Up to now the form of the u–k relationship has not been specified. Specific forms can be investigated, as well as the resulting flow-concentration models. Postulating a form for F is equivalent to defining a "fluid state".[5]

7.4.1 Greenberg's "One-Dimensional" Fluid State

Greenberg[6] has assumed a traffic fluid state *

$$\frac{du}{dt}=-\frac{c^2}{k}\frac{\partial k}{\partial x} \quad (7.32)$$

where c is a constant. (Payne[20] has extended this work to include reaction time.) Eq. 7.32 is equivalent to assuming that

* Greenberg's notation has been changed to agree with usage in this work.

$$F=-c^2/k \quad (7.33)$$

Equating the values of F from Eqs. 7.33 and 7.30 gives

$$-k\left(\frac{du}{dk}\right)^2=-\frac{c^2}{k} \quad (7.34)$$

which can be manipulated algebraically to yield

$$\frac{du}{dk}=\frac{-c}{k} \quad (7.35)$$

Eq. 7.35 can be solved as a differential equation with the boundary condition

$$\left.\begin{matrix} u=0 \\ k=k_j \end{matrix}\right\}$$

$$u/c=\pm \ln\frac{k}{k_j} \quad (7.36)$$

Note that $0\leq\frac{k}{k_j}\leq 1$; thus, $\ln\frac{k}{k_j}\leq 0$. Therefore,

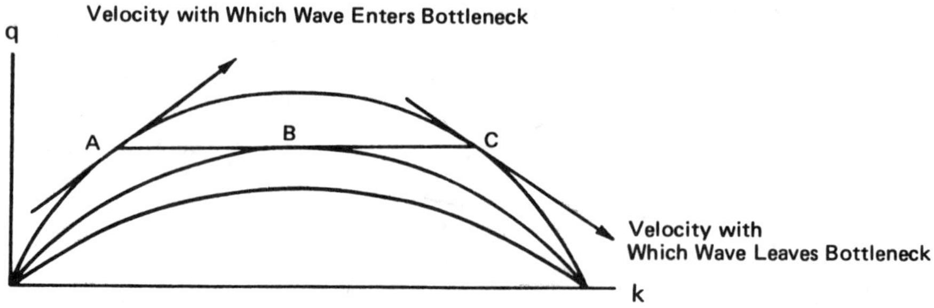

Figure 7.12 Flow-concentration curve for reflection of waves from a bottleneck as shown in Figure 7.13.[1]

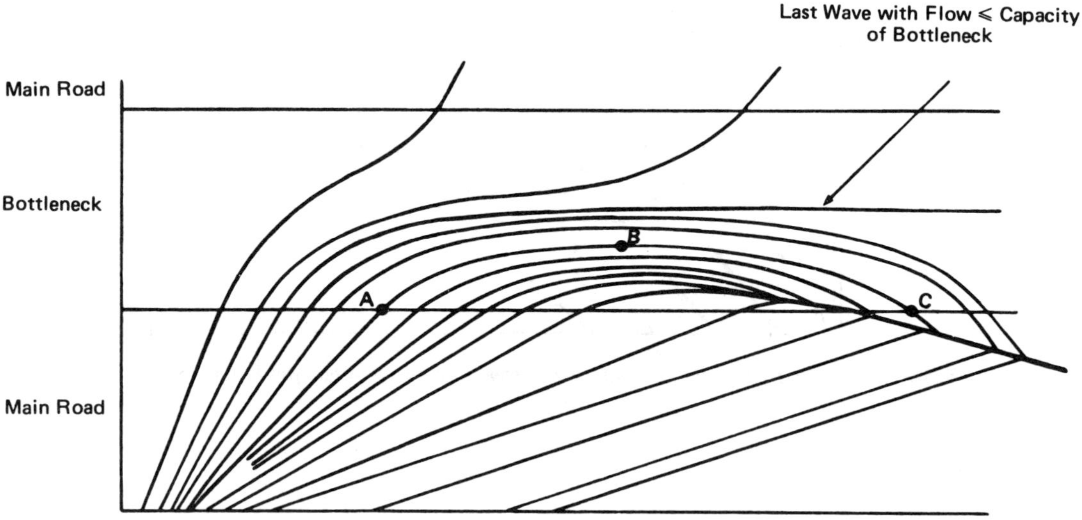

Figure 7.13 Formation of shock wave in front of a hump as it enters a bottleneck of inadequate capacity.[1]

the negative sign must be used to obtain a positive u/c:

$$u = -c\ell n \frac{k}{k_j} = c[\ell n k_j - \ell n k] \quad (7.37)$$

This is the same model as Eq. 4.2 when u_m is substituted for c. In Section 4.3.2 the maximum flow for this model was shown to be $q_m = u_m k_j/e$.

Note that the model of Eq. 7.37 is the same as that obtained in the car-following theory where $l=1$ and $m=0$ (See Table 6.5). Because of this equivalence, there is a tendency to look on this as a unifying of the microscopic (car-following) and macroscopic (fluid analogy) approaches to the theory of traffic flow. Care must be exercised, however, as to the extent to which one carries this reasoning.

7.4.2 Richards' Equation of State

Richards [7] has implied a state equation of the form

$$F = -A^2 k \quad (7.38)$$

where A is a constant having the dimensions of speed–concentration.[5] Equating the expressions for F in Eqs. 7.38 and 7.30 and performing algebraic manipulations, one obtains

$$du/dk = -A \quad (7.39)$$

Solving the differential Eq. 7.39 with the boundary conditions:

$$\left.\begin{matrix} u=0 \\ k=k_j \end{matrix}\right\} \qquad \left.\begin{matrix} u=u_f \\ k=0 \end{matrix}\right\}$$

yields

$$u=u_f(1-k/k_j) \tag{7.40}$$

Note that this is equivalent to assuming that $A=u_f/k_j$ as a traffic fluid state. Eq. 7.40, of course, is the same as Greenshields' model of Eq. 4.1.

7.4.3 Pipes' Equation of State

Pipes [4,8] has generalized Richards' work by taking an equation of state of the form

$$F=-A^2k^{2s+1} \tag{7.41}$$

where s is an integer not equal to -1. Performing the same operations on Eq. 7.41 as were performed on Eq. 7.38 yields

$$u=u_f[1-(k/k_j)^{s+1}] \tag{7.42}$$

7.5 QUANTITATIVE ANALYSIS (SIGNALIZED INTERSECTION)

A combination of wave analysis with specific flow–concentration models can be illustrated by studying queue length. The computation of queue length at a traffic signal was first discussed by Lighthill and Whitham [1] and later extended by Rorbech.[9]

Consider a one-lane approach to a traffic signal with an approach volume of q_A and an approach density k_A. The red period of the signal is t_r and the green (i.e., green plus amber) period is t_g. Figure 7.14 is a q–k curve, and Figure 7.15 is a time–space plot of the wave patterns. Figure 7.16 is a sketch of the angles used in the trigonometric computation of the queue length.

While the signal is red, traffic is stopped initially at the stopline (line SL in Figure 7.15), and a shock wave is propagated toward the rear. The waves representing the stopped traffic have a slope equal to that of the tangent at point R in Figure 7.14; the waves representing arriving traffic have a slope equal to that of the tangent at point P in Figure 7.14; the line of interference of these two waves represents the path of the shock wave and has a slope equal to that of the chord between points P and R in Figure 7.14. Point A represents the position of the shock wave (i.e., rear of queue) at the instant the signal turns green. Because the response of a traffic system is slow, traffic does not start immediately on receiving the green indication, and some cars may still be required to come to a complete stop even after the green has been displayed. Point B represents the maximum distance from the stopline for which complete stoppage is required. X_A represents the distance of point A from the stopline, and X_B represents the distance of point B from the stopline.

$$X_A=t_r \tan\alpha \qquad \tan\alpha=\frac{q_A}{k_j-k_A} \tag{7.42}$$

$$\frac{X_B}{W}=\tan\beta \qquad \frac{X_B}{t_r+W}=\tan\alpha \tag{7.43}$$

$$W=\frac{X_B}{\tan\beta} \qquad X_B=\tan\alpha\,(t_r+W)$$

$$X_B=\tan\alpha\left[t_r+\frac{X_B}{\tan\beta}\right]$$

$$\frac{X_B}{\tan\alpha}-\frac{X_B}{\tan\beta}=t_r$$

$$X_B=\frac{t_r}{\dfrac{1}{\tan\alpha}-\dfrac{1}{\tan\beta}} \tag{7.44}$$

$$\tan\beta=\left|\frac{dq}{dk}\right|_{k=k_j}=u_j$$

Where u_j is the velocity of waves representing jam density,

$$X_B=\frac{t_r}{\dfrac{k_j-k_A}{q_A}-\dfrac{1}{|u_f|}} \tag{7.45}$$

Taking Greenshields' model (linear u–k rela-

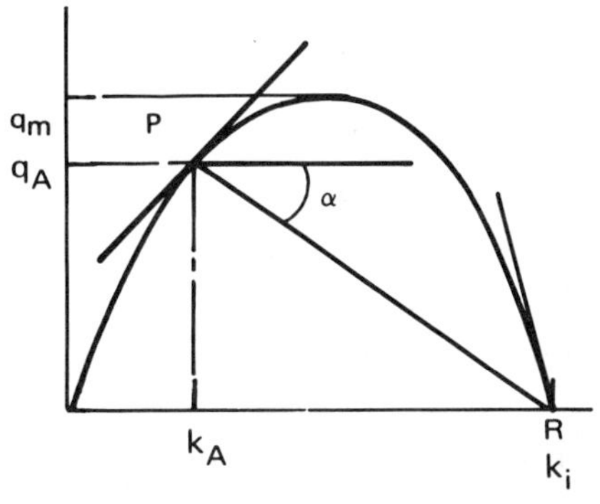

Figure 7.14 Flow–concentration curve for computation of queue length at a signal.

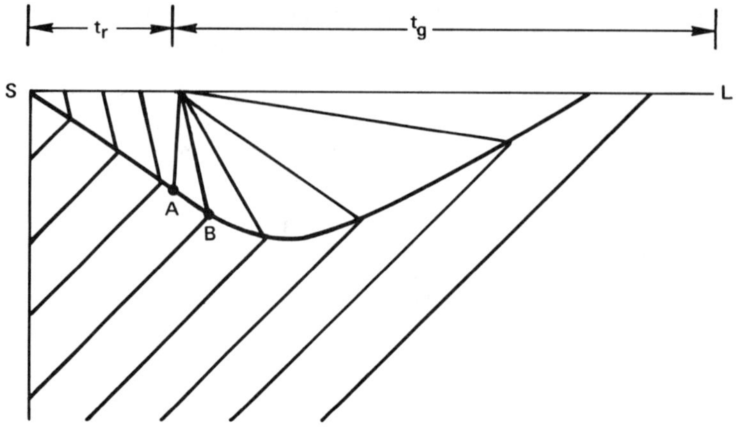

Figure 7.15 Time-space wave diagram for computation of queue length at a signal.

tionship with resulting parabolic q–k curve),

$$u = u_f\left(1 - \frac{k}{k_j}\right)$$

$$q = ku = ku_f\left(1 - \frac{k}{k_j}\right) = u_f\left(k - \frac{k^2}{k_j}\right)$$

$$\frac{dq}{dk} = u_f - \frac{2u_f k}{k_j}$$

Maximum flow (q_m) will occur at $\frac{dq}{dk} = 0$:

$$u_f - 2u_f\frac{k}{k_j} = 0$$

$$k = k_j/2$$

$$q_m = u_f\left(\frac{k_j}{2} - \frac{k_j^2}{4k_j}\right) = \frac{u_f k_j}{4} \qquad u_f = \frac{4q_m}{k_j}$$

Velocity of waves representing jam density is expressed

$$\left.\frac{dq}{dk}\right|_{k=k_j} = u_f - \frac{2u_f k_j}{k_j} = -u_f \qquad (7.46)$$

and the arrival flow is

$$q_A = pq_m$$

where p is a decimal fraction (i.e., $0 \leq p \leq 1$), from which it is possible to compute k_A.

Still using Greenshields' model,

$$q_A = k_A u_f\left(1 - \frac{k_j}{k_A}\right) = pq_m = p\left(\frac{u_f k_j}{4}\right)$$

$$\frac{q_A}{u_f} = k_A\left(1 - \frac{k_A}{k_j}\right) = p\left(\frac{k_j}{4}\right)$$

$$4k_A^2 - 4k_j k_A + pk_j^2 = 0$$

$$k_A = \frac{4k_j \pm (16k_j^2 - p16k_j^2)^{1/2}}{8}$$

$$= 0.5k_j(1 \pm \sqrt{1-p}) \qquad (7.47)$$

This gives two values of arrival density—one for flows less than maximum on the low density side and one for flows less than maximum on the high density side. The lower value is accepted and the higher is discarded. Thus,

$$k_A = 0.5k_j(1 - \sqrt{1-p})$$

$$X_A = \frac{t_r(pq_m)}{k_j - 0.5k_j[1-(1-p)^{1/2}]} = F_A\left(\frac{t_r q_m}{k_j}\right) \qquad (7.48)$$

where

$$F_A = \frac{p}{1 - 0.5[1-(1-p)^{1/2}]} = \frac{p}{0.5 + 0.5(1-p)^{1/2}} \qquad (7.49)$$

$$X_B = \frac{t_r}{\dfrac{k_j - 0.5k_j[1-(1-p)^{1/2}]}{pq_m} - \dfrac{k_j}{4q_m}}$$

$$= \left(\frac{t_r q_m}{k_j}\right)F_B \qquad (7.50)$$

where

$$F_B = \frac{1}{\dfrac{1 - 0.5[1-(1-p)^{1/2}]}{p} - \dfrac{1}{4}}$$

$$= \frac{1}{\dfrac{0.5 + 0.5(1-p)^{1/2}}{p} - 0.25} \qquad (7.51)$$

or, alternatively,

$$F_B = \frac{1}{\frac{1}{F_A} - \frac{1}{4}} \qquad (7.52)$$

Consider the case where $q_A = \frac{1}{10} q_m$; i.e., $p = 0.1$. Then $F_A = 0.103$ and $F_B = 0.105$.

Rorbech [9] has tabulated F_A and F_B for various values of p and for two flow–concentration models—Greenshields' and Greenberg's. Table 7.1 gives the results.

Consider now a numerical example, where $q_A = 0.2 q_m$, $q_m = 1{,}800$ vehicles/hr, $t_r = 30$ sec $= \frac{30}{3{,}600}$ hr, and $k_j = 200$ vehicles/mile. What distance will the queue of cars occupy at the time the signal turns green? Assuming that Greenshields' flow–concentration model has been accepted, $X_A = F_A \frac{t_r q_m}{k_j} = 0.211 \left(\frac{30}{3{,}600}\right)\left(\frac{1{,}800}{200}\right) = 0.015825$ miles $= 83.6$ ft.

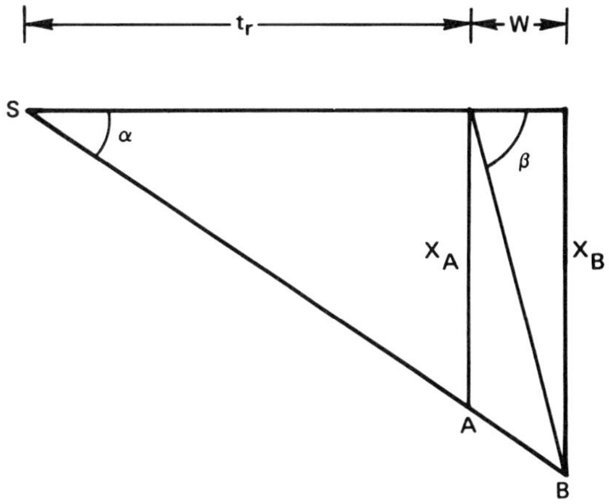

Figure 7.16 Sketch of angles for computation of queue length at a signal.

7.6 PLATOON DIFFUSION

One common objective in timing traffic signals is to synchronize them so that a platoon of cars being released from one signal arrives at the next signal at such time that it can pass through this second signal without interruption. The usual practice is to assume that all cars in the platoon move with the average speed of traffic. This simple approach breaks down because platoons do not remain in a compact state but tend to diffuse as they move away from the point of their formation. For example, Graham and Chenu [10] have experimentally found the percentage of the original number of cars in the platoon that remain in the platoon at various distances along a highway from the point of origin (Table 7.2). When platoons are represented by frequency distributions of percent of total platoon flow versus arrival time, the result is as shown in Figure 7.17. It will be noted that as distance increases, the peak of the distribution becomes lower, with an increased tendency for the distribution to "tail" out to the rear. Part of the problem lies in the fact that platoon leaders from one signal cycle to another do not travel at the same speed. Figure 7.18 illustrates the fact that arrival times of platoon leaders become more dispersed as the distance from the point of platoon formation increases.

7.6.1 Pacey's Diffusion Study

Pacey [12] has postulated that the speeds of vehicles in a platoon are normally distributed, from which he deduced the distribution of travel times. The mathematical derivation is given in Appendix F-1. The result of Pacey's method is shown in Figure 7.19.

7.6.2 Diffusion Model of Grace and Potts

By first changing to new variables and then changing back Grace and Potts [13] have shown that Pacey's model corresponds to a unidimensional fluid diffusion equation

TABLE 7.1 Values of F_A and F_B

	Greenshields' Flow–Concentration Model		Greenberg's Flow–Concentration Model	
p	F_A	F_B	F_A	F_B
0.1	0.103	0.105	0.101	0.105
0.2	0.211	0.223	0.204	0.220
0.3	0.327	0.356	0.310	0.350
0.4	0.451	0.508	0.421	0.498
0.5	0.585	0.686	0.536	0.668
0.6	0.735	0.901	0.662	0.874
0.7	0.904	1.169	0.798	1.130
0.8	1.105	1.527	0.954	1.469
0.9	1.368	2.079	1.146	1.983
1.0	2.000	4.000	1.582	3.787

$$\partial\kappa/\partial\tau = \alpha^2(\partial^2\kappa/\partial x^2) \qquad (7.53)$$

which is equivalent to

$$(\partial k/\partial t) + \bar{u}(\partial k/\partial x) = \bar{u}^2\alpha^2(\partial^2 k/\partial x^2) \qquad (7.54)$$

where

$\bar{u}$ = mean speed of vehicle speed distribution;
s = dispersion of speed distribution;
$\alpha = s/\bar{u}$ = diffusion constant;
$\tau = 1/2\, t^2$;
$\kappa = (x/\bar{u}) - t$;
x = distance from stopline at original signal; and
t = time since display of green signal.

More importantly, Grace and Potts have (1) pointed out that the establishment of the ratio α computed from stream measurements as a parameter in the kinematic diffusion theory serves to link it to stream flow and (2) described a suitable way of timing traffic signals by representing platoons as trapezoidal pulses.

TABLE 7.2 Percentage of Vehicles Remaining in a Highway Traffic Platoon [a]

Distance (miles)	Vehicles Remaining in Platoon (%)
0.25	91
0.50	85
0.75	80
1	77

[a] From Graham and Chenu.[10]

7.6.3 Platoons as Trapezoidal Pulses

When the green indication of a traffic signal is displayed, the flow does not immediately jump to its maximum value but builds up over a period of several seconds. The solid line in Figure 7.20 is a plot of flow versus time for the flow leaving a traffic signal (assuming there is sufficient demand to assure flow throughout the entire green interval). The other lines in Figure 7.20 indicate the (time) shape of the

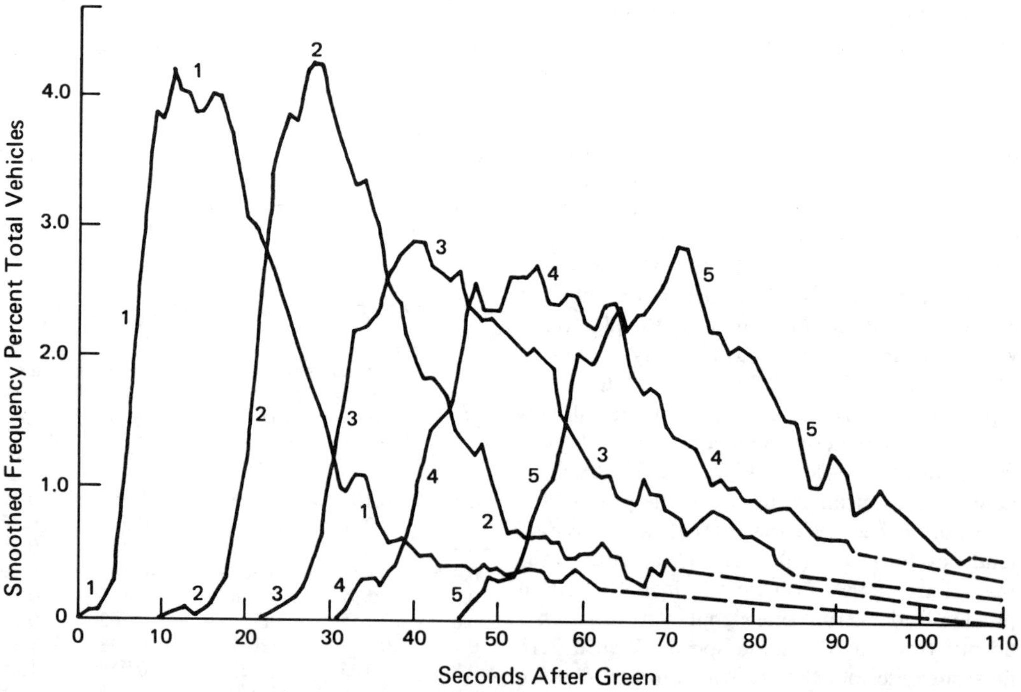

Figure 7.17 Frequency distributions of vehicle arrival at various stations.[11] Station 1, 0.03 miles from stopline of signal; station 2, 0.21 miles; station 3, 0.34 miles; station 4, 0.50 miles; station 5, 0.65 miles.

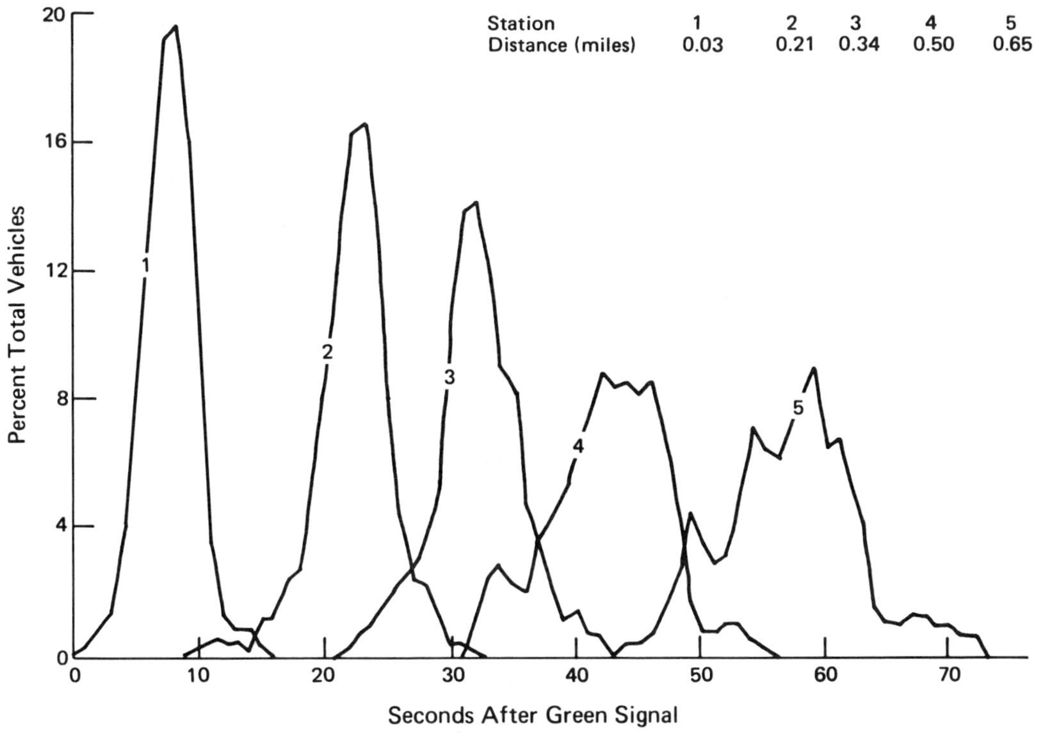

Figure 7.18 Frequency distributions of arrival times of platoon leaders at same stations as in Figure 7.17[11]

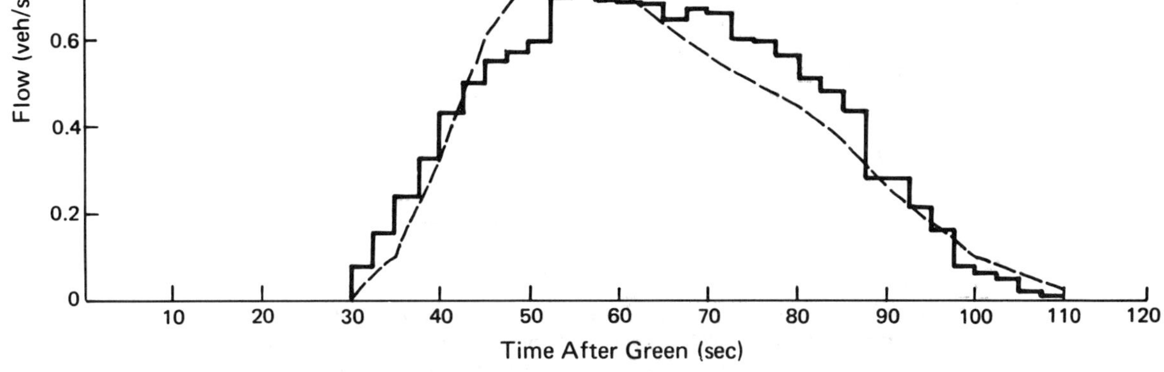

Figure 7.19 Result of Pacey's diffusion prediction.

flow pulse after it has been diffused by passage down the roadway by distances of 0.25 and 0.50 mile.

For purposes of analysis, consider the trapezoidal pulse indicated by the dotted lines of Figure 7.21. It is most convenient to treat the front and the rear of the pulse separately. Only the significant steps for the case where $b_1 = b_2$ (Figure 7.21) are given. For a detailed derivation, the reader is referred to Grace and Potts.[13]

Definitions

K= maximum density (e.g., in cars/ft) of the platoon at $t=0$;

$Q= K\bar{u}$ maximum flow (e.g., in cars/sec) of platoon at $\bar{x}=0$;

a= initial length (in ft) of a platoon;

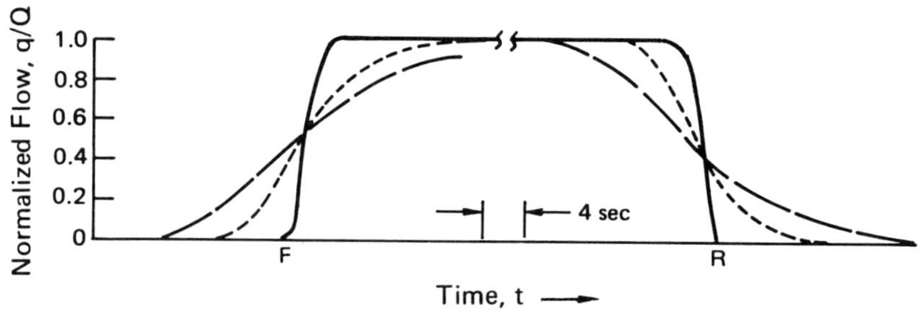

Figure 7.20 The normalized flow of a platoon plotted as a function of time at three successive points on the highway. The horizontal scale is indicated by the interval "4 sec." The initial flow builds up in approxiamtely 4 sec to its maximum value. The dotted curve represents the flow at a distance 0.25 mile down the highway; the dashed curve, at 0.5 mile. F and R indicate when the front and rear of the platoon would have reached the three points on the highway had there been no diffusion.[13]

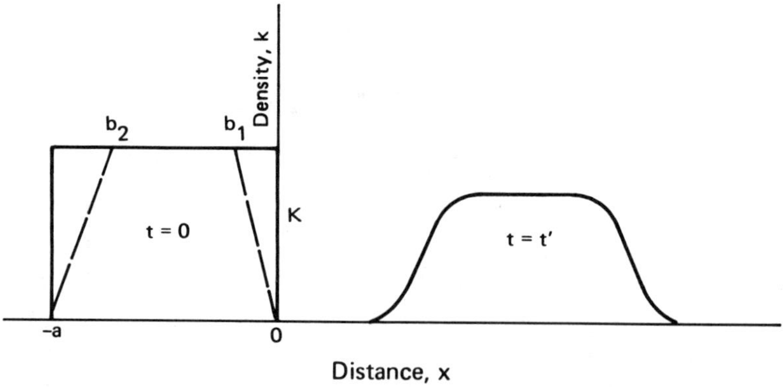

Figure 7.21 The density $k(x, t)$ of a platoon sketched (not to scale) as a function of distance x down the highway for $t=0$ and a later time $t=t'$. The solid lines indicate an initial rectangular pulse, showing how it spreads as the platoon moves down the highway. The dashed lines indicate an initial trapezoidal pulse.[13]

b_1, b_2 = length (in ft) of front (rear) of platoon in which the density builds to (falls from) its maximum value (Actually, only the case where $b_1 = b_2 = b$ has been treated);

$$erf(z) = \frac{2}{\sqrt{\pi}} \int_0^z \exp(-x^2)\,dx$$

For the Front of the Platoon

$$N_f = \tfrac{1}{2} K(\bar{u}t - x - 1/2b) + (1/2b) K \bar{u}^2 \alpha^2 t^2 \left[(x^2 + 1/2) erf(z) + (1/\pi) z \exp(-z^2)\right]_{z_1}^{z_3} \quad (7.55)$$

where N_f = number of cars that are stopped at the front of a platoon with an offset $x/\bar{u}$ sec from the previous signal and a preset (early start) of the green $(x/\bar{u} - t)$ sec;

$z_1 = (a/\bar{u} - t)/\alpha t\sqrt{2}$; and

$z_3 = [(x/\bar{u} - t) + b/\bar{u}]/\alpha t\sqrt{2}$.

For the Rear of the Platoon

$$N_r = \tfrac{1}{2} K(x - \bar{u}t - 1/2b) + (1/2b) K \bar{u}^2 \alpha^2 t^2 \left[(z^2 + 1/2)\, erf(z) + (1/\sqrt{\pi}) \exp(-z^2)\right]_{z_1}^{z_4} \quad (7.56)$$

where N_r = number of cars that are stopped at the rear of a platoon when the offset is as above and the extension (late termination) of the green is $-(x/\bar{u} - t)$; and

$z_4 = (x/\bar{u} - b/\bar{u} - t)/\alpha t\sqrt{2}$.

For various values of offset and preset or extension, the ratios of N_f/A and N_r/Q are given in Table 7.3 (for a diffusion constant $\alpha = 0.15$).

Computational Procedure

To apply these results to allow for platoon diffusion in the design of the coordination of two successive traffic lights, the following procedure is suggested:

1. Under conditions of the maximum traffic flow to be coped with, the initial flow for all lanes combined is determined from a time arrival study, from which Q (the maximum flow in cars/sec) and $b_1\bar{u}$ and $b_2\bar{u}$ (sec) (the build-up and fall-off times) are estimated.

2. The distribution of car speeds is determined and the mean speed $\bar{u}$ and dispersion $s\sigma$ (both in ft sec^{-1}) are estimated, from which the diffusion constant $\alpha = \sigma/\bar{u}$ is calculated.

3. From the distance x (ft between the traffic lights), the offset time $x/\bar{u}$ (sec) is calculated.

4. If N_f and N_r—the expected number of cars to be stopped, respectively, at the front and rear of the platoons—are decided upon, then N_f/Q and N_r/Q are calculable.

5. The appropriate values of the preset time χ (sec) and the extension time $-\chi$(sec) are then read from Table 7.3 where $\chi = (x/\bar{u} - t)$.

Numerical Example

Consider the following data: Q=two cars/sec (for a multilane facility), x=2,640 ft (the

TABLE 7.3 Expected Number of Cars Stopped,[a] Initial Trapezoidal Pulse

Offset (sec)	Preset or Extension (sec) 0	1	2	3	4	5	6	7	8	9	10
30	1.02	0.68	0.42	0.24	0.13	0.06	0.02	0.01	0	0	0
	1.02	0.77	0.58	0.44	0.33	0.24	0.18	0.14	0.10	0.08	0.06
60	2.71	2.26	1.86	1.50	1.20	0.93	0.71	0.53	0.38	0.27	0.18
	2.71	2.37	2.07	1.81	1.58	1.37	1.19	1.04	0.90	0.78	0.67
90	4.46	3.98	3.53	3.10	2.71	2.35	2.02	1.73	1.46	1.22	1.01
	4.46	4.10	3.75	3.44	3.14	2.87	2.63	2.40	2.19	1.99	1.82
120	6.24	5.74	5.26	4.80	4.36	3.95	3.57	3.21	2.87	2.55	2.26
	6.24	5.85	5.49	5.14	4.82	4.51	4.22	3.95	3.69	3.45	3.22

[a] The odd lines give the values (sec) of the ratio of the number of cars stopped at the front of the platoon to the maximum initial flow for the specified values of offset and preset. The even lines give the values (sec) of the ratio of the number of cars stopped at the rear of the platoon to the maximum initial flow for the specified values of offset and extension. The diffusion constant $\alpha = 0.15$.[13]

distance between signals), $\bar{u}=30$ mph$=44$ ft/sec (the average speed), $\alpha=s/\bar{u}=0.15$ (the diffusion constant, $b_1/\bar{u}=4$ sec (the time to build up to full-flow condition), and $b_2/\bar{u}=4$ sec (the time to decay from full-flow condition). It is a matter of engineering judgment to select the number of cars to be stopped at the front and rear of the platoon. If $N_f=2$ and $N_r=2$, the offset as computed from the data is 60 sec.

The ratio $N_f/Q=1.0$. Referring to Table 7.3, line three lists values of N_fQ for offsets of 60 sec. A ratio of 1.0 falls between a preset of 4 and 5 sec; a preset of 5 sec is selected.

The ratio $N_rQ=1.0$. Referring to Table 7.3, line four lists values of N_r/Q for offsets of 60 sec. Because the ratio 1.0 falls between 7 and 8 sec, an extension of 8 sec is selected.

Suppose now the normal duration of the green interval at the second intersection is 40 sec. The green interval as designed here would start $60-5=55$ sec after the start of the green at the first intersection in order to account for forward diffusion of platoons. Ordinarily, the green at the second intersection would terminate at $60+40=100$ sec after the start of the green at the first intersection. As designed to account for backward platoon diffusion, it will terminate $100+8=108$ sec after the start of the green at the first intersection.

7.6.4 Other Platoon Studies and Comments

Edie et al.[20] have performed experimental studies to test the kinematic model of Grace and Potts. They state: "The kinematic model well describes the spreading of platoons in medium traffic without interference."

Hillier and Rothery [21] have demonstrated that consideration of platoon phenomena can improve signal timing in the delay–difference of offset approach.

Nemeth and Vecellio [15] have concluded that Pacey's model is valid.

7.7 BOLTZMANN-LIKE BEHAVIOR OF TRAFFIC

The fact that traffic under low densities is essentially the flow of individual cars, whereas at high densities flow is by platoons, suggests an analogy between traffic and gases. The early work on this approach was performed by Prigogine.[16] Continuing work has been carried on by him, by the General Motors Research Laboratories, and by the System Development Corporation. Until now, no data have been collected under conditions prescribed by the theory; however, recent analyses based on available data demonstrate the attractiveness of the approach. Prigogine and Herman [17] discuss the development of the theory, and a brief review of assumptions, results, and some numerical applications is given in the following.

7.7.1 Velocity Distribution Function

An essential part of the theory is the introduction of the "velocity distribution function" $f(x,u,t)$, which is analogous to the probability of finding a particle in a gas at x and t with a specific momentum. In absence of any interactions, this distribution must satisfy the continuity equation for a given region*:

$$\frac{\mathrm{d}f}{\mathrm{d}t}=\frac{\partial f}{\partial t}+u\frac{\partial f}{\partial x}=0 \qquad (7.57)$$

7.7.2 Kinetic Equation

Actually, three different main features of multilane traffic are recognized and treated as separate processes in the theory. They are the relaxation process, or the speeding-up process, that expresses the attempts of drivers to achieve their own desired speeds; the interaction process, or the slowing-down process, that arises in the conflict between a faster driver and a slower driver; and the adjustment process that reduces the variance around the mean speed.[18] These processes are then expressed in a kinetic equation:

$$\frac{\mathrm{d}f}{\mathrm{d}t}=\frac{\partial f}{\partial t}+u\left(\frac{\partial f}{\partial x}\right)=\left(\frac{\partial f}{\partial t}\right)_{\text{relaxation}}+\left(\frac{\partial f}{\partial t}\right)_{\text{interaction}}+\left(\frac{\partial f}{\partial t}\right)_{\text{adjustment}} \qquad (7.58)$$

As written, this equation describes the time evolution of the speed distribution, $f[u,k(x,t),t]$, of cars on a homogeneous highway at location x and time t, where $k(x,t)$ is the concentration of cars at that point.

7.7.2.1 Relaxation. The distribution of desired speeds is defined as $f_0(x,u,t)$ and $f_0(x,u,t)\,\mathrm{d}x\mathrm{d}u$ is the number of cars at time t in

* Prigogine, Herman, et al. considered the distribution of desired speeds to be independent of density (concentration). More recent work by Andrews [22,23] discusses the case of this distribution being a function of concentration.

the interval dx with desired speeds between u and $u+du$. When the distribution f is different from f_0 it will relax toward f_0 with a time constant T. The relaxation process is described by

$$\left(\frac{\partial f}{\partial t}\right)_{\text{relaxation}} = -(f-f_0)/T \quad (7.59)$$

The desired speed distribution can exist only at low concentrations.

7.7.2.2 Interaction. The interaction process describes the effects of various cars on each other. The probability of passing (P) is related to the interaction phenomenon, which is described by

$$\left(\frac{\partial f}{\partial t}\right)_{\text{interaction}} = (1-P)k(\bar{u}-u)f \quad (7.60)$$

7.7.2.3 Adjustment. As noted in Chapter 3, as traffic density (concentration) increases the variance of speeds about the mean decreases. To account for this phenomenon, an adjustment process must be included in the model. This process is represented by

$$\left(\frac{\partial f}{\partial t}\right)_{\text{adjustment}} = \lambda(1-P)k[\delta(u-\bar{u})-f] \quad (7.61)$$

where δ is the Dirac delta function and λ is a parameter.

7.7.2.4 Computations. It is beyond the scope of this discussion to follow through the detailed mathematical steps that ensue; Prigogine and Herman[17] developed these computations. Several investigators have pursued analyses and some experimentation. Their results are as indicated in the following paragraphs.

7.7.3 Low-Density Traffic

For small concentrations, the flow becomes

$$q = \int_0^\infty uf du \approx \int_0^\infty uf_0[1-(u-\bar{u}_0)]du$$

or

$$q \approx k_j\eta[\bar{u}_0 - \gamma(\overline{u_0{}^2}\bar{u}_0{}^2)]$$

$$k_j\eta\bar{u}_0 \equiv \int_0^\infty uf\,du$$

where

$$k_j\eta\bar{u}^2{}_0 \equiv \int_0^\infty u^2 f_0 du$$

$$\frac{k}{k_j} = \eta$$

Thus, for very small values of η, q is linear with increase in concentration (see, for example, Figure 7.22); the deviations in linearity are determined by the dispersion of the desired speed distribution function.

To summarize, as $k \to 0$, $f \to f_0$; i.e., for a very small density, or light traffic, the actual speed distribution is in fact the desired distribution of each individual,

$$f = \frac{f_0}{1+Tk(1-P)(u-\bar{u})} \quad (7.62)$$

7.7.4 High-Density Traffic: Numerical Results

Several models have been examined by Anderson et al.[24]; two are reproduced in Table 7.4. The normalized flow versus concentration curves for these models are shown in Figures 7.22, 7.23, and 7.24.

Note that in Figure 7.22 the increase in flow is linear with increase in η, as mentioned in the discussion of low-density traffic. Further, notice that the higher the average speed, the lower the concentration at which the optimum flow occurs. Also in this case, collective flow does not begin until $\eta=1$ and the flow vanishes; that is, there is a "cutoff" concentration. In Figure 7.23 these cutoff concentrations are very apparent and differ from the curves in Figure 7.22 because of the use of a different model for the desired velocity distribution function.

It follows that where

$$1-Tk(1-P)\bar{u}=0 \quad (7.67)$$

leads to a "collective flow pattern," since average velocity can be seen as a function of k, T, and P only; it does not depend at all on f_0, the desired speed distribution of the individual drivers. Thus, we define the concentration at which collective flow begins as

$$k_{\text{crit}} = \frac{1}{1-T(1-P)\bar{u}} \quad (7.68)$$

Thus, for model 1 (the exponential distribution) $\eta=1$ at the initiation of collective flow.

Figure 7.24 shows the regimes of indi-

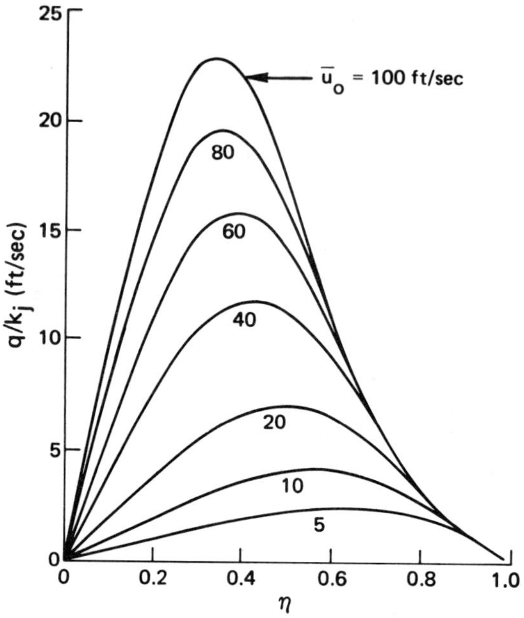

Figure 7.22 Normalized flow q/k_j vs. normalized concentration $\eta = k/k_j$ for exponential speed distribution function (model 1 in Table 7.4) for different average desired speeds and $k_j\tau = 0.1$ sec/ft.[17]

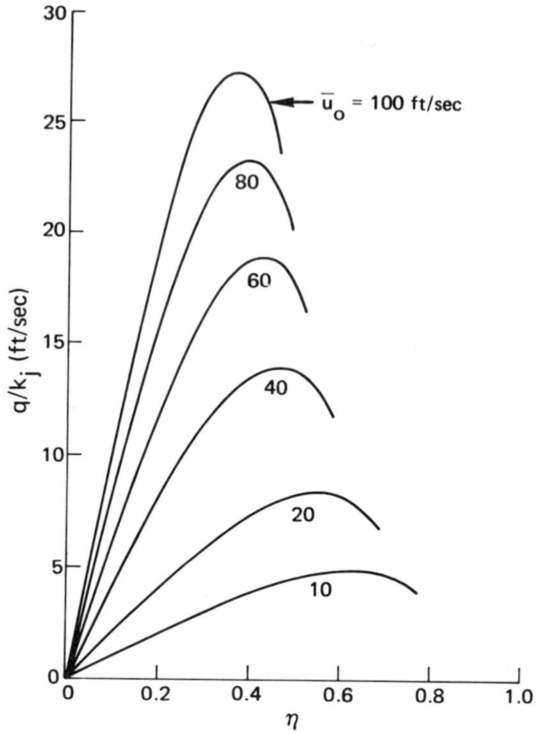

Figure 7.23 Normalized flow q/k_j vs. normalized concentration $\eta = k/k_j$ for modified exponential speed distribution function (model 2 in Table 7.4) for different average desired speeds and $k_j\tau = 0.1$ sec/ft.[17]

vidual and collective flow, where the dashed collective flow curve is given by Eq. 7.68.

In spite of these interesting results, Prigogine and Herman[17] state:

> It should be pointed out, however, that our entire discussion of the collective flow was based on the assumption of the validity of a time-independent description. This . . . is not a valid assumption, as the transition to collective flow and instabilities occur in the same range of concentration. In the quantitative description of the flow patterns at high concentrations, fluctuations will therefore play an essential role.
>
> The prediction of a transition from individual to collective flow is certainly a nice feature of the basic kinetic equation.

TABLE 7.4 Properties of Models

No.	Free-Speed Distribution Function	f_0	$\bar{u}_0$	f	λ
1	Exponential	$\left(\frac{k_j\eta}{u_f}\right)\exp\left(\frac{-u}{u_f}\right)$	u_f	$\frac{k_j\eta/u_f[\exp(-u/u_f)]}{\lambda+\gamma u}$	$(\Gamma u_f)^{-1}\exp(\Gamma u_f)^{-1}\int_{(\Gamma u_f)^{-1}}^{\infty} dx\, x^{-1}\exp(-x)$
2	Modified exponential	$\left(\frac{4k_j\eta u}{u_f}\right)\exp\left(\frac{-u}{u_f}\right)$	u_f	$\frac{(4k_j\eta u/u_f^2)[\exp(-2u/u_f)]}{\lambda+\gamma u}$	$2(\Gamma u_f)^{-1}\left\{1-2(\Gamma u_f)^{-1}\left[\exp\left(\frac{2}{\Gamma u_f}\right)\right]\int_{2(\Gamma u_f)^{-1}}^{\infty} dx\, x^{-1}\exp(-x)\right\}$

Indeed, every driver experiences that at some moment, with increasing traffic concentration, his speed drops abruptly, and he is trapped in a collective flow to which he himself contributes.

The results of the kinetic theory are nevertheless impressive, as the flow–concentration curves in Figures 7.22, 7.23, and 7.24 indicate. The material presented here is intended to give the reader the basic assumptions and conceptual formulation of the kinetic theory of traffic flow, but the interested reader may refer to the referenced literature for more details.

7.7.5 Experimental Work

A recent paper by Herman, Lam, and Prigogine [18] reports the results of a comparison of the theory and field data from two sources. Figure 7.25 contains histograms of speed for four sets of data from the Port Authority of New York and New Jersey. Figure 7.26 contains histograms for three sets of data produced by the Bureau of Public Roads (now the Federal Highway Administration) and used by Systems Development Corporation in previous work. The third and fourth moments and their ratios to the second moment (variance) were computed from the data and compared with the same values from the theory. The results are given in Tables 7.5, 7.6, and 7.7. Herman and his co-workers feel that the results are encouraging, considering the limited data available.

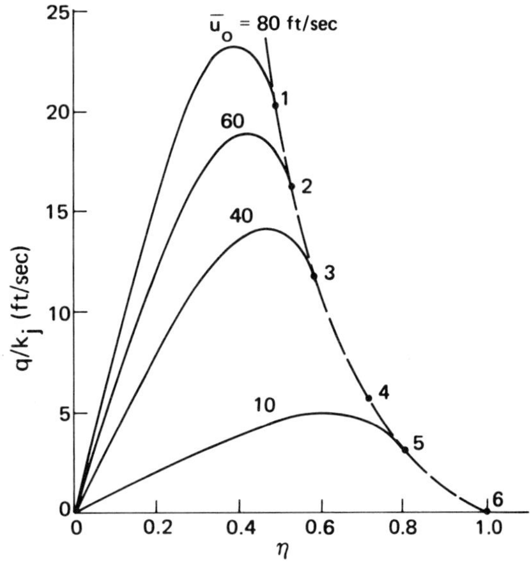

Figure 7.24 Normalized flow q/k_j vs. normalized concentration $\eta = k/k_j$ for modified exponential desired speed distribution function (model 2 in Table 7.4).[17] The solid curves represent the individual flow curves for various values of $\bar{u}_0$. The dashed curve represents the collective flow curve.

7.8 SUMMARY

1. Traffic can be analyzed by means of "waves" of constant flow and, hence, on roads of constant roadway geometry and environment, by constant speed and concentration. When there is a change in flow, resulting from changes in roadway geometry or condition, shock waves can develop. Techniques have been described by which shock waves can be plotted and used to predict performance of the traffic system. Applications to freeway bottlenecks and to traffic signals have been given.

2. The model of Greenberg was originally developed by him using a traffic "fluid state" assumption. That this model also appears in car-following gives limited confidence in the relationship between the microscopic and macroscopic theories of traffic.

3. Platoon diffusion can be represented by a model that is analogous to the diffusion of gases. Techniques of application to signal timing have been discussed.

4. The Boltzmann-like model of traffic has been introduced. This model attempts to integrate low-density flow (individual vehicles) and high-density flow (platoons) into a single model. Results achieved to the present are considered promising.

7.9 REFERENCES

1. Airborne Instruments Laboratory, Backward waves in highway traffic jams. Proc. IRE, 45 (1):2A (1957).
2. Lighthill, M. H., and Whitham, G. B., On kinematic waves: II. A theory of traffic flow on long crowded roads. Proc. R. Soc. (Lond.), Ser. A., 229 (1178):317–345 (1955); reprinted in

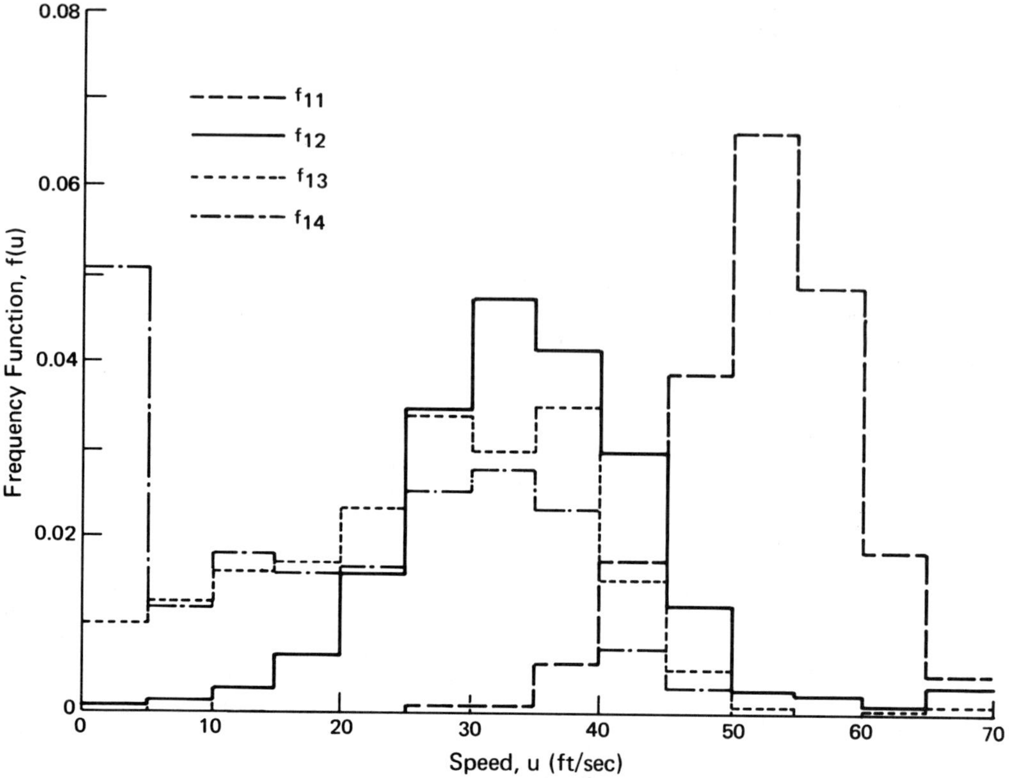

Figure 7.25 The frequency function, $f(u)$, of the observed speed distributions from the Port Authority of New York and New Jersey. The concentrations corresponding to distributions f_{11}, f_{12}, f_{13}, and f_{14} are 35.0, 62.0, 74.3, and 93.0 veh/mile, respectively.

TABLE 7.5 Moments of the Observed Speed Distributions[a]

Data Source	PANYNJ				SDC–BPR		
Distribution	f_{11}	f_{12}	f_{13}	f_{14}	f_{21}	f_{22}	f_{23}
Concentration (veh/mi)	35.0	62.0	74.3	93.0	32.7	49.5	88.4
Sample size	285	871	462	620	2709	984	1369
Mean, $\bar{u}$	52.52	34.35	26.76	20.13	48.17	47.42	41.79
Variance,[b] $m^{(2)}$	43.42	91.31	144.05	187.28	42.18	40.35	30.29
Third central moment, $m^{(3)}$	−99.22	318.23	−289.26	72.86	−19.62	−116.52	14.44
Fourth central moment, $m^{(4)}$	6487	37439	57830	57895	4379	5290	2153
$m^{(3)}/m^{(2)}$	2.285	3.485	−2.008	0.389	−0.465	−2.887	0.477
$m^{(4)}/m^{(2)}$	149.4	410.0	401.5	309.1	103.8	131.1	71.1

[a] SOURCE: Herman et al.[18]

[b] $m_0^{(2)}$ = second moment of desired speed distribution; $m^{(3)}$ = third moment of actual speed distribution; etc.

TABLE 7.6 Moment to Variance Ratios [a]

Data Source		PANYNJ					SDC–BPR	
Desired Speed Distribution (f_0)		f_{11}			f_{12}		f_{21}	
Speed Distribution		f_{12}	f_{13}	f_{14}	f_{13}	f_{14}	f_{22}	f_{23}
$m^{(3)}/m^{(2)}$	Data	3.485	−2.008	0.389	−2.008	0.389	−2.887	0.477
	Theory	−13.800	0.675	13.600	−7.144	0.515	−0.460	1.200
$m^{(4)}/m^{(2)}$	Data	410.02	401.47	309.13	401.47	309.13	131.10	71.08
	Theory	928.80	771.57	902.46	564.67	488.29	127.37	150.05
$m^{(2)}$	Data	91.31	144.05	187.28	144.05	187.28	40.35	30.29
	Theory	623.90	689.15	651.97	203.11	286.35	43.29	75.26

[a] SOURCE: Herman et al.[18]

TABLE 7.7 Calculated Moments of the Desired Speed Distribution

Data Source	PANYNJ		SDC–BPR
Distributions	f_{11},f_{12},f_{13}	f_{12},f_{13},f_{14}	f_{21},f_{22},f_{23}
$\bar{u}_0$	39.48	27.61	47.47
$m_0^{(2)}$	23.91	90.28	16.86
$m_0^{(3)}$	1712.80	−76.50	−44.88

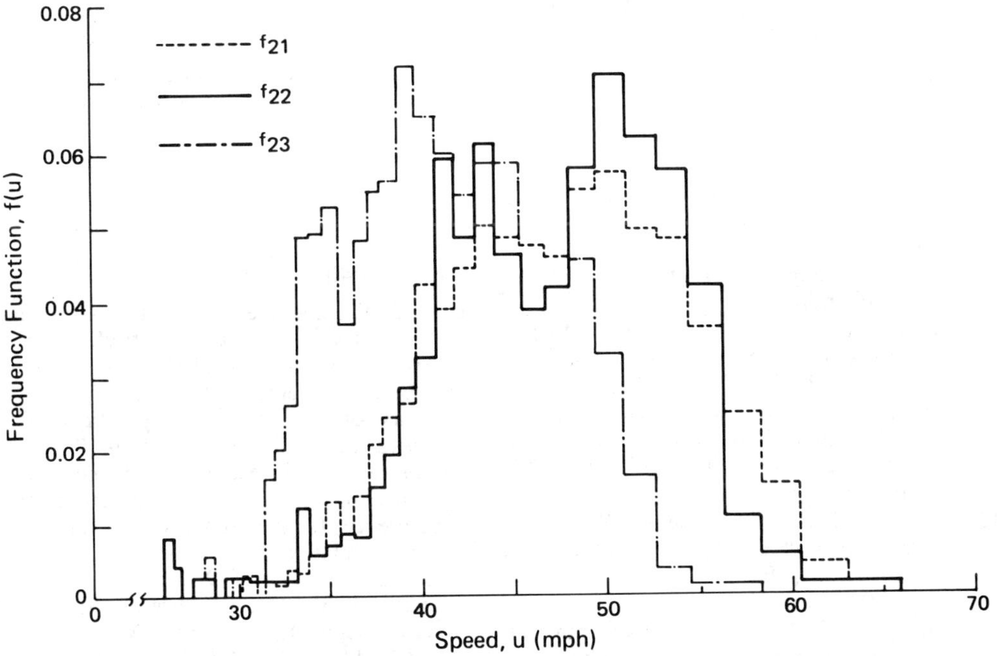

Figure 7.26 The frequency function, $f(u)$, of the observed speed distributions deduced from the Bureau of Public Roads data in Pipes.[3] The concentrations corresponding to f_{21}, f_{22}, and f_{23} are 32.7, 49.5, and 88.4 veh/mile, respectively.

Highway Research Board Special Report 79 (1964), pp. 7–35.
3. Pipes, L. A., Hydrodynamic approaches—Part I, pp. 3–5. In D. L. Gerlough and D. G. Capelle (ed) *An Introduction to Traffic Flow Theory*. Special Report 79, Highway Research Board, Washington, D.C. (1964).
4. Pipes, L. A., Vehicle accelerations in the hydrodynamic theory of traffic flow. Transp. Res., 3 (2):229–234 (1969).
5. Pipes, L. A., Topics in the hydrodynamic theory of traffic flow. Transp. Res., 2 (2):143–149 (1968).
6. Greenberg, H., An analysis of traffic flow. Oper. Res., 7 (1):79–85 (1959).
7. Richards, P. I., Shock waves on the highway. Oper. Res., 4 (1):42–51 (1956).
8. Pipes, L. A., Car following models and the fundamental diagram of road traffic. Transp. Res., 1 (1):21–29 (1967).
9. Rorbech, J., Determining the length of the approach lanes required at signal-controlled intersections on through highways—an application of the shock wave theory of Lighthill and Whitham. Transp. Res., 2 (3):283–291 (1968).
10. Graham, E. F., and Chenu, D. C., A study of unrestricted platoon movement of traffic. Traffic Eng., 32 (7):11–13 (1962).
11. Lewis, B. J., Platoon movement of traffic from an isolated signalized intersection. Highw. Res. Board Bull. 178 (1958), pp. 1–11.
12. Pacey, G. M., *The Progress of a Bunch of Vehicles Released from a Traffic Signal*. Research Note No. RN/2665/GMP. Road Research Laboratory, London (1956), mimeo.
13. Grace, M. J., and Potts, R. B., A theory of the diffusion of traffic platoons. Oper. Res., 12 (2):255–275 (1964).
14. Herman, R., Potts, R. B., and Rothery, R. W., Behaviour of traffic leaving a signalized intersection. Traffic Eng. Control, 5 (9):529–533 (1964).
15. Nemeth, Z. A., and Vecellio, R. L., Investigation of the dynamics of platoon dispersion. Highw. Res. Rec. 334 (1970), pp. 23–33.
16. Prigogine, I., A Boltzmann-like approach to the statistical theory of traffic flow. Proceedings of the [First] [International] Symposium on the Theory of Traffic Flow (Warren, Mich., 1959). Elsevier, Amsterdam, New York (1961), pp. 158–164.
17. Prigogine, I., and Herman R., *Kinetic Theory of Vehicular Traffic*. American Elsevier Publishing Co., New York (1971), pp. 17–54.
18. Herman, R., Lam, T., and Prigogine, I., Kinetic theory of vehicular traffic: comparison with data. Transp. Sci., 6 (4): 440–452 (1972).
19. Payne, H. J., Models of freeway traffic and control. In G. A. Bekey (ed.) *Mathematical Models of Public Systems*. Simulation Counc. Proc. Ser., 1 (1):51–61 (1971).
20. Edie, L. C., Foote, R. S., Herman, R., and Rothery, R., Analysis of single lane traffic flow. Traffic Eng., 33 (4):21–27 (1963).
21. Hillier, J. A., and Rothery, R., The synchronization of traffic signals for minimum delay. Transp. Sci., 1 (2):81–94 (1967).
22. Andrews, F. C., A statistical theory of traffic flow on highways. III. Distributions of desired speeds. Transp. Res., 7 (3):223–232 (1973).
23. Andrews, F. C., A statistical theory of traffic flow on highways. IV. Semi-empirical steady state theory. Transp. Res., 7 (3):233–241 (1973).
24. Anderson, R. L., Herman, R., and Prigogine, I., On the statistical distribution of traffic flow. Oper. Res., 10 (2):180–196 (1962).

7.10 RELATED LITERATURE

Hydrodynamic and Wave Models

Bick, J. H., and Newell, G. F., A continuum model for two-directional traffic flow. Q. Appl. Math., 18 (2):191–204 (1960).

De, S. C., Kinematic wave theory of bottlenecks of varying capacity. Proc. Camb. Philos. Soc., 52 (Part 3):564–572 (1956).

Foster, J., An investigation of the hydrodynamic model for traffic flow with particular reference to the effect of various speed–density relationships. Proc. Austr. Road Res. Board, 1:229–257 (1962).

Franklin, R. E., The structure of a traffic shock wave. Civil Eng. Public Works Rev., 56 (662):1186–1188 (1961).

Franklin, R. E., On the flow–concentration relationship for traffic. Proceedings of Second International Symposium on Theory of

Traffic Flow (London, 1963). O.E.C.D. (1965), pp. 120–128.
Lighthill, M. J., Dynamic instability of transport systems: the hydrodynamic analogy. Oper. Res. Q., 8 (3):109–114 (1957).
Oliver, R. M., *Fluid Flow Models of Traffic Streams*. Institute of Transportation and Traffic Engineering, University of California, Berkeley (1962), 52 pp. mimeo.
Pipes, L. A., Wave theories of traffic flow. J. Franklin Inst., 280 (1):23–41 (1965).
Wojcik, C. K., and Pipes, L. A., A study in wave theories of traffic flow. Paper presented at the National Meeting of the Operations Research Society of America, held in Santa Monica, California, May 1966.

Platoon Phenomena

Athol, P., Headway groupings. Highw. Res. Rec. 72 (1965), pp. 137–155.
Buckley, D. J., Tomlin, J. A., and Minson, W. G., Delays to traffic platoons. Proc. Austr. Road Res. Board, 3:364–376 (1966).
Daou, A., The distribution of headways in a platoon. Oper. Res., 12 (2):360–361 (1964).
Daou, A., On flow within platoons. Austr. Road Res., 2 (7):4–13 (1966).
Gerlough, D. L., Some problems in intersection traffic control, pp. 10–27. In R. Herman (ed.), Proceedings of [First] International Symposium on the Theory of Traffic Flow. Elsevier, New York (1961).
Grace, M. J., and Potts, R. B., Diffusion of traffic platoons. Proc. Austr. Road Res. Board, 1:260–267 (1962).
Kleinman, S., A mathematical model for platoon formation. Transp. Sci., 6 (2):202–204 (1972).
Newell, G. F., Platoon formation in tunnel traffic. Oper. Res., 7 (5):589–598 (1959).
Seddon, P. A., The prediction of platoon dispersion in the combination methods of linking traffic signals. Transp. Res., 6 (2): 125–130 (1972).

Boltzmann-Like Traffic Models

Andrews, F. C., A statistical theory of traffic flow on highways. I. Steady flow in low-density limit. Transp. Res., 4 (4):359–366 (1971).
Andrews, F. C., A statistical theory of traffic flow on highways. II. Three-car interactions and the onset of queueing. Transp. Res., 4 (4):367–377 (1970).
Gafarian, A. V., and Munjal, P. K., An experimental validation of two Boltzmann-type statistical models for multi-lane traffic flow. Transp. Res., 5 (3):211–224 (1971).
Herman, R., and Lam, T., On the mean speed in the "Boltzmann-like" traffic theory: analytical derivation. Transp. Sci., 5 (3):314–327 (1971).
Herman, R., and Lam, T., On the mean speed in the "Boltzmann-like" traffic theory: a numerical method. Transp. Sci., 5 (4):418–429 (1971).

7.11 PROBLEMS

1. Given a single-lane roadway as shown in the sketch and the three conditions at loca-

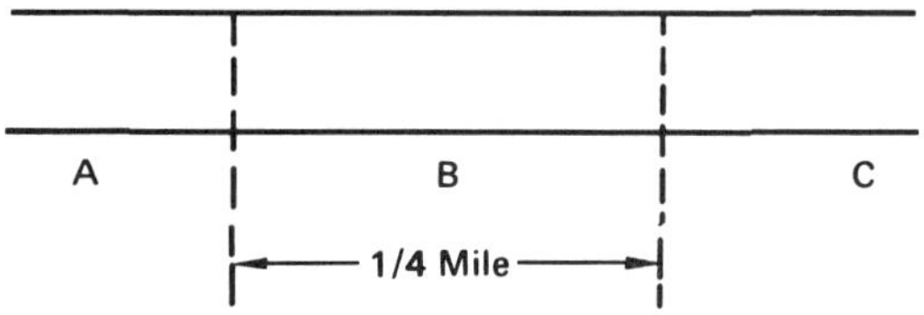

School Zone,
Operates Only for 15 Min

tions A, B, and C as: $q_A = 1{,}250$ vehicles/hr, $u_A = 50$ mph; $q_B = 1{,}000$ vehicles/hr, $u_B = 20$ mph; $q_C = 1{,}200$ vehicles/hr, $u_C = 40$ mph. Determine (a) $u_w(AB)$ (speed of shock wave AB), (b) $u_w(BC)$ (speed of shock wave BC), (c) length of queue at end of 15-min period, (d) time to dissipate queue after the end of period, and (e) distance from beginning of school zone to point of dissipation.

2. A line of traffic is moving with a free-flow speed, u_f, of 35 mph and a concentration of 40 vehicles/mile. This traffic stream is stopped for 30 sec at a signal. (a) What would be the velocity and direction of the wave of stopping? (b) What would be the length of the line of cars stopped for the signal? (c) How many cars would be stopped for the signal? (Jam concentration, $k_j = 200$ vehicles/mile.)

3. Consider a one-lane approach to a traffic signal with an approach volume of 400 vehicles/hr, a capacity volume of 1,600 vehicles/hr, a red signal time of 25 sec, and a jam density of 200 vehicles/mile. Based on the information provided in Section 7.5 and assuming Greenshield's flow–concentration model, what distance will the queue of cars occupy at the time the signal turns green? What is the distance from the stopline for which complete stoppage is required?

Chapter 8
QUEUEING MODELS*

8.1 INTRODUCTION

A desirable goal for transportation engineers is to design and operate facilities that minimize delay to the users. Delay resulting from congestion is a common phenomenon associated with many types of transportation problems. Vehicles wait in line on access ramps for an opportunity to enter a freeway; pedestrians queue up on a crosswalk in anticipation of a gap in road traffic or at a turnstile in a transit station; left-turn slots must be sufficiently long to store the maximum number of vehicles that can be expected to wait for a left-turn signal.

How long a user must wait, or what is the number of units waiting in line, or the proportion of time that a facility might be inactive (an empty parking stall, for instance)? Queueing models, employing the methods of probability and statistics, provide a means by which it is possible to predict some of these delay characteristics.

Queueing theory was first developed early in the twentieth century to deal with problems of telephone switching. Following World War II queueing was accepted for use in a wide range of situations. Adam [2] considered the problem of pedestrian delay at an unsignalized intersection in 1936. Tanner [3] expanded on the pedestrian problem in 1951, and in 1954 Edie [4] evaluated delays at toll booths by applying queueing models to an analysis of their operation. In the same year Moskowitz [5] reported on an empirical study of vehicles waiting for a gap in traffic.

The purpose of this chapter is to present some of the results of studies of probability models of traffic delay. Section 8.2 introduces some elements of queueing or waiting-line theory. The examples used in section 8.2 are concerned with delay problems that occur when all users pass through a single-channel control point, such as a left-turn slot or a single exit lane for a garage.

In section 8.3 the analysis is extended to consider several channels of service; for example, several parallel toll booths or the different stalls of a parking facility. In this section the case of a user who does not get served is also considered; for example, the person seeking a parking space but who then continues on to another destination when none is found.

Much urban traffic engineering is related to the operation of urban intersections. An understanding of delay at these intersections is necessary to obtain the greatest efficiency from existing and planned transportation systems. The analysis of delays at intersections is considered in section 8.5, beginning with an analysis of unsignalized intersections. Queueing models for more complex intersection control, such as pedestrian control or traffic-signal control, are also considered in this section.

A final application of queueing theory, the treatment of delay on roadways, is included in section 8.6. Except for the detailed development of the formulas given in section 8.2, this chapter avoids detailed mathematical development, but does present the theorists' assumptions and some results of interest. The model for service through a single channel is developed in detail because it demonstrates the relationship between probability theory and the behavior of waiting lines. Readers interested in further theoretical development of queueing models should consult textbooks such as Haight,[6] Prabhu,[7] Cox and Smith,[8] or Newell.[9]

8.2 FUNDAMENTALS OF QUEUEING THEORY

Queueing theory draws heavily on probability theory. To mathematically predict the characteristics of a queueing system, it is necessary to specify the following system characteristics and parameters:

A. Arrival pattern characteristics: (1) average rate of arrival and (2) statistical distribution of time between arrivals;

B. Service facility characteristics: (1)

* In preparing this chapter, the authors have drawn freely on the material of Cleveland and Capelle.[1] References to that material are acknowledged at this time. Specific references are noted only for the illustrations used in this text.

service time average rates and distribution and (2) number of customers that can be served simultaneously, or number of channels available;

C. Queue discipline characteristics, such as the means by which the next customer to be served is selected; for example, "first come first served," or "most profitable customer first."

To facilitate reference to these characteristics, a short notation in the form a/b/c has come into use. In this notation a letter denoting the type of arrival pattern is substituted for a; a letter denoting the type of service is substituted for b; a number designating the number of service channels is substituted for c. Symbols in the first two places are as follows:

- M = exponentially distributed (i.e., random) interarrival or service time;
- D = deterministic or constant interarrival or service time;
- G = general distribution of service times;
- GI = general distribution of interarrival times;
- E_k = Erlang distribution of interarrival or service times with Erlang parameter k.

Thus, M/G/1 designates a queue with random arrivals, general service distribution, and one server.

In some discussions it may be desirable to indicate queue length limitations and queue discipline. For such purposes the notation, M/M/1:(L/Disc) is used, where L is replaced by the maximum allowable length and Disc is replaced by a symbol for the appropriate queue discipline. The following are common disciplines:

- FIFO = first in–first out (i.e., service in order of arrival);
- SIRO = service in random order;
- LIFO = last in–first out.

Thus, E_k/D/2 (∞/FIFO) denotes a system with Erlangian arrivals, constant service, two service channels, infinite queue length (i.e., no limitation on queue length), and first come–first served discipline.

8.2.1. System State for M/M/1

The fundamental quantities characterizing a waiting line are the *states* of the system. The system is said to be in state n if it contains exactly n items (this includes all items being served as well as those waiting to be served). The value of n may be either 0 or some positive integer.

If the average arrival rate is called λ, the average interval between arrivals is $1/\lambda$. If the service rate of the system is μ, the average service time is $1/\mu$. The ratio $\rho=\lambda/\mu$, sometimes called the traffic intensity or utilization factor, determines the nature of the various states. If $\rho<1$ (that is, $\lambda<\mu$) and a sufficiently long time elapses, each state will be recurrent. This means that there is a finite probability of the queue being in any state n. If, on the other hand, $\rho\geq 1$, every state is transient and the queue length (the number in the system) will become longer and longer without limit. A fundamental theorem states that the queue will be in equilibrium only if $\rho<1$.

An understanding of the characteristics of queueing systems can be obtained from simple cases. Consider the case of a single-channel queueing system with a mean random Poisson arrival rate of λ customers per unit of time and where service times are independent and distributed exponentially with a mean rate μ. Let $P_n(t)$ be the probability that the queueing system has n items at time t. Consider the situation at time $t+\Delta t$ where Δt is so short that only one customer can enter or leave the system during this time.

Thus, for the period Δt, the following probabilities can be stated:

- $\lambda\Delta t$ = probability that one unit enters the system;
- $1-\lambda\Delta t$ = probability that no unit enters the system;
- $\mu\Delta t$ = probability that one unit leaves the system;
- $1-\mu\Delta t$ = probability that no unit leaves the system.

There are three ways in which the system can reach state n at time $(t+\Delta t)$ (when $n>0$):

1. The system was in state n at t and no customers arrived or departed in Δt. (The probability of simultaneous arrival and departure in Δt is considered to be zero.)

2. The system was in state $n-1$ at t and one customer arrived in Δt.

3. The system was in state $n+1$ at t and one customer departed in Δt.

The probability of the system being in state n at $(t+\Delta t)$ is

$$P_n(t+\Delta t)=P_n(t)[(1-\lambda\Delta t)(1-\mu\Delta t)]+P_{n-1}(t)[(\lambda\Delta t)(1-\mu\Delta t)]+P_{n+1}(t)[(1-\lambda\Delta t)(\mu\Delta t)] \quad (\text{for } n\geq 1) \quad (8.1)$$

Expanding and collecting terms,

$$P_n(t+\Delta t)-P_n(t)=-P_n(t)\ (\mu+\lambda)\Delta t+P_{n-1}(t)\lambda\Delta t+P_{n+1}(t)\mu\Delta t+\mu\lambda(\Delta t)^2\ [P_n(t)-P_{n-1}(t)-P_{n+1}(t)]$$

Neglecting terms with second-order infinitesimals and dividing by Δt,

$$\frac{P_n(t+\Delta t)-P_n(t)}{\Delta t}=\lambda P_{n-1}(t)-(\mu+\lambda)\ P_n(t)+\mu P_{n+1}(t)$$

Letting $\Delta t\to 0$,

$$\frac{dP_n(t)}{dt}=\lambda P_{n-1}(t)-(\mu+\lambda)\ P_n(t)+\mu P_{n+1}(t) \quad (8.2)$$

where $n=1,2,3\ldots.$

The probability of the system being in state 0 at time $(t+\Delta t)$ can come about in two ways: (1) There are no units in line at time t *and* none arrives in interval Δt; *or* (2) there is one unit in line at time t *and* one unit departs in interval Δt *and* none arrives in interval Δt. Expressing these relationships in terms of probabilities,

$$P_0(t+\Delta t)=P_0(t)\ (1-\lambda\Delta t)+P_1(t)[(\mu\Delta t)(1-\lambda\Delta t)]$$

Expanding, collecting terms (neglecting terms with second-order infinitesimals), and dividing by Δt,

$$\frac{P_0(t+\Delta t)-P_0(t)}{\Delta t}=\mu P_1(t)-\lambda P_0(t)$$

Letting $\Delta t\to 0$

$$\frac{dP_0(t)}{dt}=\mu P_1(t)-\lambda P_0(t) \quad (8.3)$$

When dealing with the steady state of the system (that is, when the probabilities of being in a given state do not change with time), the following results,

$$\frac{dP_n(t)}{dt}=0 \quad \text{for all } n \text{ at time } t. \quad (8.4)$$

From Eqs. 8.2, 8.3, and 8.4, it is then possible to set up systems of differential–difference equations for various steady states. The resulting equations are of the form

$$\mu P_{n+1}+\lambda P_{n-1}=(\lambda+\mu)P_n \qquad \text{for } n>0$$

and

$$\mu P_1=\lambda P_0 \qquad \text{for } n=0 \quad (8.5)$$

in which P_n is the value of $P_n(t)$ as $t\to\infty$.

The first few equations are as follows:

$$\lambda P_0=\mu P_1 \quad (8.6)$$

$$\lambda P_0+\mu P_2=(\lambda+\mu)P_1 \quad (8.7)$$

$$\lambda P_1+\mu P_3=(\lambda+\mu)P_2 \quad (8.8)$$

Recalling that $\rho=\lambda/\mu$ and noting (from Eq. 8.6) that $P_1=\rho P_0$ and substituting in Eqs. 8.7 and 8.8,

$$P_2=(\rho+1)P_1-\rho P_0=\rho^2P_0 \quad (8.9)$$

$$P_3=(\rho+1)P_2-\rho P_1=\rho^3P_0 \quad (8.10)$$

$$\vdots$$

$$P_n=\rho^nP_0 \qquad \text{for } n\geq 0 \quad (8.11)$$

Because the sum of all probabilities is 1,

$$\sum_{n=0}^{n=\infty}P_n=1$$

$$\begin{aligned}1&=P_0+\rho P_0+\rho^2P_0+\ldots\\&=P_0(1+\rho+\rho^2+\rho^3+\ldots)\\&=P_0\left(\frac{1}{1-\rho}\right) \qquad \text{for } \rho<1\end{aligned}$$

and

$$P_0=1-\rho \qquad \text{for } \rho<1 \quad (8.12)$$

Therefore, Eq. 8.11 may be written as $P_n=\rho^n(1-\rho)$.

The traffic intensity, ρ, can then be seen to express the fraction of time that the system is busy (P_0 is the probability that the system is empty and $1-P_0$ is the probability that it is occupied).

8.2.2 Average and Variance of Number of Units (Customers) in System (M/M/1)

The average number of customers in the system is

$$
\begin{aligned}
E(n) &= \sum_{n=0}^{n=\infty} nP_n \\
&= 0+P_1+2P_2+3P_3+\ldots \\
&= P_0(\rho+2\rho^2+3\rho^3+\ldots) \\
&= (1-\rho)\left[\frac{\rho}{(1-\rho)^2}\right] \qquad \text{for } \rho<1 \\
&= \frac{\rho}{1-\rho} \qquad \text{for } \rho<1 \qquad (8.13)
\end{aligned}
$$

The upper curve of Figure 8.1 illustrates this relationship. It will be noted that when the traffic intensity ρ exceeds about 0.8, the congestion (number in system) increases rapidly.

The variance of the number in the system is

$$\text{Var}(n) = \sum_{n=0}^{n=\infty} [n-E(n)]^2 P_n = \frac{\rho}{(1-\rho)^2} \qquad (8.14)$$

This relationship is plotted in Figure 8.2. The derivation of this expression may be found in a standard text on queueing theory.

8.2.3 Delay Time in the System (M/M/1)

Consider the total time a customer spends in the system (v) to be made up of two components: a time to wait before service, w (queueing time) plus a time in service, s (service time). The average number in the system, $E(n)$, is the product of the average time in the system, $E(v)$, multiplied by the arrival rate, λ, such that

$$E(n) = \lambda E(v)$$

$$E(v) = E(n)/\lambda$$

Substituting Eq. 8.13 for $E(n)$ and recalling that $\rho=\lambda/\mu$, this becomes

$$
\begin{aligned}
E(v) &= \left(\frac{\rho}{1-\rho}\right)\left(\frac{1}{\lambda}\right) \\
&= \left(\frac{\lambda}{\mu-\lambda}\right)\left(\frac{1}{\lambda}\right) = \frac{1}{\mu-\lambda} \qquad (8.15)
\end{aligned}
$$

the average time an arrival spends in the system. The expected time to wait before service (that is, the time spent waiting in a queue) is

$$E(w) = E(v) - E(s) \qquad (8.16)$$

where $E(s)$ is the average service time ($1/\mu$); thus, Eq. 8.16 may be rewritten as

$$E(w) = \frac{1}{\mu-\lambda} - 1/\mu = \frac{\lambda}{\mu(\mu-\lambda)} \qquad (8.17)$$

The average number of customers waiting to be served (the average queue length), $E(m)$, is the product of the average waiting time, $E(w)$, multiplied by the arrival rate, λ:

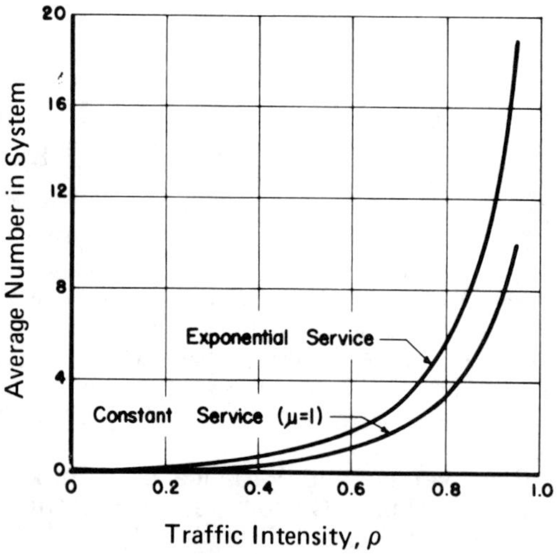

Figure 8.1 Average number in system as a function of traffic intensity.[1]

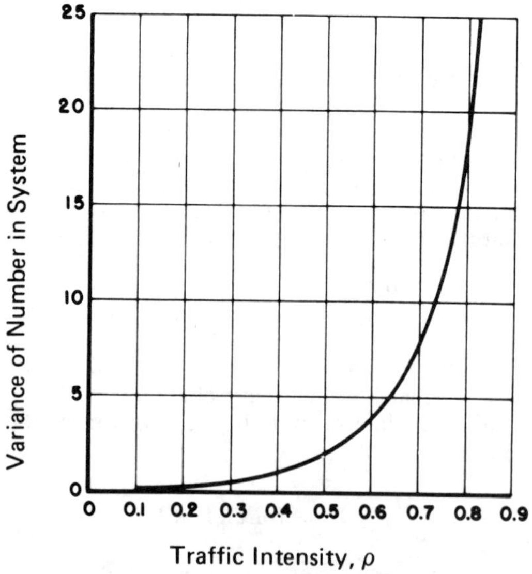

Figure 8.2 Variance of number in system as a function of traffic intensity.[1]

$$E(m)=\left[\frac{\lambda}{\mu(\mu-\lambda)}\right]\lambda=\frac{\lambda^2}{\mu(\mu-\lambda)} \quad (8.18)$$

Eq. 8.18 considers the average queue length over all time, including the periods when the queue is empty. Of interest is the average queue length, given that the queue length is greater than zero. This is defined as

$$E(m|m>0)=\frac{\text{average queue length}}{\text{prob. of nonempty queue}}=\frac{E(m)}{P(m>0)} \quad (8.19)$$

A zero queue length will occur if the system is in state zero or state one so that by Eqs. 8.11 and 8.12 the probability of a non-empty queue is

$$\begin{aligned} P(m>0) &= 1-(P_0+P_1) \\ &= 1-[(1-\rho)+\rho(1-\rho)] \\ &= 1-1+\rho-\rho+\rho^2=\rho^2=\frac{\mu^2}{\lambda^2} \quad (8.20) \end{aligned}$$

Substituting Eqs. 8.18 and 8.20 in Eq. 8.19 gives

$$E(m|m>0)=\left(\frac{\lambda^2}{\mu(\mu-\lambda)}\right)\left(\frac{\mu^2}{\lambda^2}\right)=\left(\frac{\mu}{\mu-\lambda}\right) \quad (8.21)$$

8.2.4 Example of Application of Queueing Formulas (M/M/1)

The exit from a parking garage is through a single gate where a variable fee is collected and change is made for drivers. Vehicles arrive at the gate at random at a rate, λ, of 120 vehicles/hr. The time to collect fees is exponentially distributed, with a mean duration ($1/\mu$) of 15 sec. What are the characteristics of the operation when $\lambda=120$ arrivals/hr, $\mu=4$ services/min$=240$ services/hr, and $\rho=\lambda/\mu=120/240=0.5$

(a) The probability of an idle booth (Eq. 8.12) is $1-0.5=0.5$.

(b) The probability that n vehicles will be in the system (Eq. 8.11) is

	$P_{x=n}$	$P_{x\leq n}$
$P_0=$	0.5	0.5
$P_1=$	0.25	0.75
$P_2=$	0.125	0.875
$P_3=$	0.0625	0.9375
$P_4=$	0.03125	0.96875
$P_5=$	0.015625	0.984375

If the garage operator wanted to be certain with 0.95 probability that departing vehicles would not interfere with other operations, he would need to provide space for about three vehicles, one in service, two in queue. Similarly, if he wished to be certain at the 0.99 probability level, he would have to provide space for 5 or 6 vehicles, one in service, the others in queue.

(c) The average number in the system (Eq. 8.13) is $E(n)=120/(240-120)=1$.

(d) The average number waiting in a queue (Eq. 8.18) is $E(m)=(120\times 120)/[240(120)]=0.5$.

(e) The average length of a nonempty queue (Eq. 8.20) is $E(m|m>0)=240/120=2$.

(f) The average time in the system (Eq. 8.15) is $E(v)=1/120$ hr$=0.5$ min.

(g) The average time waiting in a queue (Eq. 8.17) is $E(w)=120/[240(120)]=1/240$ hr$=0.25$ min.

8.3 THE CASE OF MULTIPLE CHANNELS WITH EXPONENTIAL ARRIVALS AND EXPONENTIAL SERVICE TIMES (M/M/N)

A parking lot (or the face of a block with on-street parking) may be considered as an example of a system with parallel service channels where the N parking slots represent the service channels. An arriving vehicle will occupy an empty slot if one is available; if not, it joins the waiting queue. The arrivals into the system are assumed to be random with rate λ and the service time per service channel (duration of parking) is also random with mean $1/\mu$. Again, ρ is defined as λ/μ. Further, ρ/N is defined as the utilization factor for the entire facility, representing the mean proportion of busy channels (full parking spaces). For the multiple-channel case the value of ρ may be greater than one but the following formulas apply only for the case where the utilization factor $\rho/N<1$.

8.3.1 Synopsis of Equations for Queues with Multiple Channels

Probability of n units in system

$$P_n=\frac{\rho^n}{n!}P_0 \qquad \text{for } n\leq N \quad (8.22)$$

$$P_n=\frac{\rho^n}{N^{n-N}N!}P_0 \qquad \text{for } n\geq N \quad (8.23)$$

Probability of no units in system

$$P_0 = \frac{1}{\sum_{n=0}^{N-1} \frac{\rho^n}{n!} + \frac{\rho^N}{N!(1-\rho/N)}} \quad (8.24)$$

Average queue length

$$E(m) = \frac{P_0 \rho^{N+1}}{N!N} \left[\frac{1}{(1-\rho/N)^2} \right] \quad (8.25)$$

Average length of nonempty queues

$$E(m|m>0) = \frac{1}{1-\rho/N} \quad (8.26)$$

Average number of units in system

$$E(n) = \rho + E(m) \quad (8.27)$$

Average time an arrival spends in the system

$$E(v) = E(n)/\lambda \quad (8.28)$$

Average waiting time in queue

$$E(w) = E(v) - 1/u \quad (8.29)$$

Average waiting time for an arrival who waits

$$E(w|w>0) = \frac{1}{\lambda} \frac{1}{1-\rho/N} \quad (8.30)$$

Probability of waiting for an empty space

$$P_{n>N} = P_0 \frac{\rho^{N+1}}{N!N(1-\rho/N)} \quad (8.31)$$

8.3.2 Example

The branch office of a customer repair service operates a fleet of vehicles that return to the office for picking up spare parts and assignments. The repair vehicles arrive at random during the day at a rate λ of four vehicles/hr. The stay at the parking lot is exponentially distributed, with a mean duration of 0.5 hr ($u=2$). There are five stalls ($=N$) in the lot set aside for the vehicles. What are the characteristics of the lot operation?

Solution

$N=5$ stalls, $\lambda=4$ arrivals/hr, $u=2$ services/hr, $\rho=\lambda/u=4/2=2$, and the utilization factor$=\rho/N=2/5=0.4$.

(a) P_0=probability of empty lot (Eq. 8.24)

$$= \frac{1}{1+\frac{2}{1!}+\frac{2^2}{2!}+\frac{2^3}{3!}+\frac{2^4}{4!}+\frac{2^5}{5!(0.6)}}$$

$$= \frac{1}{7.44444} = 0.134328.$$

(b) The probability that n vehicles will park (Eq. 8.22)* is

$P_0 = 0.134328$
$P_1 = (2/1)(0.134328) = 0.268656$
$P_2 = (2/2)(0.268656) = 0.268656$
$P_3 = (2/3)(0.268656) = 0.179104$
$P_4 = (2/4)(0.179104) = 0.089552$
$P_5 = (2/5)(0.089552) = 0.035821$

(c) The probability that a vehicle will have to wait on arrival (Eq. 8.31) is $P(n>5) = 0.134328 \frac{2^6}{5!5(0.6)} = 0.023880$, which is the same as $1-(P_0+P_1+P_2+P_3+P_4+P_5)$.

(d) The average number of vehicles waiting for an empty slot (Eq. 8.25) is $E(m) = 0.134328 \frac{2^6}{5!5(0.6)^2} = 0.0398$ vehicles.

(e) The average number of waiting vehicles when the lot is full (Eq. 8.26) is $E(m|m>0) = 1/0.6 = 1.67$ vehicles.

(f) The average number of vehicles parked and waiting (Eq. 8.27) is $E(n) = 2+0.0398 = 2.0398$ vehicles.

(g) The average time a vehicle spends parking and waiting (Eq. 8.28) is $E(v) = 2.0398/4 = 0.50995$ hr.

(h) The average time spent in waiting for an empty slot (Eq. 8.29) is $E(w) = 0.50995 - 1/2 = 0.00995$ hr.

(i) The average time a waiting vehicle waits for an empty slot (Eq. 8.30) is $E(w|w>0) = \frac{1}{4} \frac{1}{0.6} = 0.417$ hr.

8.3.3 System with Infinite Stalls: An Example of the M/M/∞ Case

As the number of stalls becomes very large ($N \to \infty$), Eq. 8.24 gives

$$\lim_{N\to\infty} P_0 = \frac{1}{e^\rho} = e^{-\rho} \quad (8.32)$$

and the expected number of parked vehicles is

$$E(n) = \sum_{n=0}^{\infty} nP(n) = \rho P_0 \left(1+\rho+\frac{\rho^2}{2!}+\ldots \right) = \rho \quad (8.33)$$

* Computation can be simplified by observing that $P_n = (\rho/n)P_{n-1}$.

8.3.4 The System with Loss

A more realistic type of operation for a parking lot occurs when vehicles unable to park go away instead of waiting in line; that is,

$$P_n = 0 \qquad \text{for } n > N$$

For this model the probability that n vehicles will be parked is

$$P_n = \frac{\rho^n/n!}{\sum_{i=0}^{N} \rho^i/i!} \qquad \text{for } n = 0,1,2,\ldots,N \qquad (8.34)$$

The probability of an empty lot is

$$P_0 = \frac{1}{\sum_{i=0}^{N} \rho^i/i!} \qquad (8.35)$$

and the probability that a car cannot park is the probability that there are N slots occupied, so that

$$P_N = \frac{\rho^N/N!}{\sum_{i=0}^{N} \rho^i/i!} \qquad (8.36)$$

Eq. 8.36 is called Erlang's Loss Formula, $L_N(\rho)$, the probability that an incoming unit is "lost" to the system.

Finally, the average number of vehicles in the parking lot may be developed as follows:

$$E(n) = \sum_{n=0}^{N} n\,P_n$$

$$= \frac{\sum_{n=0}^{N} n\frac{\rho^n}{n!}}{\sum_{n=0}^{N} \frac{\rho^n}{n!}}$$

$$= \frac{\left[\frac{1\rho}{1} + \frac{2\rho^2}{2\cdot 1} + \frac{3\rho^3}{3\cdot 2!} + \frac{4\rho^4}{4\cdot 3!} + \cdots + \frac{N\rho^N}{N\cdot(N-1)!}\right]}{\sum_{n=0}^{N} \frac{\rho^n}{n!}}$$

$$= \frac{\rho\left[1 + \rho + \frac{\rho^2}{2!} + \frac{\rho^3}{3!} + \cdots + \frac{\rho^{N-1}}{(N-1)!}\right]}{\sum_{n=0}^{N} \frac{\rho^n}{n!}}$$

$$E(n) = \rho \frac{\sum_{n=0}^{N-1} \frac{\rho^n}{n!}}{\sum_{n=0}^{N} \frac{\rho^n}{n!}} \qquad (8.36)$$

If Eq. 8.36 is multiplied by $e^{-\rho}/e^{-\rho}$, the expression becomes

$$E(n) = \rho \frac{\sum_{n=0}^{N-1} \frac{e^{-\rho}\rho^n}{n!}}{\sum_{n=0}^{N} \frac{e^{-\rho}\rho^n}{n!}} \qquad (8.37)$$

in which case it is possible to use tabulations of the Poisson distribution function to get the desired answer.

A more elegant discussion of the relationship between the Poisson distribution and the queueing formulas developed here is given by Kometani and Kato [10] and by Haight and Jacobson.[11]

8.3.5 Example of a Queueing System Operating with Loss

Assume the same data given in example 8.3.2, except that vehicles do not wait for an empty space. Again, $N=5$, $\lambda=4$ arrivals/hr, $u=2$ service/hr, $\rho=\lambda/u=4/2=2$, and utilization factor $=\rho/N=2/5=0.4$.

The probability of an empty slot (Eq. 8.35) is

$$P_0 = \frac{1}{1+\frac{2}{1}+\frac{2^2}{2}+\frac{2^3}{3!}+\frac{2^4}{4!}+\frac{2^5}{5!}} = \frac{1}{7.2667}$$

$$= 0.137614$$

$$P_1 = (2/1)P_0 = 0.275228$$
$$P_2 = (2/2)P_1 = 0.275228$$
$$P_3 = (2/3)P_2 = 0.183485$$
$$P_4 = (2/4)P_3 = 0.091743$$
$$P_5 = (2/5)P_4 = 0.036697$$ (probability that a car cannot park)

The average number of vehicles parked (Eq. 8.37) is $E(n) = 2\frac{0.947}{0.983} = 1.93$.

8.4 SYSTEM M/D/1 BUSY PERIOD

In this section the number of units, n, that will be served before the system will again become empty is considered, given r units in the system at a given time. For example, there may be five vehicles (r) in a queue at the start of the green interval on a traffic signal. It is desired to find the number n of vehicles that will pass through the intersection before the queue is dissipated; i.e., the system is again empty. The solution to this particular problem has been accomplished by assuming the M/D/1 queueing model; that is, random arrivals into the single queue with a uniform service time for each of the "customers."

In this system vehicles arrive randomly at a rate λ, there is one server, and each vehicle is "served" for exactly B units of time, so that r vehicles in line will require rB units of time for service. The distribution of the number of units served, n, before the system again becomes empty in a busy period starting with an accumulation of r units is given by the Borel–Tanner distribution:

$$P(n|r) = \frac{r}{n}\,\frac{e^{-\lambda nB}(\lambda nB)^{n-r}}{(n-r)!} \qquad (8.38)$$

$$n = r, r+1, \ldots$$

The development of Eq. 8.38 may be found in Prabhu.[7] This equation can be rewritten by noting that $\rho = \lambda B$

$$P(n|r) = \frac{r}{n}\,\frac{e^{-n\rho}n\rho^{n-r}}{(n-r)!} \qquad (8.39)$$

$$n = r, (r+1), (r+2), \ldots .$$

Tabulated values for a limited range of n and r, given $\rho = 0.2$, are presented in Table 8.1.

For example, if there are three units in a queue, including the one being served, the probability that no more units will arrive during the service for these three is 0.549, but the probability that exactly five will be served before the queue empties again is 0.110. Haight [12] used the Borel–Tanner distribution to analyze delay at a signalized intersection (see section 8.5.3), whereas Tanner [13, 14] has applied the Borel–Tanner distribution to a model for delays to vehicles on two-way, two-lane roads.

TABLE 8.1 State Probabilities for Borel–Tanner Distribution ($\rho = 0.2$)

n \ r→	1	2	3	4	5
1	0.819				
2	0.134	0.670			
3	0.033	0.220	0.549		
4	0.010	0.072	0.270	0.449	
5	0.003	0.025	0.110	0.294	0.368
6	0.001	0.009	0.043	0.145	0.301
7	a	0.003	0.017	0.064	0.172
8	a	0.001	0.007	0.028	0.086
9	a	a	0.003	0.012	0.040
10	a	a	0.001	0.005	0.018
11	a	a	a	0.002	0.008
12	a	a	a	a	0.003

[a] Probability $(n|r)$ less than 0.001.

8.5 DELAYS AT INTERSECTIONS

Applications of queueing theory to problems represented by traffic situations are more complex than those developed in sections 8.2 and 8.3. For example, the time a vehicle waits in line at a stop-sign-controlled intersection is a function of a combination of gap acceptance characteristics, the passage of gaps in the main stream and the characteristics of the waiting stream of traffic.

Treatment of intersection problems may be categorized by two elements: (1) the type of control (stop sign, yield sign, fixed-time signal, or traffic-actuated signal) and (2) the element controlled (vehicles or pedestrians).

At the stop-sign-controlled intersection, it is assumed that the side-street traffic waits for an adequate gap in the main-street traffic before crossing.

The problem of crossing the main street will be considered for both pedestrians and vehicles. There is a fundamental difference between these two cases. Pedestrians arrive at the crossing and accumulate at the curb until an opportunity to cross presents itself. The entire group then crosses together, independent of the number of pedestrians waiting. On the other hand, later vehicular arrivals cannot cross the main stream until the first vehicle in line has departed. If side-street vehicular flow is so low that two or more vehicles will rarely be waiting, the calculation of delays to individual vehicles will be similar to those used for individual pedestrians.

The problem of pedestrians crossing at a pretimed signalized intersection is tractable when conflicts with cross-street turning traffic are ignored. Under these conditions delays to these pedestrians can be easily determined from knowledge of the pedestrian arrival distribution and the traffic signal timing. The unsignalized and signalized intersection delay problems are treated in sections 8.5.2 and 8.5.3, respectively.

8.5.1 Blocks, Gaps, Intervals, and Lags—Some Definitions

A stream of traffic can be considered as the passage of a succession of vehicles, or a succession of gaps, or a succession of blocks.

Raff[15] and Oliver[16] have considered the problem of defining the intervals between vehicles that might be considered as acceptable to a vehicle crossing or merging with a stream of traffic.

Consider a stream of traffic as shown in Figure 8.3 where the time of arrival of main-street vehicles is shown on a time scale (events 2 through 9).

Intervals 2 through 9 are time intervals (h) between the arrival of main-street vehicles at the projected path of the crossing side-street vehicle. The first interval (a lag) is defined by the time from the arrival of the side-street vehicle at the point of crossing (event number 1) to the passage of the next main-street vehicle. If h is greater than the critical headway τ, the waiting driver or pedestrian will cross, otherwise he will wait.

Intervals 2, 6, 7, and 9 are all greater than τ, but it is evident that only a portion of these four intervals is available for crossing. That time during which no crossing is possible has been defined by Raff as a "block"; conversely, the rest of the time is defined as "antiblocks."

Oliver defined any time interval ($h>\tau$) as a gap and the remaining intervals as nongaps, as shown in Figure 8.3. The length of a gap is seen to be the length of a block less the critical headway (for $h>\tau$). Time intervals 2, 6, 7, and 9 are each a gap; time intervals 3, 4, and 5 are grouped into a single nongap; and

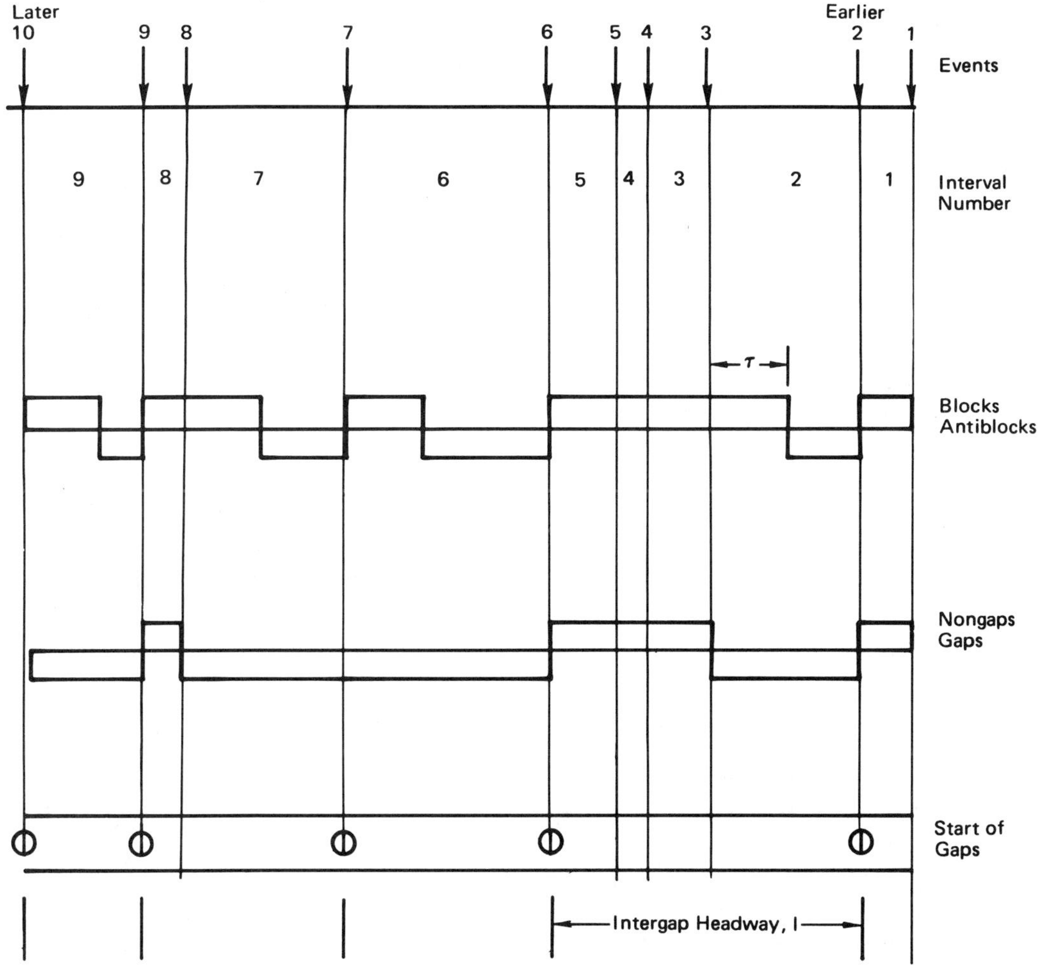

Figure 8.3 Definitions of time interval for stream flow.

intervals 1 and 8 are also nongaps. It is of interest to note that the number of blocks, antiblocks, and gaps are all equal. The number of vehicles in an intergap headway is here defined as the number occurring just after the vehicle (or event) defining the start of a gap but including the last vehicle that defines the end of the nongap. For example, the first intergap headway in Figure 8.3 contains four vehicles (numbers 3, 4, 5, and 6); the last intergap headway shown contains one vehicle (number 10). If event 3 were to represent the arrival of a pedestrian, he would be delayed until events 4, 5, and 6 (passage of main-street vehicles) had taken place.

8.5.2 Unsignalized Control—Pedestrian Delay

Delays are discussed with the assumption that the delayed person is in a position to accept or reject a gap without the added delay of waiting for the person in front to enter or cross the traffic stream. A group of pedestrians waiting to enter a crosswalk is an example of this case.

Pedestrian delay at an unsignalized intersection was first treated by Adams [2] in 1936 in one of the earliest theoretical traffic papers. He assumed that pedestrian and vehicle arrivals are random and made field observations that generally justified the assumption. If it is assumed that the main-street flow is q (vehicles/sec) and that an interval τ (the critical gap in seconds) is required between successive arrivals on the main street for a pedestrian to cross safely, several delay relationships can be derived.

By the equations of Chapter 3, the probability that a pedestrian will pass without delay is

$$P(h>\tau)=e^{-q\tau} \tag{8.40}$$

The probability that pedestrians will be delayed is

$$P_d=1-e^{-q\tau} \tag{8.41}$$

which is plotted in Figure 8.4 with main-street flow expressed in vehicles per minimum acceptable gap.

Of particular interest is the average duration of blocks, antiblocks, gaps, and nongaps as defined in Figure 8.3 and the time a pedestrian must wait (block time) for an appropriate gap in order to cross the roadway. Roadway events are considered relative to the passage of vehicles over an elapsed time t, during which the number of events is the accumulated volume, qt. Further, the mean headway $(1/q)$ is defined as T.

Each time interval $(h>\tau)$ is the beginning of an antiblock and therefore also marks the end of a block, so that in the elapsed time t the number of time intervals

$$\begin{aligned}(h>\tau) &= \text{number of antiblocks}\\ &= \text{number of blocks}\\ &= \text{number of gaps}\\ &= qt\,e^{-q\tau}\end{aligned} \tag{8.42}$$

the time spent in antiblocks is

$$=te^{-q\tau} \tag{8.43}$$

and the time spent in blocks is

$$t-te^{-q\tau}=t(1-e^{-q\tau}) \tag{8.43a}$$

The total time spent in gaps (Figure 8.3) is the sum of antiblock time $+\tau\times$ (number of antiblocks)

$$\begin{aligned}&=te^{-q\tau}+qte^{-q\tau}\ \tau\\ &=(t+qt\tau)e^{-q\tau}\end{aligned} \tag{8.44}$$

and the proportion of time spent in gaps (Eq. 8.44/t)

$$=(1+q\tau)e^{-q\tau} \tag{8.45}$$

which was first proposed by Adams in his 1936 study of delay to pedestrian traffic.

The average time duration (sec) of all gaps = (total gap time, Eq. 8.44)/(number of gaps, Eq. 8.42)

$$=\frac{(t+qt\tau)e^{-q\tau}}{qt\,e^{-q\tau}}=T+\tau \tag{8.46}$$

which is equation III of Adams.

The average time duration for all intervals $(h<\tau)$ = average length of nongaps

$$\begin{aligned}&=(\text{total nongap time})/\\ &\quad(\text{number of intervals } h<\tau)\\ &=[t-(t+qt\tau)e^{-q\tau}]/qt(1-e^{-q\tau})\\ &=T-\frac{\tau e^{-q\tau}}{1-e^{-q\tau}}\end{aligned} \tag{8.47}$$

which is Eq. IV of Adams.

The average time duration of blocks is (total block time, Eq. 8.43a)/(number of blocks, Eq. 8.42)

$$\begin{aligned}&=\frac{t(1-e^{-q\tau})}{qte^{-q\tau}}=\frac{t}{qte^{-q\tau}}-\frac{te^{-q\tau}}{qte^{-q\tau}}\\ &=\left(\frac{1}{q}\right)\left(\frac{1}{e^{-q\tau}}\right)-T=(T/e^{-q\tau})-T\end{aligned} \tag{8.48}$$

while the average time duration of antiblocks (Eq. 8.43/Eq. 8.42)

$$= \frac{te^{-q\tau}}{qte^{-q\tau}} = 1/q = T$$

This last equation can be compared with the length of gaps expressed in Eq. 8.46, demonstrating the relationship between gaps and antiblocks.

In considering the problem of delays, reference is again made to Figure 8.3, recalling that the beginning of a gap can be defined by the arrival of a side-street vehicle or pedestrian, as well as by the passage of a main-street vehicle. A pedestrian may arrive in two positions: (a) as the event defining a gap (no delay) or (b) during the nongap interval. In the latter instance, the arrival must wait for the remaining vehicles within the gap to pass before he can cross the stream.

The average number of vehicles between the start of gaps (see Figure 8.3 for relationship between start of gaps and the start of blocks)

$$= \frac{\text{volume}}{\text{number of blocks}} = \frac{qt}{qte^{-q\tau}} = \frac{1}{e^{-q\tau}} \qquad (8.49)$$

A delayed vehicle or pedestrian has to wait for one less vehicle than is given in Eq. 8.49, or

$$\frac{1-e^{-q\tau}}{e^{-q\tau}} \qquad (8.50)$$

From this, it is noted that the expected delay, $E(t)$, is found by multiplying the average number of waits (Eq. 8.50) by the average length of gap ($h<\tau$), Eq. 8.47; that is,

$$E(t) = \left(\frac{1-e^{-q\tau}}{e^{-q\tau}}\right) \times \left(T - \frac{\tau e^{q\tau}}{1-e^{-q\tau}}\right)$$
$$= \frac{1}{qe^{-q\tau}} - T - \tau \qquad (8.51)$$

Eq. 8.51 is plotted in Figure 8.5 with delay in terms of the minimum crossing gap required. For example, if there is one vehicle per minimum gap and the minimum required gap is 5 sec (equivalent flow=720 vehicles/hr), the delay will be $(0.7 \times 5) = 3.5$ sec. Similarly, for a 10-sec value of τ (equivalent flow=360 vehicles/hr) the expected delay will be $(0.7 \times 10) = 7.0$ sec.

The amount of delay for all delayed pedestrians is equal to the average delay/proportion delayed:

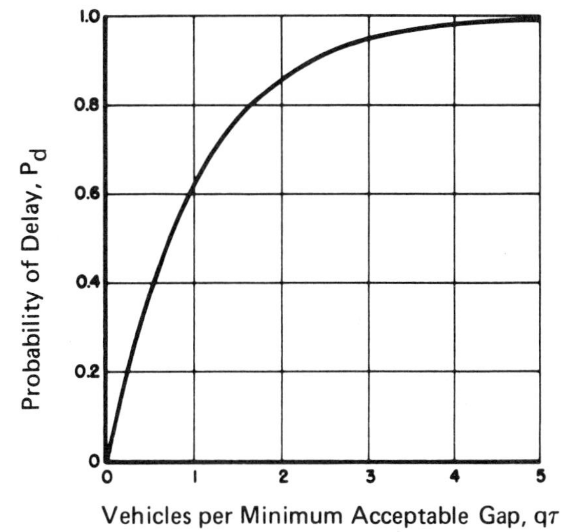

Figure 8.4 Probability of pedestrian delay.[1]

$$E_d(t) = \left(\frac{1}{qe^{-q\tau}} - \frac{1}{q} - \tau\right) \times \left(\frac{1}{1-e^{-q\tau}}\right)$$
$$= \frac{1}{qe^{-q\tau}} - \frac{\tau}{1-e^{-q\tau}} \qquad (8.52)$$

which is also plotted in Figure 8.5.

Underwood,[17] in applying the formulas for pedestrian delay to determine warrants for installation of pedestrian crossings, considered three levels of pedestrian treatment: (a) no treatment zone; (b) a "walking legs" sign zone (pedestrian sign and crosswalk markings in U.S. practice); and (c) a traffic control signal zone. His proposed method requires that three values be established: (a) minimum vehicular volume warrant; (b) minimum pedestrian volume warrant; and (c) maximum pedestrian volume warrant.

If the volume is less than that required by warrants (a) or (b), no treatment is required. In the case of warrant (a), the delay to pedestrians would be acceptable; in the case of warrant (b), because there are so few pedestrians delay would be acceptable. The "walking legs" sign zone (includes painted crosswalk) is used if warrants (a) and (b) are exceeded. If warrant (c) is also exceeded, a traffic control signal is justified.

The minimum vehicular volume warrant is determined from the application of Eq. 8.52. which is plotted in Figure 8.6 for values of τ at 9, 12, 15, and 18 sec. Line AA represents a point such that below the line delays increase

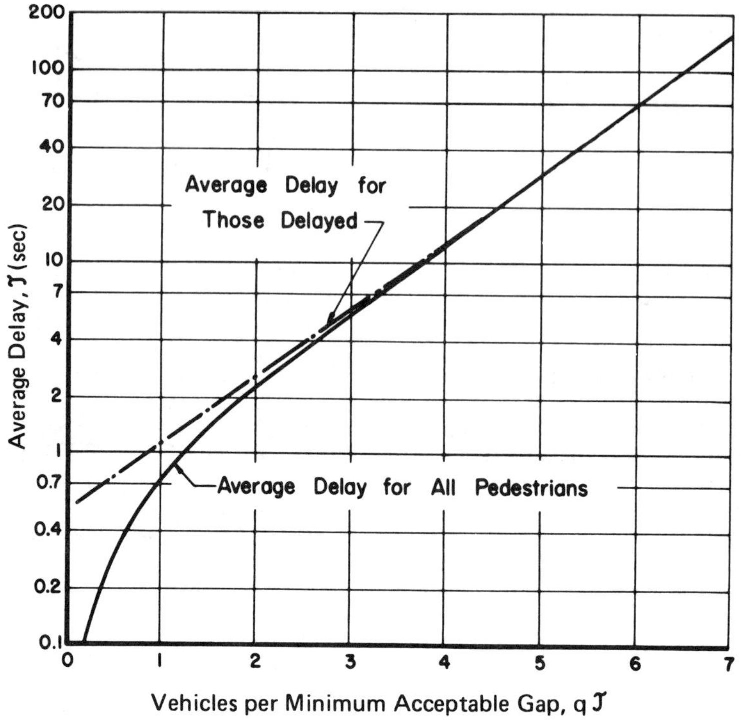

Figure 8.5 Average delay to pedestrians.[1]

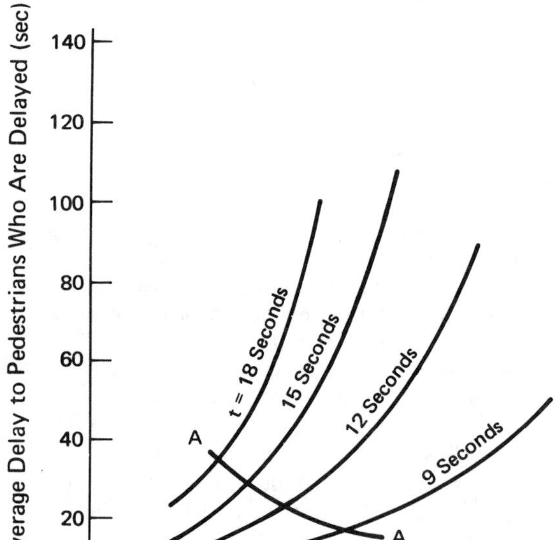

Figure 8.6 Pedestrian delay vs. vehicular volume.[17]

gradually; above it, increase at an accelerating rate. In each case the value $V \times \tau$ is approximately 6,000, where V=hourly vehicular volume (=3,600 q) and τ is the intervehicular spacing (sec) required by pedestrians. To attain the minimum,

$$V = \frac{6{,}000}{\tau} \tag{8.53}$$

an approximate formula for τ as given by Underwood is

$$\tau = \frac{R \times S}{30} + \frac{W}{v} + 2$$

where

R= perception time (sec; 2 or 3 sec);
S= speed limit (mph);
W= width of roadway (ft); and
v= crossing speed (ft/sec); about 4 ft/sec.

The minimum pedestrian volume is taken to be an average of one pedestrian per antiblock. The number of pedestrians delayed at an average of one delayed pedestrian per antiblock is $Ve^{-q\tau}$ (from Eq. 8.42) and the proportion of pedestrians delayed is $(1-e^{-q\tau})$; thus, the total number of pedestrians who may cross the traffic stream such that, on the average, not more than one pedestrian is waiting at any particular time is

$$P_{(\min)}=\frac{Ve^{-q\tau}}{(1-e^{-q\tau})}\text{pedestrians/hr} \qquad (8.54)$$

which is the minimum pedestrian warrant.

For example, at $V=720$ vehicles/hr ($q=0.2$ vehicles/sec) and $\tau=9$ sec, $P_{(\min)}=\frac{720\times e^{-1.8}}{(1-e^{-1.8})}=\frac{720\times 0.165}{0.835}=142$ pedestrians/hr.

Finally, Underwood reasoned that motorists who were required to give way to pedestrians in the crosswalks should have at least 60 percent of the available time. If less than 60 percent of the time is available, traffic signals are warranted to assure adequate time to the motorists.

The proportion of nondelayed vehicular traffic (from Eq. 8.40) is $e^{-Pg/3,600}$ where P is the hourly pedestrian volume and g is the time interval between pedestrians in the crosswalk such that there is a minimum safe distance between pedestrians.

Underwood takes the minimum safe distance between pedestrians as equal to the width of a vehicle plus 6-ft clearance on either side, or, for an 8-ft vehicle, a spacing of 20 ft. The time interval between pedestrians is given by $g=20/v$, where v is the crossing speed of pedestrians (ft/sec), so that

$$\begin{aligned}\text{Probability (no delay)} &= e^{-20P/3,600v}\\ &= e^{-P/180v} \qquad (8.55)\end{aligned}$$

For 60 percent probability $e^{-P/180v}=0.60$ $P/180v=0.51$ and $P\approx 90v$.

If $v=4$ ft/sec, a signal is warranted when pedestrian volume exceeds 360 pedestrians/hr.

Underwood prepared a family of curves based on Eqs. 8.53, 8.54, and 8.55, as shown in Figure 8.7. Areas to the left of and below the dashed curve associated with each minimum gap represent the domain of combined pedestrian and vehicular volumes for which no treatment is required. To the right of and above the same curves "walking legs" sign zones are required. Traffic control signals are required when the pedestrian volume exceeds 360 persons per hour; for example, if the pedestrian volume is 100/hr and the minimum acceptable gap is 10 sec, no control is required for vehicular volumes less than 800/hr. For the same minimum acceptable gap and a pedestrian crossing velocity of 4 ft/sec, a signal will be required for pedestrian volumes over 360/hr if the vehicular volume exceeds 600 vehicles/hr.

In 1951, Tanner [18] published the results of a comprehensive study of pedestrian crossing delays. His work is an extension of the delay relationships developed by Adams.[2] He assumed random arrivals of both main-street vehicles and crossing pedestrians and presented three proofs of the crossing delay distribution. (Details of these proofs are beyond the scope of this presentation but may be found in the original paper.) Tanner also developed a method of considering varying values of gap acceptance for different pedestrians and gave some attention to the problem of groups of pedestrians crossing the street.

Of particular interest to those designing pedestrian controls are the distribution of the size of pedestrian groups crossing together and the distribution of the number of pedestrians waiting at a random time.

The average size of a group crossing together is

$$E(n_c)=\frac{pe^{p\tau}+qe^{-q\tau}}{(p+q)e^{(p-q)\tau}} \qquad (8.56)$$

in which p is pedestrian flow and q is vehicular flow. Figure 8.8 shows this relationship. The average number waiting to cross is

$$E(n_w)=\frac{p}{q}(e^{q\tau}-q\tau-1) \qquad (8.57)$$

which is plotted in Figure 8.9.

As an example, consider a crossing with a vehicular volume of 720 vehicles/hr and a pedestrian volume of 360 persons/hr. The minimum acceptable gap, τ, is constant at 10 sec. For this example, $q=0.2$, $p=0.1$.

The probability that a pedestrian will be delayed (Eq. 8.41) is $P_d=1-e^{-2}=0.865$. The mean delay for all pedestrians (Eq. 8.51) is $E(t)=\frac{1}{0.2e^{-2}}-\frac{1}{0.2}-10=21.95$ sec; the mean delay per delayed pedestrian (Eq. 8.52) is $E_d(t)=\frac{1}{0.2e^{-2}}-\frac{10}{0.865}=25.38$ sec; the average number of pedestrians crossing together (Eq. 8.56) is $E(n_c)=\frac{0.1e^{1}+0.2e^{-2}}{e^{(0.1-0.2)10}(0.1+0.2)}=2.71$ pedestrians; the average number waiting to cross at

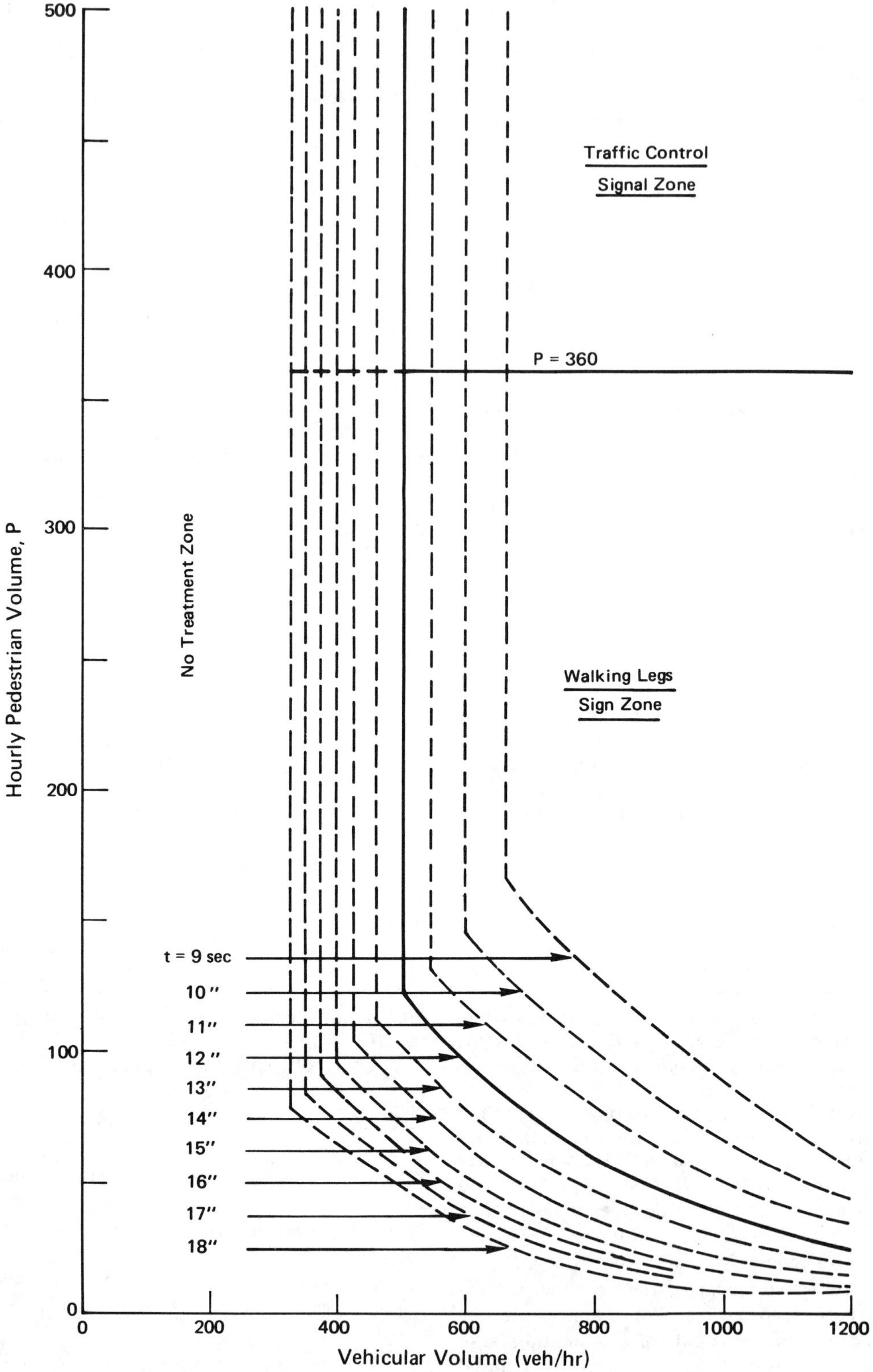

Figure 8.7 Warrants for pedestrian controls.[17]

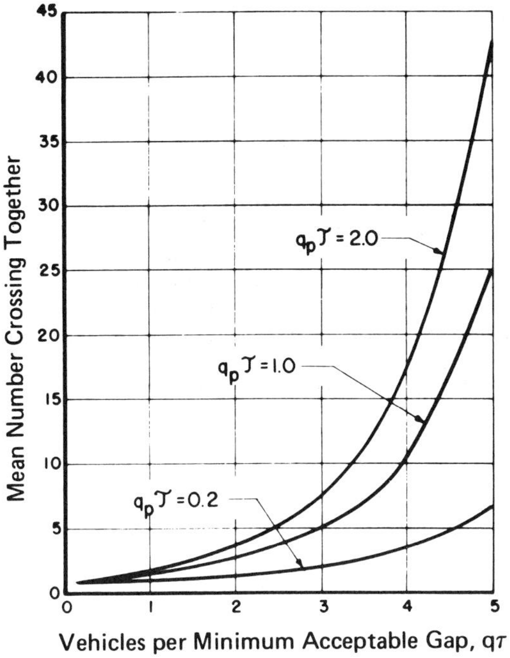

Figure 8.8 Average number of pedestrians crossing together.[1]

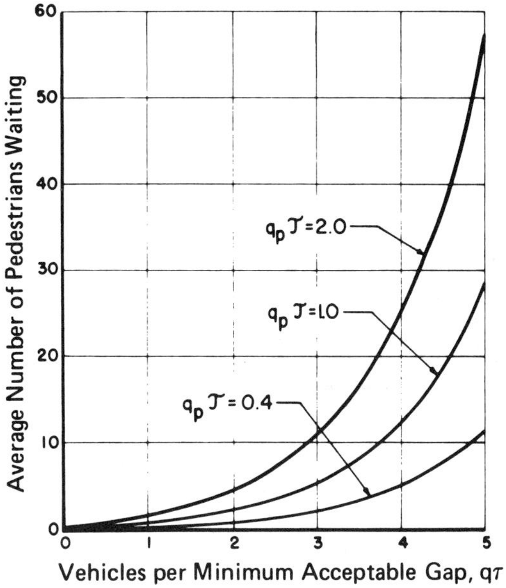

Figure 8.9 Average number of pedestrians waiting to cross street.[1]

a random time (Eq. 8.57) is $E(n_W) = \frac{0.1}{0.2}(e^2 - 2 - 1) = 2.19$ pedestrians.

Tanner also compared the delay to pedestrians crossing the entire roadway at one time with the delay to those stopping in the middle at a refuge island when necessary. His field studies indicated that pedestrians crossing the street where no island exists look for a gap of at least the critical lag in both directions of movement rather than for some combination of near- and far-stream gaps. The average delay, assuming the same volumes as given in the previous example, is 21.95 sec. When the pedestrian can stop in the middle of the street at a refuge island, assuming the vehicular volume is equal in both directions and that the critical gap required to cross one-half of the street is reduced to $\tau/2$, the formula for average delay is

$$E(t_s) = \frac{4}{qe^{-(q\tau/4)}} - \frac{4}{q} - \tau \qquad (8.58)$$

In the present example the delay with a pedestrian refuge would be $\frac{4}{0.2e^{-0.5}} - \frac{4}{0.2} - 10 = 2.97$ sec. These delays are plotted in Figure 8.10.

In the previous discussion of this section all of the authors assumed that distributions of main-street headways were negative exponential; but, as noted in Chapter 3, this distribution is unrealistic in describing small time headways between successive vehicles moving in a single roadway lane. Mayne [19] generalized Tanner's results to include an arbitrary distribution of independent main-street headways. He also considered the effects of introducing refuge islands on a wide crossing, showing that for the same average delay the pedestrian flow must be at least four times as great when an island is present as when there is no island.

Jewell [20] obtained the distribution, mean, and variance of waiting times for arbitrary mainstream headway distributions and for several main-street situations at the time a side-street vehicle presents itself. His relationships were developed for a critical lag τ and extended for other gap acceptance criteria. He obtained results for the number of minor-street vehicles that can be discharged during a fixed time period when only one side-street vehicle can cross during each acceptable main-street gap. He also showed that the mean delay for the side-street vehicle increases in proportion to the second or higher power of the critical gap and to the first or higher power of flow. The

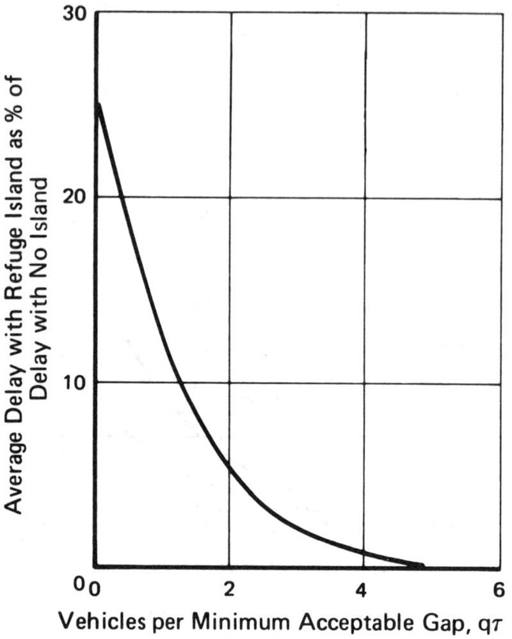

Figure 8.10 Pedestrian delays with and without refuge island.[1]

variance of delay increases in proportion to the third or higher power of the critical gap.

Weiss and Maradudin [21] and Herman and Weiss [22] developed several generalizations of the crossing delay problem studied by earlier investigators. The approach is based on renewal theory. A renewal process in time is the occurrence of random spacings from a known gap distribution. With their technique, it is possible to deal with a general independent distribution of main-street gaps and a general gap acceptance distribution. This makes it possible to consider the "yield sign" delay problem where the side-street vehicle has a different critical lag, depending on whether it is moving or stopped. It is also possible to develop delay functions for the impatient driver whose probability of accepting a given gap in the main street increases with the passage of main-street vehicles.

Weiss and Maradudin expressed delay characteristics for several gap and gap acceptance distributions. Herman and Weiss fitted shifted exponential constants experimentally. For Poisson main-street traffic with mean headway T and shifted exponential gap acceptance, the mean delay to side-street traffic as developed by Herman and Weiss [22] is

$$\mathrm{E}(t)=\frac{\mathrm{e}^{q\tau}-1}{q}-\tau+\frac{1}{b}\Bigg[\mathrm{e}^{q\tau}-1-q\tau+\left(\frac{q}{q+b}\right)^{2}\times(1+q\tau+b\tau)(1-\mathrm{e}^{-q\tau})+\mathrm{e}^{-q\tau}\left(\frac{q}{q+b}+q\tau\right)\Bigg] \qquad (8.59)$$

in which τ is the minimum acceptable gap, q is the main-street flow, and b is the parameter of the shifted exponential gap acceptance distribution, which equals $1/(T-\tau)$. Eq. 8.59 may be compared with Eq. 8.51, the delay with a constant gap acceptance value. The probability, $F(t)$, of accepting a gap of t sec in the main-street flow is

$$F(t)=0 \qquad (t\leq\tau)$$
$$F(t)=1-\exp[-b(t-\tau)] \qquad (t\geq\tau) \qquad (8.60)$$

The upper curve of Figure 8.11 presents a graph of this relationship for Herman and Weiss's constants, $\tau=3.3$ sec and $b=2.7$ sec^{-1}. The lower curve shows the results of assuming that all drivers have an acceptable gap of 3.3 sec.

Weiss and Maradudin also considered the yield sign problem, which differs from the pedestrian delay problem in acceptable gap between moving and stopped vehicles. If a moving vehicle requires a gap of τ_1 and a stopped vehicle requires a gap of $\tau_2(\tau_1\leq\tau_2)$, the mean delay is

$$\mathrm{E}(t)=\frac{\mathrm{e}^{q\tau_2}}{q}(1-\mathrm{e}^{-q\tau_1})+\mathrm{e}^{-q\tau_1}(\tau_2-\tau_1)-\tau_2 \qquad (8.61)$$

As an example, assume $\tau_2=3.3$ sec and $\tau_1=2.0$ sec. These values substituted into Eq. 8.61 yield a plot, as shown in Figure 8.12, that compares the average side-street vehicle delay (at a yield sign) with a stop sign situation where all side-street drivers are required to stop and wait for a main-street gap of 3.3 sec.

Weiss [23] further demonstrated that the delay to a single vehicle crossing or merging with a traffic stream is practically independent of the velocity distribution of the mainstream flow, for flow rates less than 1,600 vehicles/hr.

A different approach to the distribution of gaps in the main-street flow is given by Miller,[24] who postulated that bunches of vehicles (nongaps in Figure 8.3) are randomly distributed on a highway. Letting the flow of queues (bunches) be q/unit time and defining λ as the parameter of the exponential distribution of intervals between queues, $\lambda e^{-\lambda t}$, Miller relates q to λ as $1/q=1/\lambda+\bar{t}$, where $\bar{t}$ is the average length of a

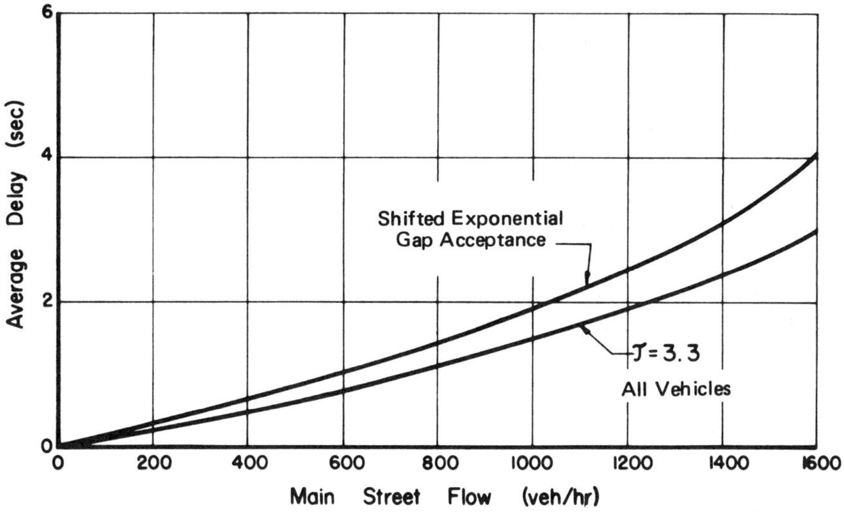

Figure 8.11 Average delay crossing a street.[1]

queue (Eq. 8.47 if traffic flow is random). For example, if the average length of a queue is 10 sec ($\bar{t}=10$) and $\lambda=0.025$, $1/q=1/0.025+10=50$ and $q=0.02$ queues/sec = 72 queues/hr.

One can arrive at an intuitive feeling for λ by observing that in one hour there will be 72 queues of 10-sec average duration $\bar{t}$, or a total of 720 sec of queues (nongap time). The remaining gap time ($3{,}600-720=2{,}880$ sec) is distributed among the 72 gaps (intervals between queues) so that the average gap length $=2{,}880/72=40$ sec$(1/\lambda)$. This is equivalent to a flow rate λ of 0.025 units/sec.

Miller derived an expression for the mean waiting time for a pedestrian or side-street vehicle as

$$E(t)=\lambda(\bar{t}+\tau)^2/2, \qquad (8.62)$$

where τ is the minimum acceptable gap and λ and $\bar{t}$ are defined as previously. For example, if it is assumed that a pedestrian needs a time gap of at least 10 sec ($\tau=10$), that λ is equivalent to 90 events/hr ($\lambda=1/40$), and that it takes on the average 10 sec ($\bar{t}=10$) for a queue to pass, $E(t)=1/2\times 1/40\ (10+10)^2=5$ sec.

The probability that a side-street vehicle can cross immediately is given by

$$P_0=(1-q\bar{t})e^{-\lambda\tau} \qquad (8.63)$$

Substituting appropriate values,

$$P_0=(1-10/50)e^{-0.25}=0.622.$$

Miller made a limited comparison of the average side-street delay and frequency of undelayed crossings predicted by the random bunches model with those produced by the random vehicle model. He found little difference in average waiting time for crossing vehicles. The random bunches model predicted the opportunities for immediate crossing better

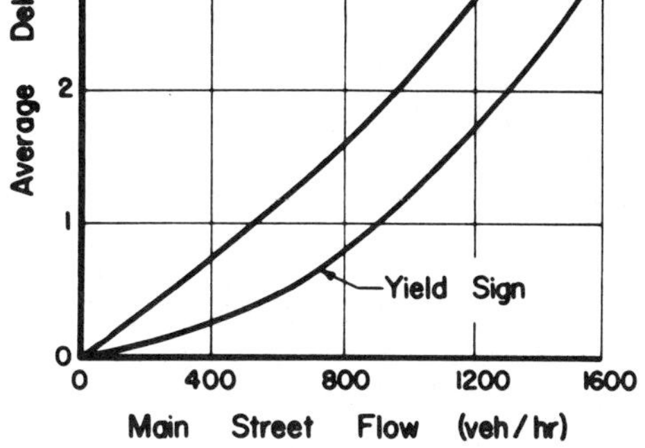

Figure 8.12 Average delay for side-street vehicles.[1]

than did the random vehicles model. Figure 8.13 gives the observed values for immediate crossing opportunities and the values predicted by the two theoretical models for 14 sets of observations, each set representing one hour of data collection.

8.5.3 Unsignalized Control—Vehicular Delay

Because the equations in section 8.5.2 make no allowance for queueing on the minor street, it is necessary to introduce a delay that reflects the time required for a second-in-line vehicle to get into position to accept or reject a gap. The case of a queue of vehicles waiting on a side road or entrance ramp before merging or crossing traffic on a main highway has been discussed by Evans, Herman, and Weiss,[25] Major and Buckley,[26] and Ashworth,[27] among others.

Consider a single inexhaustible queue of vehicles on the side street under the following conditions: when a main-street highway is less than τ, no vehicle enters; when a main-street headway is between τ and 2τ, one vehicle enters; when a main-street headway is between 2τ and 3τ, two vehicles enter, etc. The number of vehicles N that can enter from the side street, developed by Major and Buckley,[26] is found as follows:

Size of Headway	Number of Vehicles Entering Headway	Number of Headways This Size/Unit Time
$<\tau$	0	$q(1-e^{-q\tau})$
τ–2τ	1	$q(e^{-q\tau}-e^{-2q\tau})$
2τ–3τ	2	$q(e^{-2q\tau}-e^{-3q\tau})$
3τ–4τ	3	$q(e^{-3q\tau}-e^{-4q\tau})$
etc.		

The number of vehicles that will enter or cross the main-street flow per unit time (capacity of cross flow) is

$$N=q(e^{-q\tau}-e^{-2q\tau})+2q(e^{-2q\tau}-e^{-3q\tau})+3q(e^{-3q\tau}-e^{-4q\tau})+\dots$$
$$=qe^{-q\tau}+qe^{-2q\tau}+qe^{-3q\tau}+\dots$$

from which

$$N=\frac{qe^{-q\tau}}{1-e^{-q\tau}} \tag{8.64}$$

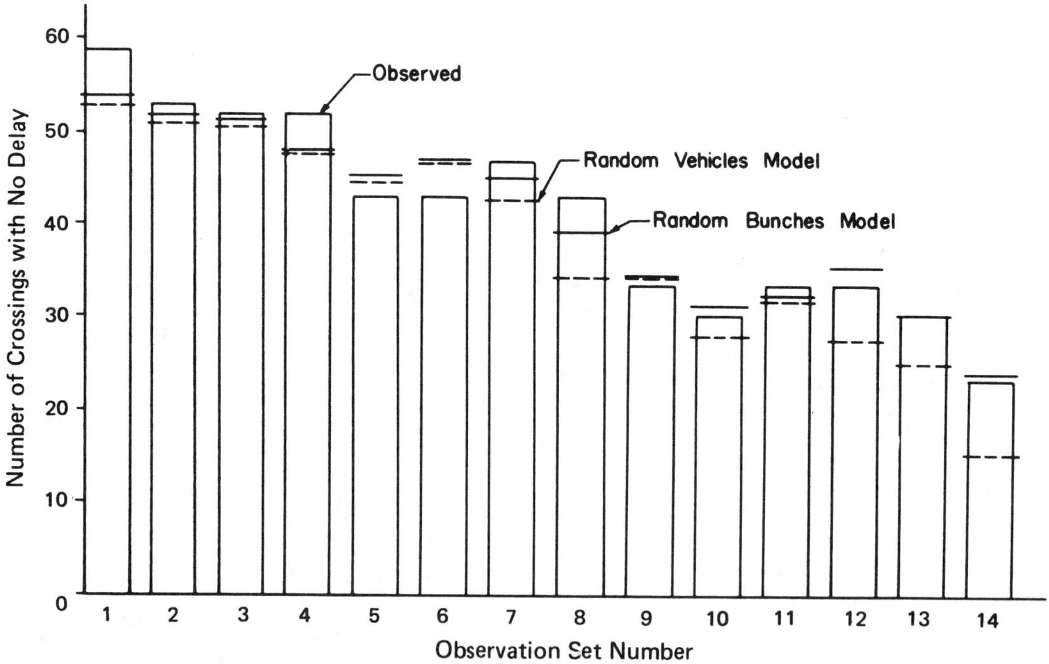

Figure 8.13 Comparison of undelayed crossing opportunities.[1] Each set represents 1 hr of observations. (Adapted from Miller,[24] Table 3.)

The more realistic assumption can be made that the headway for following vehicles is β_2 so that two vehicles require a headway of $\tau+\beta_2$ and three vehicles require $\tau+2\beta_2$, etc. Provided $\beta_2 \leq \tau$, Eq. 8.64 becomes

$$N=\frac{qe^{-q\tau}}{1-e^{-q\beta_2}} \tag{8.65}$$

If it is assumed that only one vehicle enters during each antiblock, the capacity of the cross street (from Eq. 8.4) is equal to $qe^{-q\tau}$ vehicles/unit time.

Ashworth[27] modified the approach used by Major and Buckley by assuming that the critical gap of the driver is τ_1, whereas that of the second driver is τ_2 sec. Further, he assumed a move-up time equal to a constant β_2 sec, during which time—following the departure of the vehicle at the head of the queue—the second vehicle moves into the front position but is unable to take advantage of any suitable gap offered. With these assumptions the author gives the average waiting delay at the head of the queue for those vehicles actually delayed:

$$E_d(t)=\ell+(e^{q\tau_2-1})\left(T-\right)\frac{\tau_2 e^{-q\tau_2}}{1-e^{q\tau_2}} \tag{8.66}$$

where

$$\ell=\frac{1+q(\tau_1-\beta_2)-e^{q(\tau_1-\tau_2-\beta_2)}(1+q\tau_2)}{q(1-e^{q(\tau_1-\tau_2-\beta_2)})}$$

and $T=1/q$.

The average waiting delay for all vehicles (provided $\tau_2>\tau_1-\beta_2$) is

$$E(t)=\tau_1-\tau_2-\beta_2+T(e^{q\tau_2}-e^{q(\tau_1-\beta_2)}) \tag{8.67}$$

Eqs. 8.67 and 8.66 may be compared with Eqs. 8.51 and 8.52, respectively, where the last two equations consider delay for the pedestrian case.

The influence of an acceleration lane on the merging problem is treated by Blumenfeld and Weiss.[28] In their model, vehicles on the acceleration lane of length L are assumed to travel at a constant velocity ν and vehicles on the main road travel at a constant velocity $V(>\nu)$. The model further assumes that the merging driver continues to move along the acceleration lane at a velocity ν until he either finds a suitable gap or continues to the end of the lane, at which point the velocity goes instantaneously to 0. The time T_L is the time for the vehicle to reach the end of the lane$=L/\nu$. Suppose a merge occurs at some time $t<T_L$; if the driver had not been merging, the distance traveled would have been Vt instead of νt actually traveled. This delay is defined as

$$D=\left(1-\frac{\nu}{V}\right)t=\beta t$$

where $\beta=1-\nu/V$.

If the time to merge is greater then T_L (say $t=T_L+t_1$, where $t_1>0$) the delay is defined by

$$D=t-\frac{\nu T_L}{V}=\beta T_L+t_1$$

The total delay to a vehicle in the acceleration lane can be written as the sum of two random variables, D_m+D_s, where D_m is the time spent traveling on the lane and D_s is the delay while stopped. The formula for the expected delay (developed in detail by Blumenfeld and Weiss[28]) is awkward, but the authors have found numerical solutions to relate the probability of reaching the end of the ramp without merging (Figure 8.14) and the expected delay to merging vehicles (Figure 8.15) as a function of the length of the acceleration lane.

8.5.4 Signalized Intersections

There are several models that may be used in the investigation of queues and delays at signalized intersections. In section 8.5.4.1 a continuum or fluid model is considered in which various measures of queue length and delay are developed. In section 8.5.4.2 various probability models are compared and different assumptions implicit to several types of models are noted.

In estimating delay at intersections, the traffic is considered as consisting of identical passenger car units (PCU). A truck, for example, may be considered as 1.5 or 2 PCU and a turning vehicle may be assigned some value depending on the type of maneuver that is made.

The following notation, after Allsop,[29] is used. Let

$c=$ the cycle time (sec);
$g=$ the effective green time (sec);
$r=$ the effective red time (sec);
$q=$ the average arrival rate of traffic on the approach (PCU/sec);

$$I = \frac{\text{variance of number of PCU arriving in one signal cycle}}{\text{mean number of PCU arriving in one signal cycle}};$$

s = the saturation flow on the approach (PCU/sec);
d = the average delay to PCU on the approach (sec);
Q_0 = the overflow (PCU);
$\lambda = g/c$ (i.e., the proportion of the cycle that is effectively green;
$y = q/s$ (i.e., the ratio of average arrival rate to saturation flow); and
$x = qc/gs$ (i.e., the ratio of average number of arrivals/cycle to the maximum number of departures/cycle).

Thus, $r+g=c$ and $\lambda x = y$. The ratio x is called the degree of saturation of the approach, and y is called the flow ratio of the approach.

The effective green time is the portion of the cycle time during which PCUs are assumed to pass the signal at a constant rate s, providing vehicles are waiting on the approach. Greenshields et al.,[30] for example, reported that the total time for a queue of n stopped vehicles to pass a signal can be given by

$$\text{Total time} = 14.2 + 2.1(n-5) \text{ sec} \qquad \text{for } n \geq 5$$

Had all of the vehicles departed at the saturation rate s (1/2.1), the first five vehicles would have required 10.5 sec; that is, the effective green is the signal green time less 3.7 sec. In most studies it is assumed that a waiting queue of vehicles will take advantage of the yellow clearance interval, although the effective green time may be adjusted to reflect particular operating conditions.

The meaning of arrival time and departure time for a PCU on an approach can be demonstrated by reference to Figure 8.16, in which distance–time curves are plotted for each of four vehicles. AB represents the passage of an undelayed vehicle, where the line PQ represents the stopline at which the first vehicle waits when there is a queue. CDEF represents the trajectory of the first vehicle that is delayed by a signal. The straight portions of CD and EF are parallel to AB and projected to meet PQ

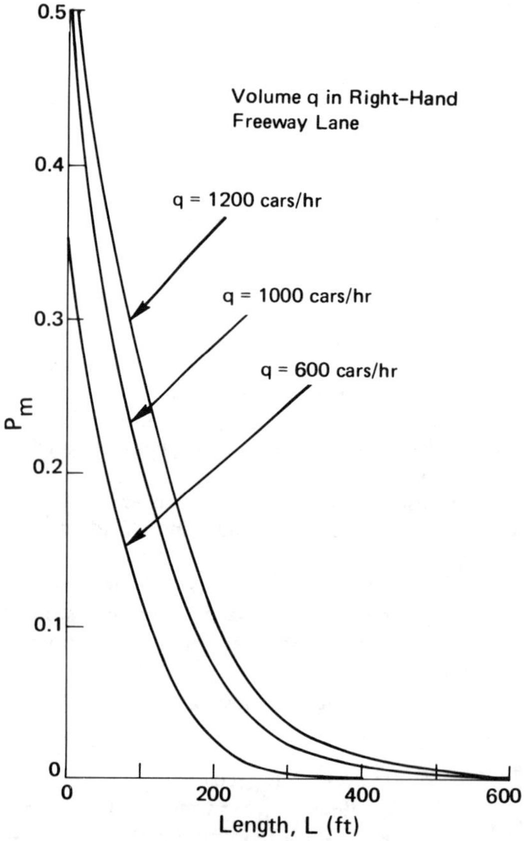

Figure 8.14 Probability of reaching the end of the acceleration lane without merging as a function of length.[28]

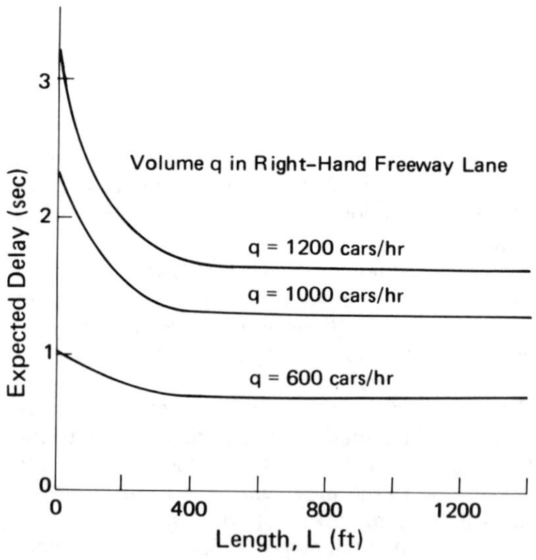

Figure 8.15 Expected delay as a function of acceleration lane length.[28]

at X and Y so that the length XY is the delay to the first vehicle. Similarly, X′Y′ and X″Y″ represent the delays for the next two vehicles. XX′ and X′X″ represent the arrival headways.

In a number of the studies it has been convenient to assume that events such as the arrival of a vehicle may occur only at certain instants that are equally and closely spaced in time. It is convenient to choose the interval in seconds between two events such that $\Delta t = 1/s$ and to define time in multiples of Δt. If the time origin $t=0$, events occur at time $t = n\Delta t$ $(n = 1,2,3, \ldots)$.

Let $c = C\Delta t$; $r = R\Delta t$; and $g = G\Delta t$, where C, R, and G are integers, and let α = average number of vehicles in time $\Delta t = q\Delta t = y$. These definitions are useful when using a binomial model of vehicular arrivals as used in section 8.5.4.2.

8.5.4.1 Continuum Model for Pretimed Signal. A representation of a continuum model proposed by May [31] is given in Figure 8.17. The vertical axis represents the cumulative arrivals qt and the horizontal axis the time t.

Case I represents the behavior when the capacity of the green interval exceeds the arrival during the green+red time. Case II is concerned with the instance when the discharge during the green phase is equal to the arrivals during the green+red period. In Figure 8.17 the vertical distance ca represents the number of vehicles that have accumulated since the signal entered the red phase. The horizontal distance ab represents the total time from arrival to departure for any given vehicle.

May developed the following measures of queue behavior:

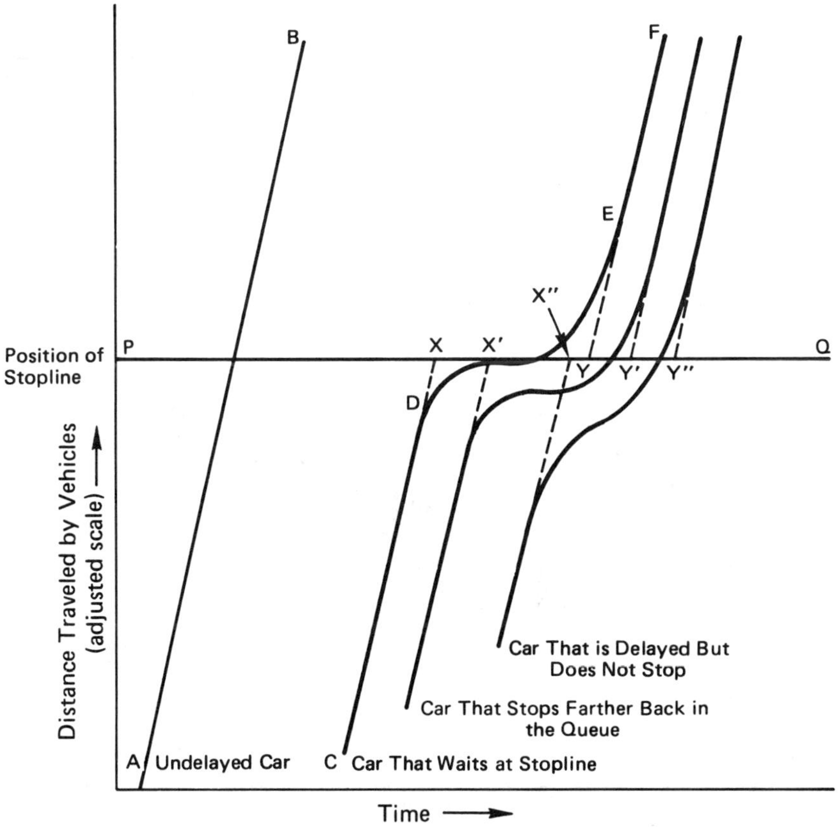

Figure 8.16 Diagram to illustrate the definitions of supposed arrival and departure times.[20]

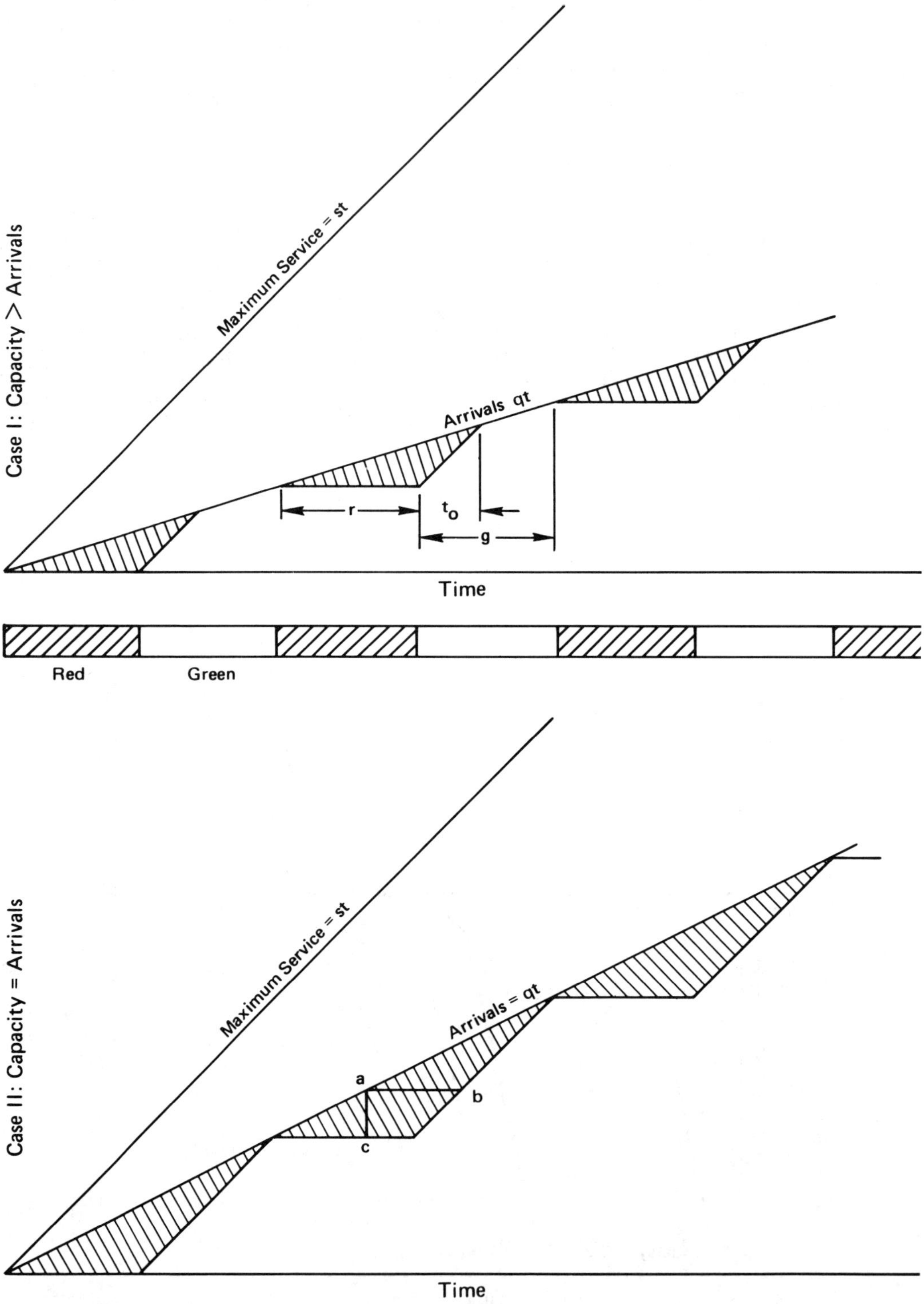

Figure 8.17 Representation of queuing at a signalized intersection.

1. Time after start of green that queue is dissipated (t_0);
2. Proportion of cycle with queue (P_q);
3. Proportion of vehicles stopped (P_s);
4. Maximum number of vehicles in queue (Q_m);
5. Average number of vehicles in queue ($\bar{Q}$);
6. Total vehicle-hours of delay per cycle (D);
7. Average individual vehicle delay (d);
8. Maximum individual vehicular delay (d_m).

The formulas for these two cases are developed from simple geometric relationships:

1. For any given cycle it is evident that at time t_0 after the start of green, the arrivals equal the discharge:

$$q(r+t_0)=st_0$$

letting $y=q/s$,

$$t_0=yr/(1-y) \quad (8.68)$$

2. Proportion of cycle with queue is equal to queue time/cycle length

$$P_q=(r+t_0)/c \quad (8.69)$$

3. Proportion of vehicles stopped is equal to vehicles stopped/total vehicles per cycle

$$P_s \quad q(r+t_0)/q(r+g)=t_0/(yc) \quad (8.70)$$

4. The maximum vehicles in queue will be seen by inspection to be the height of the triangle at r units after start of red:

$$Q_m=qr \quad (8.71)$$

5. The average number of vehicles in the queue, over the total length of cycle (c) is

$$\bar{Q}=\frac{(qr/2)r+(qr/2)t_0+0(g-t_0)}{r+t_0+g-t_0}$$

which yields

$$\bar{Q}=[(r+t_0)/c](qr/2) \quad (8.72)$$

6. The total vehicle-time of delay is given by the area of the triangle

$$D=(qr/2)(r+t_0)=(qr/2)[r/(1-y)]$$
$$=qr^2/[2(1-y)] \quad (8.73)$$

7. The average individual delay is given by dividing the total delay by the number of vehicles, or

$$d=\left[\frac{qr^2}{2(1-y)}\right]\frac{1}{qc}=\frac{r^2}{2c(1-y)} \quad (8.74)$$

8. The maximum individual vehicular delay will be seen from Figure 8.16 to be

$$d_m=r. \quad (8.75)$$

If the departures sc are less than the arrivals qc, the queue grows with each successive cycle and the foregoing formulas are no longer applicable.

Consider the following example (after May) of the behavior of a queue at a signalized intersection. Assume that the green phase g is 40 sec, the red phase r is 20 sec, the discharge rate s is 1,200 vehicles/hr, and the input rate q is in one case 600 vehicles/hr and in another case 800 vehicles/hr. The results are given in Table 8.2.

8.5.4.2 Probability Models for Pretimed Signals. This section owes much to the paper by Allsop [29] in which he critically reviews the various theoretical models of delay at fixed-time signals. The reader should refer to this paper and those of the individual authors cited for a more detailed analysis than that presented here.

Several models have been used to describe the arrival of vehicles at intersections. The simplest involves regular arrivals, as discussed in section 8.5.4.1. Winsten and co-workers [32] were the first to use the binomial model in analyzing delays at pretimed signals.

An approach has binomial arrivals if for some fixed Δt and α, P (1 PCU arrives at time $n\Delta t$) $=\alpha$ and P (no PCU arrives at time $n\Delta t$) $=1-\alpha$ for each n, independent of any other instant, and no arrivals can occur at other times. The average arrival rate is $\alpha/\Delta t$; if A_N

TABLE 8.2 Queueing Characteristics at Fixed-Time Signalized Intersection

Queueing Characteristic	Value for Input Rate (q) of	
	600 vph	800 vph
t_0	20 sec	40 sec
P_q	0.67	1.0
P_s	0.67	1.0
Q_m	3.3 veh	4.4 veh
$\bar{Q}$	1.1 veh	2.2 veh
D	66.7 veh-sec	133.3 veh-sec
d	6.67 sec	10.0 sec
d_m	20 sec	20 sec

is the number of PCU in a period containing N instants of $n\Delta t$, then A_N has a binomial distribution. (See section 3.2.2.)

For this distribution the ratio of variance to mean (I), equal to $1-\alpha$, is less than 1; for urban roads, however, the value of I is often observed to exceed 1, as reported by Miller.[33]

The Poisson distribution (section 3.2.1) has been used by Adams,[2] Webster,[34] and Wardrop.[35] The value of I will equal 1 and the model is most appropriate when an approach is lightly loaded in relation to the capacity of the approach.

Newell [36] used a model in which the arrival headways were assumed to have a shifted exponential distribution (section 3.5.1). This assumption imposes a minimum headway of $1/s$. Other models of arrivals cited by Allsop include those of Darroch,[37] Kleinecke,[38] McNeil,[39] and Miller.[40]

Models for the departure of PCU from a queue are simpler than the arrival models. Most models assume departures at equal time intervals $1/s$ providing a queue exists and the first departure is at the start of the effective green time. For the discrete time assumption, Δt is taken as $1/s$. The first PCU departs at the start of the effective green time $n\Delta t$, and one PCU departs at each succeeding $n\Delta t$ until the queue clears or the green time ends. Other departure models have been proposed, but variations in assumptions for departure models do not have the same impact on delay as the variations in arrival models.

Allsop demonstrates that for an approach with regular arrivals at intervals $1/q$ (arrivals would be plotted as a step function rather than as a straight line in Figure 8.17), the average delay/PCU (sec) is

$$d=\left[\frac{1}{2c(1-y)}\right]\left(r-\frac{1}{2s}\right)^2+\frac{y(2-y)+\theta(1-y)^2}{12q^2} \qquad (8.76)$$

where θ is shown by Allsop to have the range $-\frac{1}{3\times 3^{1/2}}<\theta<\frac{2}{3\times 3^{1/2}}$, and c, q, r, s, and y are as previously defined.

The first term of Eq. 8.76 is the same as that developed by Wardrop [35]:

$$\mathrm{d}=\left(r-\frac{1}{2s}\right)^2/2c(1-y) \qquad (8.77)$$

If q is allowed to increase indefinitely (the flow ratio $y=q/s$ remaining constant) in Eq. 8.76, the average delay per PCU will tend to $d=r^2/[2c(1-y)]$, the result given in Eq. 8.74, based on an assumed continuum model.

Winsten et al.[32] have demonstrated that for a traffic signal with binomial arrivals the average delay to a PCU passing through the approach is

$$d=\frac{R}{(1-\alpha)}\left[\frac{\mathrm{E}(Q_0)}{\alpha}+\frac{R+1}{2}\right]\Delta t \qquad (8.78)$$

Although the Winsten group was not able to develop the probability distribution of the overflow (the number of vehicles failing to clear the intersection during a given cycle), Dunne [41] and Potts [42] have developed the probability-generating function for the total delay in a cycle when binomial arrivals are assumed and the green time is long enough for the queue to clear.

Newell [43] was able to develop an estimate of the mean value of the overflow $\mathrm{E}(Q_0)$, which, as $\alpha\rightarrow g/c$ (i.e., the ratio x approaches 1), can be approximated by

$$\mathrm{E}(Q_0)\approx\frac{Rg}{2c(g-\alpha c)} \qquad (8.79)$$

One of the better-known formulas for delay is that developed by Webster [34] using data resulting from computer simulation of intersection operations:

$$d=\frac{c(1-\lambda)^2}{2(1-\lambda x)}+\frac{x^2}{2q(1-x)}-0.65\left(\frac{c}{q^2}\right)^{1/3}x^{(2+5\lambda)} \qquad (8.80)$$

Because $c(1-\lambda)=r$ and $\lambda x=y$, the first term is the same as that obtained by assuming continuum flow (Eq. 8.74). Allsop [29] shows that the second term is obtained by assuming that a queue with constant service $1/\lambda s$ is interposed between the signal and the arriving traffic. The mean waiting time in the interposed queue is $x^2/2q(1-x)$.

The third term, developed by Webster from regression analysis of data generated by simulated signal behavior, is a correction term representing from 5 to 15 percent of the total mean delay. From this, Allsop suggests that the average delay may be taken as

$$d=\frac{9}{10}\left[\frac{c(1-\lambda)^2}{2(1-\lambda x)}+\frac{x^2}{2q(1-x)}\right] \qquad (8.81)$$

Miller [44] made no assumptions about the distribution of arrivals, except that the queue on the approach was in statistical equilibrium and that the numbers of arrivals in successive red and green times were independently distributed. With this assumption he found an approximation for the mean number of overflow vehicles

$$E(Q_0) = \frac{e^{-1.33(\lambda cs)^{1/2}(1-x)/x}}{q(1-x)} \quad (8.82)$$

which when applied to the model for delay with Poisson arrivals gives

$$d = \frac{1-\lambda}{2(1-y)}\left[c(1-\lambda) + \frac{e^{-1.33(\lambda cs)^{1/2}(1-x)/x}}{q(1-x)}\right] \quad (8.83)$$

Miller[44] also developed a delay formula in which he relaxed the assumption of Poisson arrivals and applied the same general model of arrivals used in developing Eq. 8.82. The resulting equation for delay, when $x > 1/2$ is

$$d = \frac{1-\lambda}{2(1-\lambda x)}\left[c(1-\lambda) + \frac{(2x-1)I}{q(1-x)} + \frac{I+\lambda x-1}{s}\right] \quad (8.84)$$

When $x < 1/2$, the middle term in the brackets is replaced by zero. Miller found that his equation and Webster's (Eq. 8.80) gave similar results with I near to 1, and that Eq. 8.84 gave better results when I was appreciably greater than 1.

Newell[45] considered the continuum model developed in section 8.5.4.1 and the further delay caused by the overflow Q_0. His work leads to the expression for delay

$$d = \frac{c(1-\lambda)^2}{2(1-\lambda x)} + \frac{IH(\mu)x}{2q(1-x)} + \frac{I(1-\lambda)}{2s(1-\lambda x)^2} \quad (8.85)$$

where $H(\mu)$ is a function of the spare capacity of the approach and

$$\mu = (sg-qc)/(Isg)^{1/2} = (1-x)/(sg/I)^{1/2} \quad (8.86)$$

(See Figure 8.18.)

Hutchinson[46] made numerical comparisons of several of the delay expressions that had been discussed by Allsop,[29] as well as analyzed the following expressions: (a) Webster's full expression (Eq. 8.80), (b) Miller's first expression (Eq. 8.84), (c) Miller's second expression (Eq. 8.83), (d) Newell's expression (Eq. 8.85), (e) a modification of the simplified

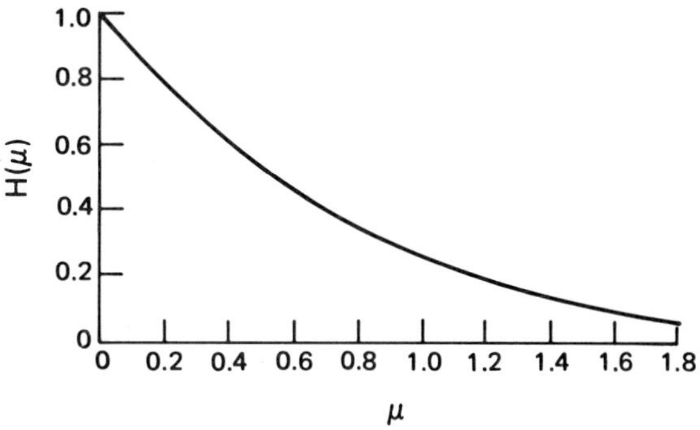

Figure 8.18 Value of function $H(\mu)$.[45]

Webster expression (Eq. 8.81), introducing the variable I, such that

$$d = \frac{9}{10}\left[\frac{c(1-\lambda)^2}{2(1-\lambda x)} + \frac{Ix^2}{2q(1-x)}\right] \quad (8.86)$$

Because Webster's full expression has been subject to the most extensive field testing, it has been taken as the standard for the numerical comparisons made by Hutchinson.

Curves of Webster's average delay (Eq. 8.80) versus degree of saturation x and proportion of cycle effectively green are shown in Figures 8.19 and 8.20, respectively.

The influence of the parameter I (variance/mean) for the several expressions for delay is presented in Figure 8.21.

Hutchinson also calculated the percentage differences between various estimates of average delay and the delay given by Webster's full expression as a function of the flow ratio y. Figure 8.22 shows the results obtained when $s=0.5$, $c=90$, and $\lambda=0.5$. Similar results are presented by Hutchinson for different combinations of s, c, and λ. The heavy pair of curves that nearly envelope the other curves are the limiting values of the percentage differences that result as $y \to 0$ $(x \to 0)$ and as $y \to \lambda$ $(x \to 1)$.

The percentage differences related to the proportion of cycle effectively green λ, as calculated by Hutchinson, are shown in Figure 8.23. As a result of his analysis, the author concluded that the results are sufficiently similar that convenience and simplicity of calculation may dictate the choice of model. For example, if one were required to differentiate

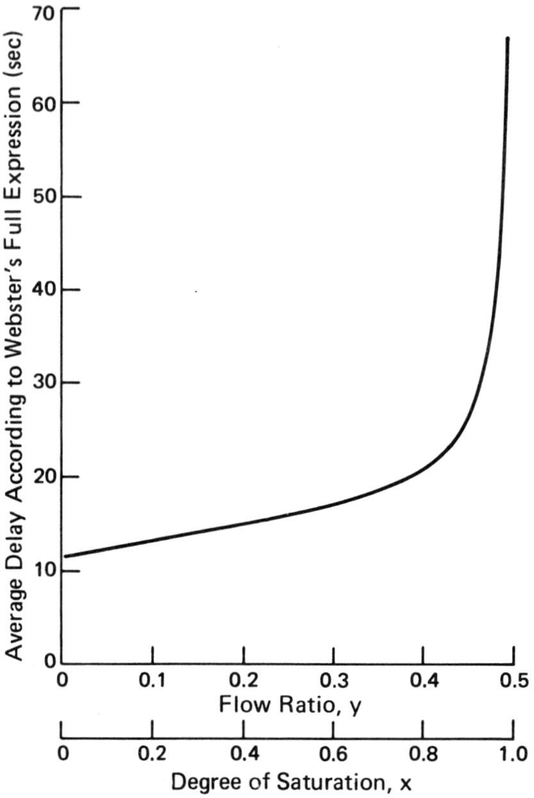

Figure 8.19 Average delay according to Webster's full expression as a function of flow ratio for an approach with $s = 1.0$, $c = 90$, and $\lambda = 0.5$ (so $g = 45$).[46]

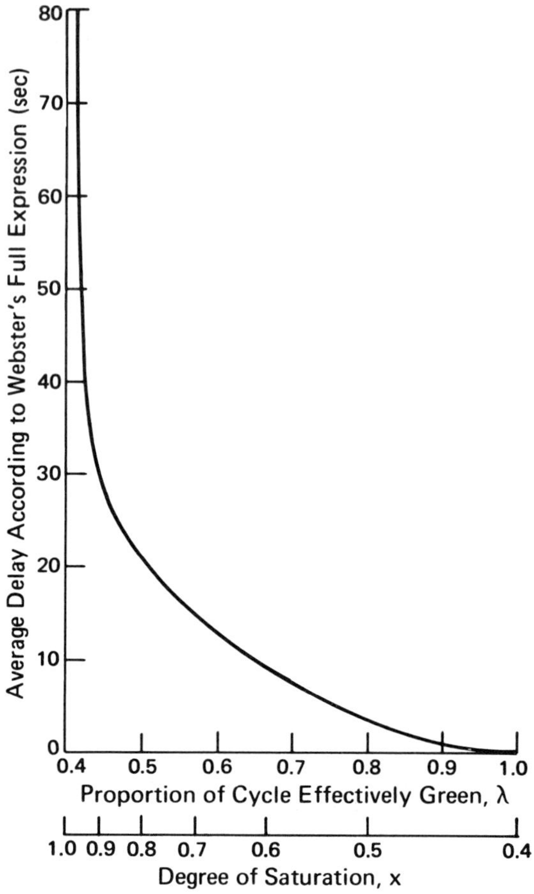

Figure 8.20 Average delay according to Webster's full expression as a function of the proportion of the cycle effectively green for an approach with $s = 1.0$, $c = 90$, and $y = 0.4$ (so $q = 0.4$).[46]

the expressions in determining optimal signal setting parameters, Eqs. 8.81, 8.83, 8.84, or 8.86 would be easiest to handle.

In this section only fixed-time signals have been considered. Recent analyses of traffic behavior at traffic-actuated signals have been reported by Little,[47] Newell,[48] and Newell and Osuna.[49] Readers may refer to these papers for a discussion of queueing behavior.

The foregoing discussion has considered only the case of isolated intersections, where vehicle arrivals are not influenced by adjacent traffic control devices. When groups of intersections are considered, either in succession along an arterial or as a cluster in a downtown street network, the assumptions made for isolated intersections no longer hold. A review of much of the literature related to signal operation of systems on urban arterials will be found in the work by Wagner, Gerlough, and Barnes.[56] The same authors also report [57] on a review of literature related to signal networks. The reader is referred to these for an introduction to the appropriate literature.

8.6 QUEUEING MODELS FOR ROADWAYS

Queueing theory methods have been applied to flow analysis on roadways. For instance, it is possible to calculate the behavior of a queue of cars at a bottleneck, where the

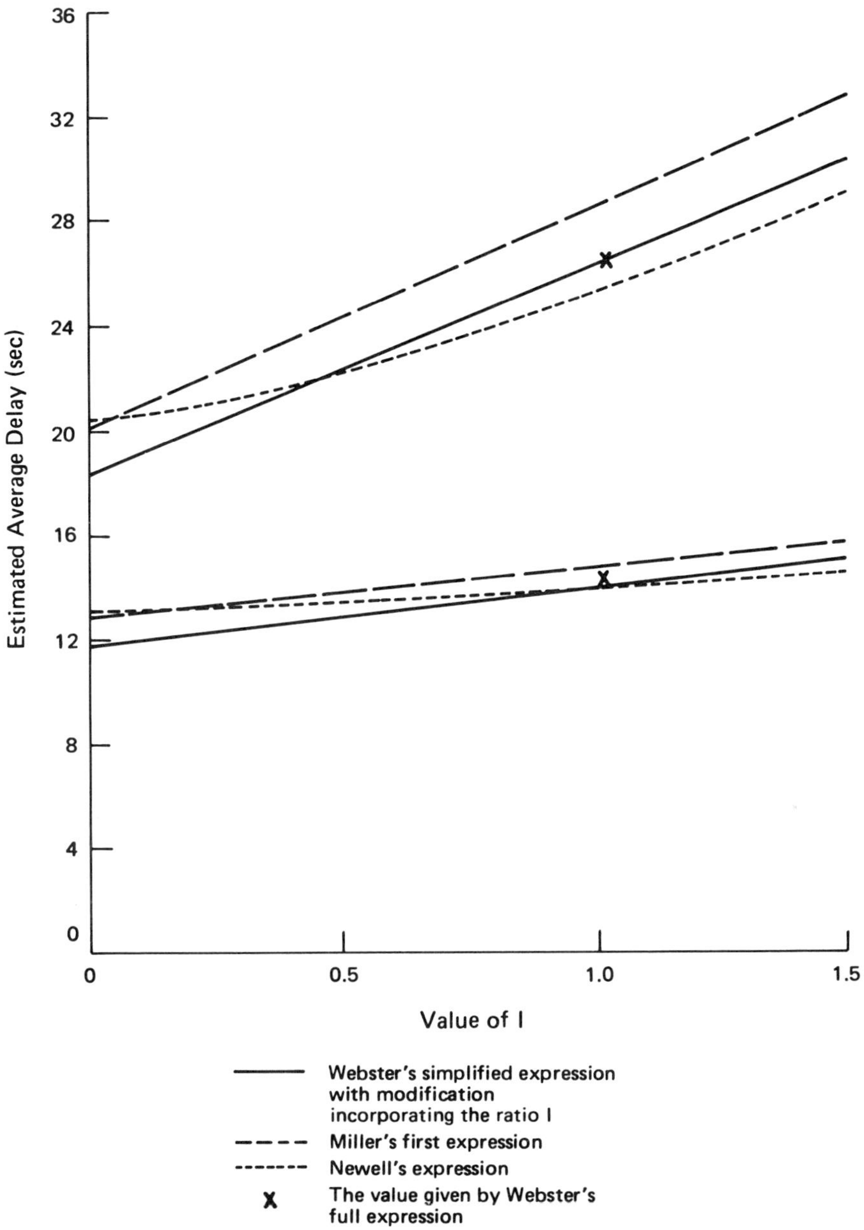

Figure 8.21 Examples of the effect of *I* on the estimates of average delay by various expressions. For the upper set of curves $s = 1.0$, $c = 90$, $\lambda = 0.5$, and $y = 0.45$ (so $x = 0.9$); for the lower set, $s = 1.0$, $c = 90$, $\lambda = 0.6$, and $y = 0.45$ (so $x = 0.75$).[46]

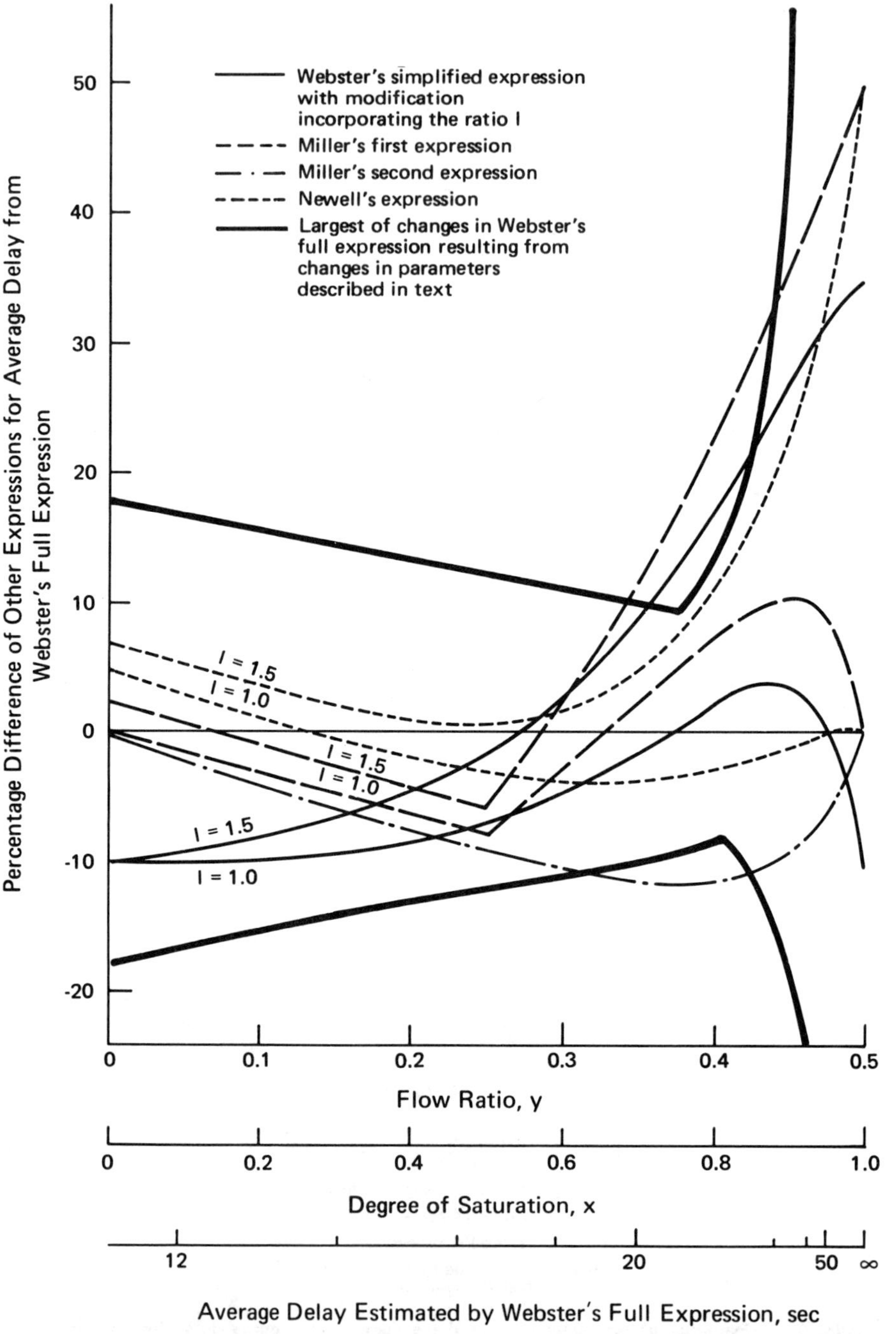

Figure 8.22 Percentage difference between various estimates of average delay and that given by Webster's full expression as a function of flow ratio for an approach with $s = 0.5$, $c = 90$, and $\lambda = 0.5$ (so $g = 45$).[46]

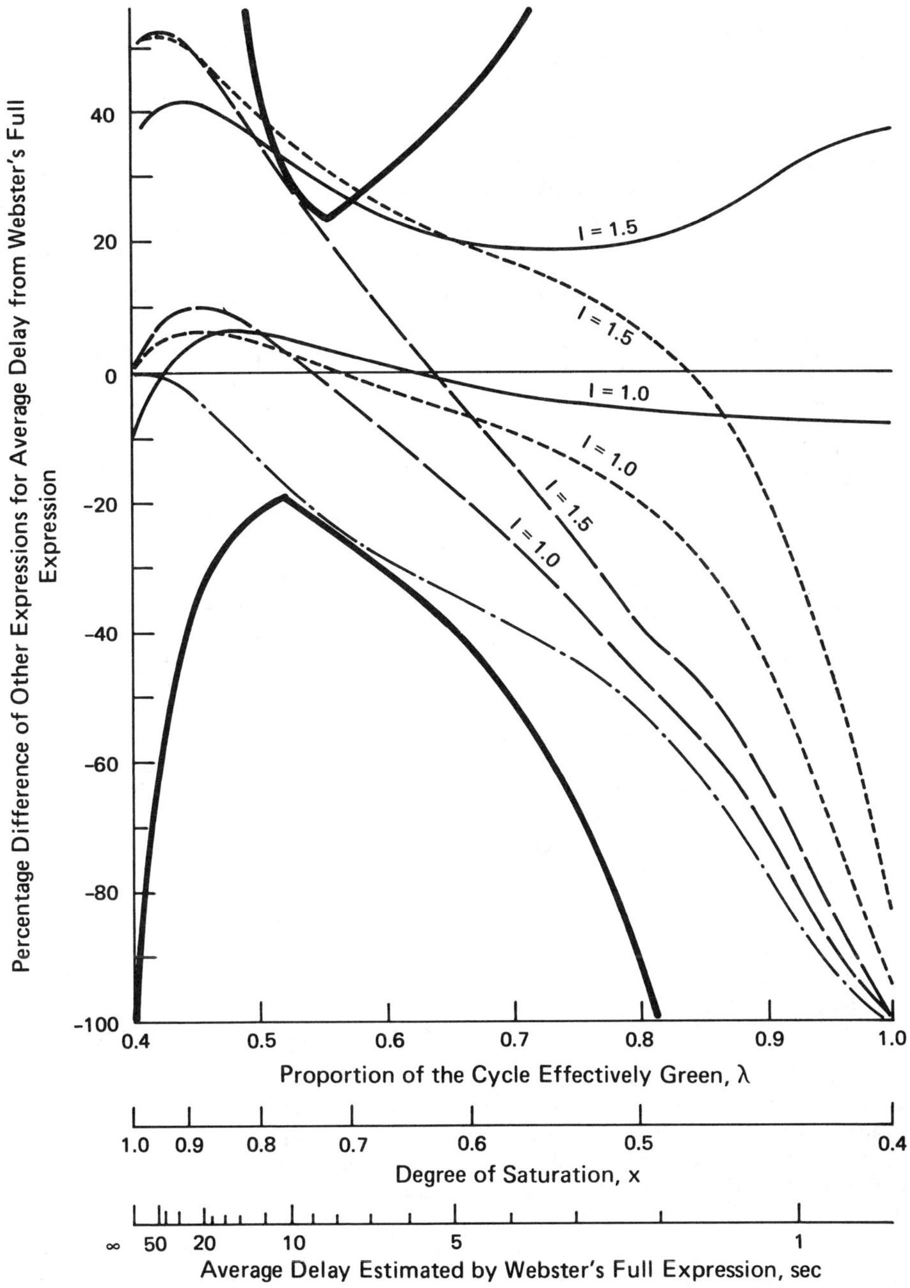

Figure 8.23 Percentage difference between various estimates of average delay and that given by Webster's full expression as a function of the cycle effectively green for an approach with $s = 0.5$, $c = 30$, and $y = 0.4$ (so $q = 0.2$). (See Figure 8.22 for key.)[46]

bottleneck may be considered the server, and vehicles waiting to be served (i.e., to pass the bottleneck) is the observed behavior. This topic is discussed in section 8.6.1. Another application of queueing techniques is describing uninterrupted vehicular flow. In models of this type vehicles are assumed to be operating on a roadway at some desired speed until they overtake a slower vehicle, join a queue behind it, and await an opportunity to pass. Models of this type are discussed in section 8.6.2.

8.6.1 Queue Behavior at Bottlenecks

May [31] applied the technique of continuous flow to the problem of the temporary bottleneck (i.e., a roadway blocked by a rail crossing or a single lane blocked by an accident). The problem may be represented by the behavior of a queue during one cycle of a traffic signal (Figure 8.17), where the duration of the delay or blockade is equivalent to the length of the red interval r and the time for the queue to dissipate after removal of the blockade is the same as the time t_0 in Figure 8.17.

The following notation, similar to that used in section 8.5.4, is used for May's roadway model. Let

q = average arrival rate of traffic (vehicle/min) upstream of the bottleneck;
s = saturation flow rate or capacity (vehicles/min) of uninterrupted flow;
s_r = flow rate (vehicles/min) at bottlenecks during blockade ($s_r < q < s$);
r = duration of blockade (min);
t_0 = time (min) for queue to dissipate after blockade is removed;
t_q = total elapsed time (minutes from start of blockade until free flow resumes) $= r + t_0$.

The value of s_r may be zero when the roadway is completely blocked, as for an at-grade railroad crossing, or some value ($s_r < q$) when the roadway is partially blocked by a disabled vehicle. May developed the following set of relationships:

$$\text{Duration of queue, } t_q = r\left(\frac{s - s_r}{s - q}\right)$$

Number of vehicles affected, $N = qt_q$

Maximum number of vehicles in queue, $Q_m = r(q - s_r)$

Average number of vehicles in queue, $\bar{Q} = Q_m/2$

Total vehicle-minutes of delay, $D = \dfrac{r(q - s_r)t_q}{2}$

Average minutes vehicle is delayed, $d = \dfrac{r}{2}(1 - s_r/q)$

Maximum minutes of individual delay, $d_m = r(1 - s_r/q)$.

May and Keller [50] have applied a similar technique to analyze the behavior of queueing vehicles during rush-hour traffic. For this problem, it is assumed that although the capacity flow of the roadway s remains constant, the demand q varies from some value less than s, to equals s, and then to a maximum rate $q_2 > s$. The authors consider two cases for the shape of the demand curve: (a) trapezoidal (i.e., demand increases at constant rate to maximum q_2, remains at q_2 for some fixed time, and then decreases at a constant rate to constant post-peak demand) and (b) triangular (i.e., demand increases at constant rate to q_2 and then immediately decreases at constant rate to constant post-peak demand).

By an analysis similar to that used for bottlenecks, it is possible to calculate values of delay, queue length, and duration of queue.

McNeil [51] has considered the bottleneck problem by treating the bottleneck as the M/G/1 queueing situation (i.e., random arrivals, general service function, and one queue). The service time for the first vehicle in the queue b_1 is the delay time s_1 plus the time a_1 for its effective length (front bumper to front bumper of following vehicle) to clear the block point. The time s_1 could vary from a few seconds in the event of a short delay to several minutes in the event of a vehicle breakdown.

Each subsequent vehicle in the queue is assumed to have a service time b_i that is made up of the departure period s_i plus the time a_i for each vehicle to clear its own effective length. The departure period s_i ($i = 1, 2, 3, \ldots$) is a sequence of independent, identically distributed variables with a mean value that corresponds to the headway at the maximum flow rate. The value of a_i is constant for all vehicles, including the first, and is the effective length over the mean velocity V. The first and subsequent vehicles are each assumed to instantly resume the speed V after passing the point of blockage.

McNeil notes that if the vehicles are assumed to have zero length (a_i will be zero), then the number of vehicles delayed by the initial blockage of duration s_1 is the number of customers served in the server's busy period for the queue M/G/1. These delays must be modified to allow for the effective length of the vehicle. Some of the results of McNeil's model are presented in the following. Let

$E(b) = E(a) + E(s)$ = mean service time for all vehicles after the first;
$E(b_1) = E(a_1) + E(s_1)$ = mean service time for the first delayed vehicle;
q = normal arrival rate at the bottleneck (vehicles/min);
$\rho = qE(b)$;
$\rho_1 = qE(b_1)$; and
$\rho_A = qE(a)$.

Then

$$E(N) = 1 + \frac{\rho_1}{1-\rho}$$

= total number of vehicles affected;

$$E(W) = E(s_1) + \frac{1}{2}qE(s_1b_1) + \frac{1}{2}\rho_1[(2+\rho_1+\rho_1C_1^2)E(s) + q\mathrm{cov}(s,b)](1-\rho)^{-1} + \frac{1}{2}\rho\rho_1E(s)(1+\rho C_b^2)(1-\rho)^{-2}$$

= total vehicle delay time (vehicle-min);

where

$$C_1^2 = \mathrm{var}(b_1)/E^2(b_1);$$
$$C_b^2 = \mathrm{var}(b)/E^2(b);$$

$\mathrm{cov}(s,b)$ = the covariance between the headway s and the service time b; and

$$E(t_q) = E(s_1) + E(s)E(n-1)$$

= duration of queue (min).

McNeil also develops the variance of the measures listed in the foregoing and provides a numerical example for the solution of the expected total vehicle delay time.

8.6.2 Delays on Roadways

The delay experienced by vehicles while traveling on two-lane roads in accordance with postulated rules has been of particular interest to the traffic flow theorist. If not interrupted, each vehicle will travel at its own desired speed. When a slower vehicle or group of vehicles is overtaken, passing without delay will occur if there is an acceptable gap in the opposing stream of vehicles. If an acceptable gap for passing is not available in the opposing stream, the faster vehicle will be required to assume the speed of the slower vehicle or queue of vehicles and to await an opportunity to pass.

Several authors have applied principles of queueing to the problems of passing on two-lane roads. For example, Tanner's model [13,14] makes use of the Borel–Tanner distribution (section 8.4) in determining the size of queues. If a vehicle with velocity u is to pass a vehicle with velocity u_1, the passing maneuver requires a time of

$$t = \frac{A_1}{u-u_1} \qquad (8.87)$$

and a distance of

$$x = \frac{A_1u}{u-u_1} \qquad (8.88)$$

in which A_1 is a parameter describing the distance required for the passing vehicle relative to the vehicle being passed.

Tanner's model deals with vehicles traveling in both directions along a two-lane road and can be extended to one-way flow on a two-lane facility. In Figure 8.24, the flow in one direction is q_1 vehicles/unit time. All vehicles travel at the same constant speed u_1, except the vehicle under study that travels at some greater desired speed u, or at speed u_1 if it is unable to pass. The minimum spacing of vehicles in this stream is S_1. In the opposite direction the flow is q_2, with all vehicles traveling at the same constant speed u_2 and with no spacing less than S_2. The Borel–Tanner distribution is assumed for the number of vehicles n in the "bunches." The distribution of gaps is a modification of random arrivals, which requires that no spacing be less than the minimum.

The delays experienced by the single vehicle traveling at speed u in the q_1 flow direction is the problem for which Tanner offered a model. To solve this problem, the vehicle is assumed to act in accordance with the following rules:

1. A group of n vehicles in the q_1 stream traveling at their minimum separation S_1 is overtaken in a single maneuver. The overtaking vehicle can only reenter the q_1 lane between two groups and cannot break into any one group or approach the rear vehicle of any group by a distance less than S_1.

2. When the overtaking vehicle reaches the tail of any group of n vehicles and there is

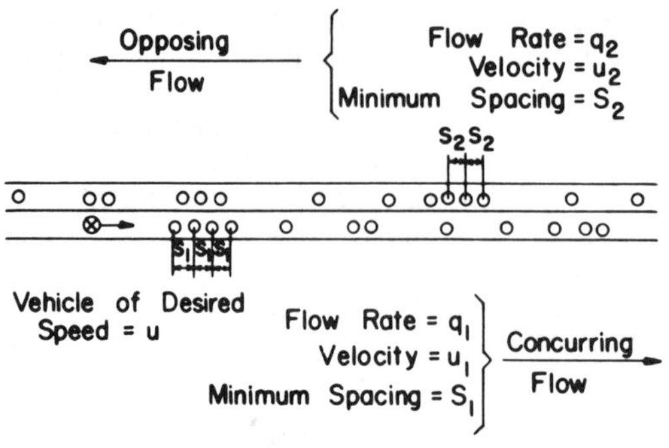

Figure 8.24 Two-lane roadway, showing the corresponding assumed terms in the Tanner model.[1]

a distance of at least $d_n = d + nS_1(u+u_2)/(u-u_1)$ in the q_2 stream, the vehicle passes without decelerating. [d is defined as the least acceptable clear distance between the u vehicle and the passed traffic as the u vehicle clears the bunch in passing. It can be expressed by $d = A_1 u/(u-u_1)$, with A_1 being some distance between 50 and 100 ft.]

3. If the required distance d_n is not available, the vehicle decelerates instantaneously to u_1, follows as closely as possible behind the vehicle ahead, waits for a clear distance of at least $D_n = d_n + t(u_1+u_2)$ in the q_2 stream, waits a further time t, accelerates instantly to u, and passes. The additional time t required for the overtaking vehicle to remain in the q_1 stream because of having slowed down is used to compensate for the assumed instantaneous accelera-

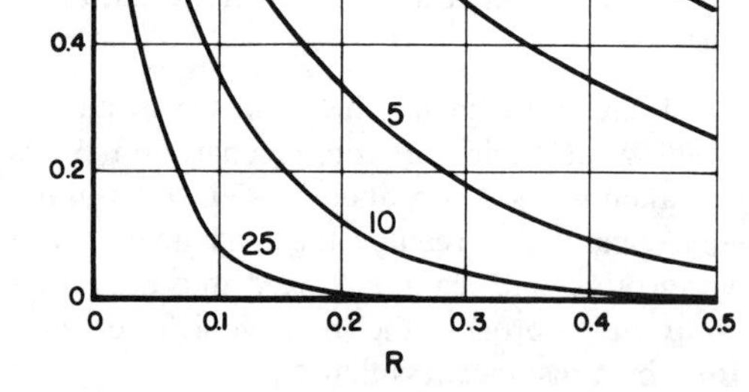

Figure 8.25 Relationship between *K* and the parameters *R* and *C/G*.[1]

tion and could be expressed as $t = A_2(u-u_1)/a$, where a is the constant acceleration of the overtaking vehicle and A_2 is approximately one.

Tanner's major objective was to determine the average speed E(u) of a single vehicle desiring to travel at a velocity u over an infinitely long trip. He was able to express the average speed E(u) in terms of the average waiting time behind all vehicles, E(t_w), which included zero waits. The expression

$$E(u) = \frac{uu_1^2 + q_1(u-u_1)(u_1 - s_1 q_1)u_1 E(t_w)}{u_1^2 + q_1(u-u_1)(u_1 - s_1 q_1)E(t_w)} \tag{8.89}$$

was developed; thus the problem became one of solving for E(t_w). Algebra involved in the computation of E(t_w) is formidable:

$$E(t_w) = \left[\frac{1}{(u_1+u_2)(1-R)}\right]\left\{\left(\frac{Ke^{-cd}}{c+G-GK}\right) \times \left(1 - \frac{c\exp[G(u_1+u_2)t]}{c+G}\right) + (\exp[Gt(u_1+u_2)+Gd]) \times \left[\frac{N}{G} - \frac{c}{G(c+G)}\right] - \frac{1-R}{c(1-n)}\right\} \tag{8.90}$$

in which $g = q_1/u_1$, $G = q_2/u_2$, $r = S_1 g$, $R = S_2 G$, $c = q_1(u-u_1)/u_1(u+u_2)$, K = root between 0 and 1 of $K = \exp[R(K-1-c/G)]$, and N = the smaller real root of $N = \exp[n(N-1+G/c)]$ (which exists only when $r \exp(1-n+nG/c) \leq 1$).

Limited solutions for K and N have been included in Figures 8.25 and 8.26, respectively.

Substitution of the values of E(t_w) in Eq. 8.89 gives an expression for the average speed E(u) in terms of the desired speed u, the velocity of the q_1 stream u_1, and the flow rate q_1 of the stream. Limited solutions of this equation were made by Tanner using specific values of the various parameters. Figure 8.27 shows the effect of traffic flow when $q_1 = q_2$ for various values of u and $u_1 = 30$ mph. This model indicates that for a total flow $(q_1 + q_2)$ of more than 800 vehicles/hr a vehicle will have to assume very nearly the velocity of the q_1 stream, regardless of its own desired velocity.

The effect of varying proportions of q_1 and q_2 on the average speed E(u) is shown in Figure 8.28, which shows that in this model the average speed E(u) is least when one-half to three-fourths of the total traffic is traveling in the opposite direction of flow q_2 (one-half

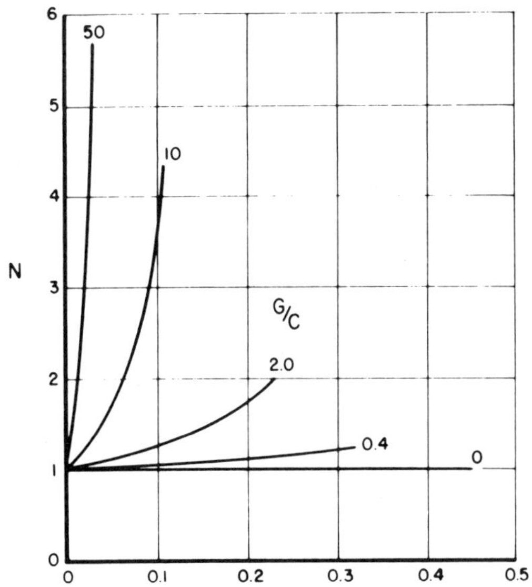

Figure 8.26 Relationship between *N* and the parameters *r* and *G/C*.[1]

being applicable at low volumes and the value of three-quarters as one approaches higher volumes).

It is worthwhile to point out that the delay implied by $E(u)$ is the only delay involved; all other vehicles are, by the assumptions, not delayed. The u vehicle would ultimately pass all q_1 vehicles, and no passing would occur among q_1 or q_2 vehicles.

Morse and Yaffe[52] have developed a queueing model for the two-way two-lane situation at low volumes. The model assumes a sequence of *free* or *lead* cars driving at various speeds, some of which are followed by a queue of trapped cars that are traveling at the same speed as the lead car. No assumption is made of the desired speed of the trapped vehicle, but some are able to escape the queue by passing and becoming lead cars. The authors develop a formula for the length of queue trapped behind a lead vehicle with given velocity v, the mean velocity of the stream, and the mean time for a faster vehicle to pass a slower vehicle. Tables that aid in the solution of the resulting equations are included.

Queueing models on multilane roadways differ from the two-lane situation by virtue of

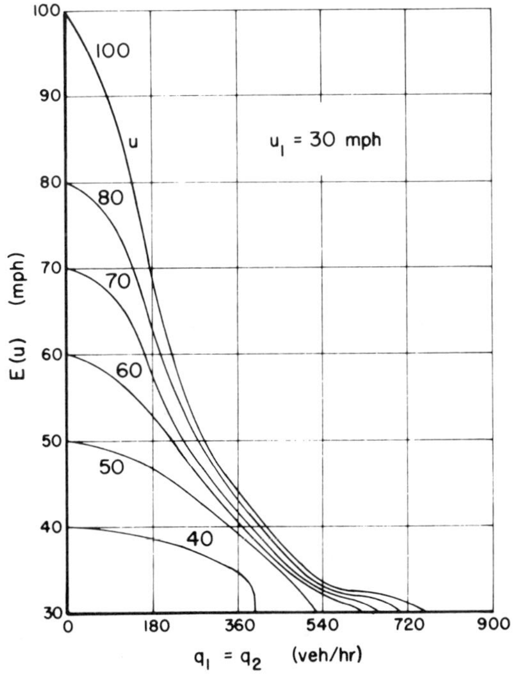

Figure 8.27 Effect of traffic when equally divided between directions for various values of *u* (two-way traffic).[1]

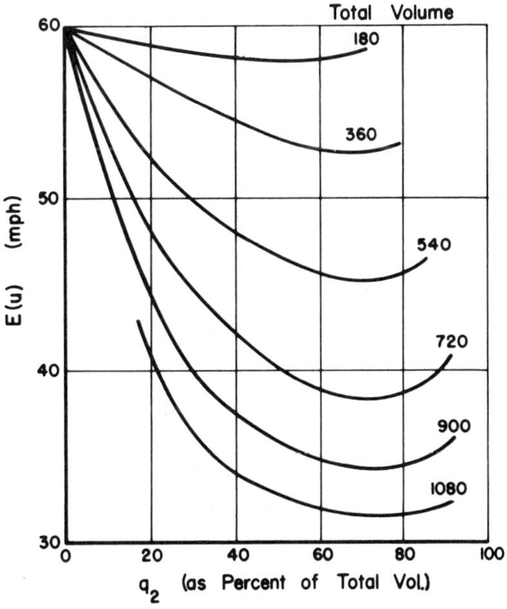

Figure 8.28 Effect of varying proportion of opposing traffic for various levels of total traffic.[1]

the fact that a vehicle is free to use any of the two or more lanes available in his direction of travel and is restricted from passing only during the period when another vehicle is in the adjacent passing lane. Miller[53] proposes an empirical model, which is an extension of the "product density" method of analysis. Miller develops and criticizes the product density approach and then develops a technique that for want of a better term, he calls the "termination rate method," where this latter method is analogous to the statistical model of mortality rates for humans. Miller provides some verification of his model with data collected on a multilane freeway.

Schach[54] gives an example of the Markov process as it might be applied to a model of multilane flow. Some quantities that he is able to calculate are the average use of each lane, the average speed of cars in the system, and the expected number of lane changes per unit time. Input parameters are N, the density of vehicles on a roadway, intensities of weaving from lane 1 to lane 2 or lane 2 to lane 1 (λ, μ), and a parameter r that the author sets equal to 2.

Holland[55] considers the behavior of three lanes of traffic, all moving in the same direction. He allows the special behavior of each lane—traffic in the right lane is free to exit or move to the middle lane; traffic in the middle lane may move to either of the two adjacent lanes; traffic in the left lane may enter the middle lane only. The solution technique is iterative (a flow chart of the computational process is included), and his results are plotted against previously published data.

Drew[58] and Worrall, Bullen, and Gur[59] have used the Markov process to describe the lane-changing process on a multilane freeway. The interested reader will find information on application of this process to traffic analysis in these two references.

8.7 REFERENCES

1. Cleveland, D. E., and Capelle, D. G., Queueing theory approaches, pp. 49–98. In D. L. Gerlough and D. G. Capelle (ed.) *An Introduction to Traffic Flow Theory*. Special Report 79. Highway Research Board, Washington, D.C. (1964).
2. Adams, W. F., Road traffic considered as a random series.: J. Inst. Civil Eng., 4: 121–130 (1936).
3. Tanner, J. C., The delay to pedestrians crossing a road. Biometrika, 38 (Parts 3 and 4):383–392 (1951).
4. Edie, L. C., Traffic delays at toll booths. Oper. Res., 2 (2):107–138 (1954).
5. Moskowitz, K. "Waiting for a gap in a traffic stream. Proc. Highw. Res. Board, 33:385–395 (1954).
6. Haight, F. A., *Mathematical Models for Traffic Flow*. Academic Press, New York (1963).
7. Prabhu, N. U., *Queues and Inventories*. Wiley, New York (1965).
8. Cox, D. R., and Smith, W. L., *Queues*. Chapman and Hall, London (1971).
9. Newell, G. F., *Applications of Queueing Theory*. Chapman and Hall, London (1971).
10. Kometani, E., and Kato, A., On the theoretical capacity of an off-street parking space. Memoirs, Faculty of Engineering, Kyoto University, 18 (4):315–328 (1956).
11. Haight, F. A., and Jacobson, A. S., Some mathematical aspects of the parking problem. Proc. Highw. Res. Board, 41: 363–374 (1962).
12. Haight, F. A., Overflow at a traffic light. Biometrika, 46:420–424 (1949).
13. Tanner, J. C., A simplified model for delays in overtaking on a two-lane road. J. R. Stat. Soc., Ser. B, 20 (2):408–414 (1958).
14. Tanner, J. C., Delays on a two-lane road. J. R. Stat. Soc., Ser. B, 23 (1):38–63 (1961).
15. Raff, M. A., *A Volume Warrant for Urban Stop Signs*. Eno Foundation for Highway Traffic Control, Saugatuck, Connecticut (1950), pp. 62–75.
16. Oliver, R. M., Distribution of gaps and blocks in a traffic stream. Oper. Res., 10 (2):197–217 (1962).
17. Underwood, R. T., Tentative warrants for the installation of pedestrian crossings. Res. Memo. No. 13. Country Roads Board, Victoria (1957), p. 18.
18. Tanner, J. C., The delay to pedestrians crossing a road. Biometrika, 38:383–392 (1951).
19. Mayne, A. J., Some further results in the theory of pedestrians and road traffic. Biometrika, 41:375–389 (1954).
20. Jewell, W. S., *Waiting for a Gap in Traffic*. Res. Rep. No. 6. Operations Research Center, University of California (1961).

21. Weiss, G. H., and Maradudin, A. A., Some problems in traffic delay. Oper. Res., 10:1, 74–104 (1962).
22. Herman, R., and Weiss, G., Comments on the highway-crossing problem. Oper. Res., 9:828–840 (1961).
23. Weiss, G. H., The intersection delay problem with gap acceptance function depending on speed and time. Transp. Res., 1 (4):367–371 (1967).
24. Miller, A. J., A queueing model for road traffic flow. J. R. Stat. Soc., Ser. B, No. 23 (1961), pp. 64–75.
25. Evans, D. H., Herman, R., and Weiss, G. H., The highway merging and queueing problem. Oper. Res., 12 (6):832–857 (1964).
26. Major, N. G., and Buckley, D. J., Entry to a traffic stream. Proc. Austr. Road Res. Board (1962), pp. 206–228.
27. Ashworth, R., The capacity of priority-type intersections with a non-uniform distribution of critical acceptance gaps. Transp. Res., 3 (2):273–278 (1969).
28. Blumenfeld, D. E., and Weiss, G. H., Merging from an acceleration lane. Transp. Sci., 5 (2):161–168 (1971).
29. Allsop, R. E., Delay at a fixed time traffic signal. I: Theoretical analysis. Transp. Sci., 6 (3):260–285 (1972).
30. Greenshields, B. D., Schapiro, D., and Erickson, E. L., Traffic performance at urban intersections. Bureau of Highway Traffic, Tech. Rep. No. 1. Yale University Press, New Haven, Conn. (1947).
31. May, A. D., Jr., Traffic flow theory—the traffic engineer's challenge. Proc. Inst. Traffic Eng. (1965), pp. 290–303.
32. Beckmann, M., McGuire, C. B., and Winsten, C. B., *Studies in the Economics of Transportation*. Yale University Press, New Haven, Conn. (1956).
33. Miller, A. J., Settings for fixed-cycle traffic signals. Proc. Austr. Road Res. Board, 2:342–365 (1964).
34. Webster, F. V., *Traffic Signal Settings*. Road Research Tech. Paper No. 39. Great Britain Road Research Laboratory, London (1958).
35. Wardrop, J. G., Some theoretical aspects of road traffic research. Proc. Inst. Civ. Eng. 1:325–362 (1952).
36. Newell, G. F., Statistical analysis of the flow of highway traffic through a signalized intersection. Q. Appl. Math., 13:353–369 (1956).
37. Darroch, J. N., On the traffic light queue. Ann. Math. Stat., 35:38–88 (1964).
38. Kleinecke, D. C., Discrete-time queues at a periodic traffic light. Oper. Res., 12: 809–814 (1964).
39. McNeil, D. R., A solution to the fixed-cycle traffic light problem for compound Poisson arrivals. J. Appl. Probab., 5: 624–635 (1968).
40. Miller, A. J., Settings for fixed-cycle traffic signals. Oper. Res. Q., 14:373–386 (1963).
41. Dunne, M. C., Traffic delay at a signalized intersection with binomial arrivals. Transp. Sci., 1:24–31 (1967).
42. Potts, R. B., Traffic delay at a signalized intersection with binomial arrivals. Transp. Sci., 1:126–128 (1967).
43. Newell, G. F., Queues for a fixed-cycle traffic light. Ann. Math. Stat., 31:589–597 (1960).
44. Miller, A. J., Some operating characteristics of fixed time signals with random arrivals. Institute of Highway and Traffic Research, University of New South Wales (1969).
45. Newell, G. F., Approximation methods for queues with applications to the fixed-cycle traffic light. SIAM Rev., 7 (2): 223–240 (1965).
46. Hutchinson, T. P., Delay at a fixed time signal. II: Numerical comparisons of some theoretical expressions. Transp. Sci., 6:286–305 (1972).
47. Little, J. G., Jr., Queueing of side street traffic at a priority type vehicle-actuated signal. Transp. Res., 5:295–300 (1971).
48. Newell, G. F., Properties of vehicle actuated signals. II. One-way streets. Transp. Sci., 3:30–52 (1969).
49. Newell, G. F., and Osuna, E. E., Properties of vehicle actuated signals: II. Two-way streets. Transp. Sci., 3:99–125 (1969).
50. May, A. D., Jr., and Keller, H. E. M., A deterministic queueing model. Transp. Res., 1:117–128 (1967).
51. McNeil, D. R., Growth and dissipation of a traffic jam. Transp. Res., 3:115–121 (1969).
52. Morse, P. M., and Yaffe, H. J., A queueing model for car passing. Transp. Sci., 5:48–63 (1971).
53. Miller, A. J., An empirical model for multilane road traffic. Transp. Sci., 4: 164–186 (1970).

54. Schach, S., Markov models for multi-lane freeway traffic. Transp. Res., 4:259–266 (1970).
55. Holland, H. J., A stochastic model for multilane traffic flow. Transp. Sci., 1: 184–205 (1967).
56. Wagner, F. A., Gerlough, D. L., and Barnes, F. C., Improved criteria for traffic signal systems on urban arterials. National Cooperative Highway Research Program Rep. 73. Highway Research Board, Washington, D.C. (1969), 55 pp.
57. Wagner, F. A., Barnes, F. C., and Gerlough, D. L., Improved criteria for traffic signal systems in urban networks. National Cooperative Highway Research Program Rep. 124. Highway Research Board, Washington, D.C. (1971), 86 pp.
58. Drew, D. R., *Traffic Flow Theory and Control.* McGraw-Hill, New York (1968).
59. Worrall, R. D., Bullen, A. G. R., and Gur, Y., An elementary stochastic model of lane-changing on a multilane highway. Highw. Res. Record 308 (1970), pp. 1–12.

8.8 RELATED LITERATURE

Allan, R. R., On the highway crossing problem. Proc. Austr. Road Res. Board, 4:529–548 (1968).

Allan, R. R., Queueing on the minor road. Proc. Austr. Road Res. Board, 4:549–566 (1968).

Allsop, R. E., An analysis of delays to vehicle platoons at traffic signals, pp. 98–104. In *Strassenbau und Strassenverkehrstechnik 1969* (IVth International Symposium on the Theory of Traffic Flow). Karlsruhe (1968).

Andrews, F. C., A statistical theory of traffic flow on highways. I. Steady-state flow in low density limit. II. Three-car interactions and the onset of queueing. Transp. Res., 4: 359–377 (1970).

Andrews, F. C., A statistical theory of traffic flow on highways. III. Distributions of desired speeds. IV. Semi-empirical steady state theory. Transp. Res., 7:223–242.

Ashton, N. R., Buckley, D. J., and Miller, A. J., Some aspects of capacity and queueing in the vicinity of slow vehicles on a rural two-lane road. Proc. Austr. Road Res. Board, 4:595–612 (1968).

Ashworth, R., The capacity of priority-type intersections with non-uniform distribution of critical acceptance gaps. Transp. Res., 3: 273–278 (1969).

Blumenfeld, D. E., and Weiss, G. H., On the robustness of certain assumptions in the merging delay problem. Transp. Res., 4: 125–140 (1970).

Blumenfeld, D. E., and Weiss, G. H., On queue splitting to reduce waiting times. Transp. Res., 4:141–144 (1970).

Blunden, W. R., and Pretty, R. L., On the theory of deterministic cyclic traffic flows in networks, pp. 287–299. In *Vehicular Traffic Science.* Proceedings Third International Symposium on the Theory of Traffic Flow, New York, 1967.

Buckley, D. J., Leong, Y. F., Ong, C. K., Some results for a deterministic model of an element of a traffic signal system. *Strassenbau und Strassenverkehrstechnik 1969* (IVth International Symposium on the Theory of Traffic Flow). Karlsruhe, 3:364–376 (1968).

Darroch, J. N., Newell, F. F., and Morris, R. W. J., Queues for a vehicle-actuated traffic light. Oper. Res., 12:882–895 (1964).

DeSmit, J. H. A., The transient behaviour of the queue at a fixed cycle traffic light. Transp. Res., 5:1–14 (1971).

Dunne, M. C., Traffic delay at a signalized intersection with binomial arrivals. Transp. Sci., 1:24–31 (1967).

Dunne, M. C., and Buckley, D. M., Delays and capacities at unsignalized intersections. Proc. Austr. Road Res. Board, 6:345–362 (1972).

Dunne, M. C., and Potts, R. B., Algorithms for traffic control. Oper. Res., 12:870–881 (1964).

Dunne, M. C., and Potts, R. B., Control of traffic at an intersection. Proc. Austr. Road Res. Board, 2:265–273 (1964).

Fry, A. T., and Buckley, D. J., Delay and interference in combined lanes at signalized intersections. Proc. Austr. Road Res. Board, 5: 15–39 (1970).

Gaver, D. P., Jr., Time-dependent delays at traffic merges. Oper. Res., 14: 812–821 (1966).

Gazis, D. C., Newell, G. F., Warren, P., and Weiss, G. H., The delay problem for crossing an n-lane highway, pp. 267–279. In *Vehicular Traffic Science.* Proceedings of the

Third International Symposium on the Theory of Traffic Flow, New York, 1967.

Hauer, E., and Templeton, J. G. C., Queueing in lanes. Transp. Res. 6:247–259 (1972).

Iyer, K. S. S., and Maheswaran Nambudripad, D. B., On a stochastic study of highway traffic. *Strassenbau und Strassenverkehrstechnik 1969* (IVth International Symposium on the Theory of Traffic Flow). Karlsruhe (1968), pp. 67–71.

Kleinecke, D. C., Discrete time queues at a periodic traffic light. Oper. Res., 12:809–814 (1964).

Leong, H. J. W., Some aspects of urban intersection capacity. Proc. Austr. Road Res. Board, 2:305–338 (1964).

Martin-Lof, A., Computation of an optimal control for a signalized traffic intersection. Transp. Sci., 1:1–5 (1967).

Miller, A. J., Analysis of bunching in rural two-lane traffic. Oper. Res., 11:236–247 (1963).

Miller, A. J., Queueing in rural traffic, pp. 122–137. In *Vehicular Traffic Science*. Proceedings of the Third International Symposium on the Theory of Traffic Flow, New York, 1967.

Pearce, C. E., Queueing systems with transport service processes. Transp. Sci., 1:218–223 (1967).

Pretty, R. L., and Blunden, W. R., On the computer simulation of a single channel queueing facility for a wide range of arrival and departure distributions. Proc. Austr. Road Res. Board, 2:248–260 (1964).

Rosenshine, M., Queues with state-dependent service times. Transp. Res., 1:97–104 (1967).

Shaw, L., On traffic jam queues. Transp. Res., 4:281–292 (1970).

Weiss, G. H., The intersection delay problem with correlated gap acceptance. Oper. Res., 14:614–618 (1966).

Weiss, G. H., Pedestrian queueing at an *n*-lane intersection, pp. 280–286. In *Vehicular Traffic Science*. Proceedings Third International Symposium on the Theory of Traffic Flow, New York, 1967.

8.9 PROBLEMS

1. A parking authority has found that for every day an automatic ticket dispenser is out of service \$50 is spent to assign operators to that station. Ticket dispensers fail at an average rate (λ) of three/day. The authority has the option of using repair system A that costs \$40/day and is capable of repairing equipment at a rate (μ) of four/day. A more expensive system, B, capable of repairs at a rate (μ) of six/day, will cost \$60/day. The breakdowns occur at random (Poisson), and the length of time to repair a dispenser is exponentially distributed. Which of the two systems, A or B, will minimize the total daily cost of repairs plus breakdowns? (Answer: A = \$190, B = \$110).

2. Vehicles arrive at a toll plaza at random at a rate of 10/min. Calculate the queue characteristics for each of the following arrangements: (a) a single toll booth where the service is random at an average rate of 20 vehicles/min; (b) two toll booths where at each the service is random at an average rate of 10 vehicles/min.

3. Pedestrians require a minimum headway τ of 6.0 sec to cross a roadway. Using the equations of section 8.5.2, find the mean delay per pedestrian and the mean delay per delayed pedestrian for flow rates varying from 100 to 1,200 vehicles/hr (by 100-vehicle/hr increments). Plot the results.

4. The cycle length c at an intersection is 90 sec. Effective green time g is 60 sec. Vehicles arrive at a rate of 720 vehicles/hr and the saturation flow s is 1/3 vehicle/sec. (All values are passenger car units.) The ratio of variance:mean, I, is 1.10. Find the mean delay by each of the methods discussed in section 8.5.4.

5. The peak-hour demand volume on a freeway is 4,500 vehicles/hr and the capacity for uninterrupted flow is 5,700 vehicles/hr. In the event of a vehicle breakdown on the freeway, the capacity is reduced to 4,200 vehicles/hr. It requires 15 min to clear the breakdown. Use the technique discussed by May (section 8.6.1) to determine the queue and delay characteristics. What happens if the breakdown can be reduced to 10 min?

Chapter 9

SIMULATION OF TRAFFIC FLOW

9.1 INTRODUCTION

As any discussion of theory proceeds, one usually has a growing desire to conduct experimentation. In some cases there is a desire to test a particular model; in others, a desire to evaluate a parameter or constant. In still other situations there is a desire to investigate situations that have not yet yielded to theoretical treatment. At times there is a need for experimentation simply to gain enough knowledge of a system to begin modeling. The conduct of traffic experiments on operating facilities has many difficulties. The experimenter must find a suitable site, prepare suitable instrumentation, and then wait for the appropriate traffic condition to occur. If the condition lasts only a short time, tests may have to be conducted on several days or weeks; as a result, it may not be possible to repeat a problem in the field. Some traffic situations may not occur at all on an operating facility. Some experimental runs may imply conditions that would be hazardous. Some experiments might require the construction of expensive facilities. Test tracks, such as that maintained by the (British) Transport and Road Research Laboratory in Crowthorne, England, enable execution of a certain range of tests neither requiring elaborate construction nor entailing hazard. But even experiments on a test track can be quite expensive, requiring the provision of vehicles, drivers, etc.

Since the development of the high-speed automatic computer, there has been a growing tendency to use digital computer simulation as a method of conducting a variety of experiments, especially those concerning systems having important stochastic features.

9.2 NATURE OF SIMULATION

Over the past 20 years various definitions and interpretations have been given to the term "simulation." Generally, present-day usage refers to an experiment performed on an artificial model of a real system. The preponderance of current simulations makes use of a digital computer to implement a model. Such is the case with traffic flow simulations. Although some early investigators examined the possible use of analog computers for traffic flow simulation, and a few recent investigators have proposed applications of hybrid computers for traffic flow simulation (see, for instance, Green et al.[43]), the important traffic flow simulations remain those performed using a digital computer.

The reasons for simulation include (but need not be limited to):

1. The need to test the behavior of a new system or operating procedure prior to its actual construction:
   (a) The construction of the new system may be very expensive and/or time consuming.
   (b) Experimentation with the real system may entail considerable risk (such as traffic accidents).

2. The need to test alternate systems under identical conditions. (For instance, it is never possible to exactly reproduce a specific traffic condition in the field; in simulation it is quite routine to submit the same traffic conditions to several alternative systems.)

Any simulation may be divided into the following steps:

1. Formulation of a model.
2. Reduction of the model to a language acceptable by the computer.
3. Program checkout and internal verification of the model.
4. Experiment planning and design:
   (a) Design of an experiment that will yield the desired information.[1]
   (b) Determination of how each of the test runs will be executed.
5. Performance of experiment(s):
   (a) Model validation.
   (b) Simulation of new conditions.
6. Interpretation of results.

Each of these steps is discussed in the following paragraphs.

9.3 HISTORICAL NOTE ON TRAFFIC FLOW SIMULATION

Although suggestions for the simulation of traffic flow took place as early as 1949,* it was not until several years later that there were published papers discussing possible techniques,[2,3] and it was about 1954 before traffic simulations were actually run on computers in the United States.[4–6,] † Many studies were concerned with how to conduct traffic simulation. By about 1960 it became generally accepted that traffic simulation was possible and feasible, and efforts were directed to the development, validation, and use of large-scale simulation programs. The work described by Wagner et al.[8] is typical of these efforts.

9.4 GENERATION OF RANDOM INPUTS

One of the most important features of simulating traffic is the ability to generate random events. Such generation takes place in two steps: First, a random number following a uniform (rectangular) distribution is generated. Second, this random number is treated as a probability to substitute into an appropriate distribution function in order to solve for the associated event.

9.4.1 Generation of Random Numbers

Many programming systems have "built-in" random number generators. When using a system having such a generator, the investigator can simply call for random numbers, using the appropriate command for that system. Even though most present-day simulation programmers have a random number generator available to them, it is interesting and worthwhile to be familiar with the fundamentals of random number generation.

Most correctly, the term "pseudorandom numbers" should be used rather than random numbers; the procedures used to generate numbers are highly nonrandom, but the numbers generated, when subjected to statistical tests, do not exhibit any nonrandomness. There are many algorithms that can be used to generate pseudorandom numbers,[9] but the one that remains the most common by virtue of its reliability and ease of implementation is the simple multiplicative procedure.[10] This method may be described as follows:

$$R_m = \rho R_{m-1} \text{Mod } b^n \qquad (9.1)$$

where

R_m = the mth random number;
ρ = the multiplier;
n = number of digits in a normal word on the particular computer used;
b = number base of computer;
Mod b^n = instruction to use only the low-order or less significant half of the full ($2n$-digit) product (the remainder after dividing the product by b^n the maximum integral number of times); and
R_0 = any odd number selected as a starting number.

This multiplicative procedure is a special case of a general "congruential" algorithm, which may be stated [11]:

$$R_m = [\rho R_{m-1} + c] \text{ Mod } b^n$$

Thus, the multiplicative procedure is a congruential method with $c=0$.

The results of such generation algorithms may be used as a series of fractions following a uniform (rectangular) distribution. As an example of the multiplicative procedure, let $R_0 = 3$, $\rho=97$, and $b^n=100$ (i.e., $b=10$, $n=2$).

Then

$$03 \times 97 = 0291 \qquad R_1 = 91$$
$$91 \times 97 = 8827 \qquad R_2 = 27, \text{ etc.}$$

The results of this sample generation process are given in Table 9.1, in which it may be seen that the "random series" consists of 20 values, after which the series is repeated.* This illustrates one of the problems in the use of such pseudorandom generation procedures (i.e., a repeating period). However, if the word size of the computer is sufficiently large and the values of R_0 and ρ are properly selected,

* Suggestions by M. Asimow in private discussion with D. L. Gerlough.

† It appears that there may have been some computer runs of traffic simulation in Great Britain as early as 1953 (D. W. Davies of British National Physical Laboratory in conversation with D. L. Gerlough). Webster [7] made use of such work in preparing his classic work on traffic signal settings.

* It will be noted that all numbers are odd. This is a result of selecting an odd number as the starting value and a multiplier that is a power of a prime number.

the period is usually sufficiently long for most simulation purposes. It should be noted that by saving the last random number, it is possible to restart the series without returning to the beginning with each start. Alternatively, if it is desired for some particular experiment to provide, say, the same series of traffic arrivals to each part of the experiment, using the same starting value for the random generation will provide this feature.†

The value of ρ may be selected by taking a base that is prime relative to the number base of the computer and raising it to the highest power that can be held by one word of the computer. If it is necessary to design a generator for the longest possible period, the procedures discussed by Hull and Dobell [12] may be used.

TABLE 9.1 Example of Random Numbers

m	R_m
0	03
1	91
2	27
3	19
4	43
5	71
6	87
7	39
8	83
9	51
10	47
11	59
12	23
13	31
14	07
15	79
16	63
17	11
18	67
19	99
20	03

9.4.2 Production of Desired Random Variate

Production of the desired random variate drawn from a specified distribution function may be explained by use of Figure 9.1, which shows the cumulative probability distribution of a variable X. Although this distribution curve would usually be thought of as representing the probability as a function with the value of X as the argument, in the present application the roles of the variables are reversed. The random number (fraction) generated as described in section 9.4.1 is interpreted as a probability and is used as an argument to enter the distribution, giving the value of X as the function. [For this procedure it is necessary to use the (cumulative) probability distribution rather than the probability density function.]

When an analytic expression for the cumulative distribution is known, the conversion can be accomplished by a simple equation-solving procedure. For instance, in obtaining variates distributed according to the negative exponential distribution, the solution proceeds as follows:

$$P(h<t)=1-e^{-t/T}$$

which simplifies to

$$P=1-e^{-t/T}$$
$$t=-T\ln(1-P) \qquad (9.2)$$

† If random numbers are used for several functions (beside random arrival), separate generation routines should be provided if the same traffic is to be submitted to several experimental conditions.

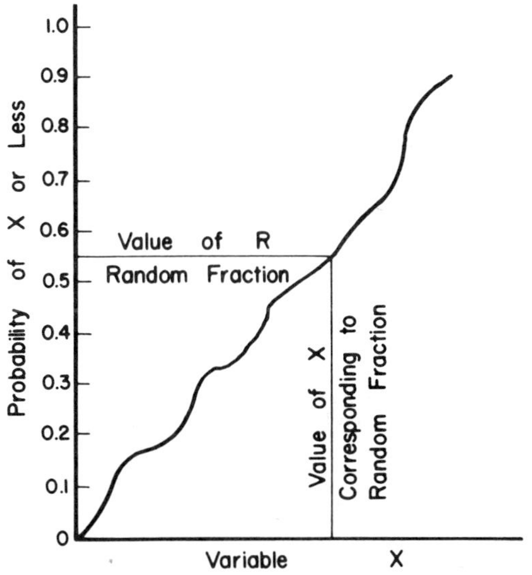

Figure 9.1 Use of cumulative probability distribution to convert random numbers from uniform distribution to desired distribution.[10]

A random number may be substituted for P or $(1-P)$ and the resulting value of t used as the desired headway.

For the shifted negative exponential the relationship is:

$$t=(T-\tau)[-\ln(1-P)]+\tau \qquad (9.3)$$

If, however, a counting distribution such as the Poisson distribution is required, a step-by-step solution must be used. This procedure for the Poisson distribution is illustrated by Figure 9.2 and described as follows: First, a random fraction, R, is generated as previously described. The cumulative Poisson distribution is then formed, term by term, using Eq. 3.6. At each step the cumulation is compared with R. When the first value $P(x)$ satisfying the relationship $P(x) \geq R$ is found, the corresponding value of x is taken as the random variate (number of arrivals). A flow diagram for accomplishing this process on a computer is given by Gerlough.[10]

An alternate approach is to generate and store in advance a table with probability as an argument and time as a function. This may be done by assuming various values of t and computing P. When in operation, random fractions are used to search the P values of the table. When a match is found, the corresponding t-value is used as the arrival headway.

For the production of random variates from a normal distribution the following special procedure may be used:

1. Draw 12 random fractions from a uniform distribution.

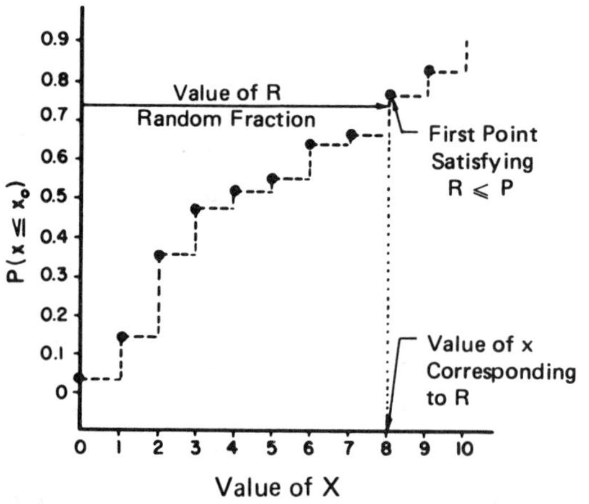

Figure 9.2 Production of random variates following counting distribution, using step-by-step calculation.[10]

2. Take the sum of these fractions, giving the first normal variate.
3. Repeat using 12 different random fractions, giving the second normal variate.

The results will be variates from a distribution having a mean of 6 and a variance of 1. (This result may be explained by the central limit theorem.)

9.5 MODEL FORMULATION

The formulation of a simulation model starts with the definition or selection of the following items:

1. The traffic situation to be simulated.
2. The "figure of merit" or "measure of effectiveness" to be used.
3. The degree of complexity to be included.
4. The traffic generation (arrival) model to be used.
5. Model for processing traffic through the simulated situation.
6. Computer language to be used (see section 9.6).
7. Computer to be used (which is usually the one most readily available).

At first it may seem inappropriate to include items 6 and 7 under model generation. However, in many cases these items influence the selection of other items, especially items 3 and 5.

Rather than attempt to describe model formulation in general terms, the procedures are illustrated by selected examples.

9.5.1 Simulation Example: Intersection Load Factor *

(NOTE: Intersection load factor has been defined as "a ratio of the total number of green signal intervals that are fully utilized by traffic during the peak hour to the total number of green intervals for that approach during the same period.")

Once it has been decided that a study of load factor as it relates to capacity should be conducted by simulation, the following steps of model formulation are implemented:

1. Situation to be simulated: Intersection approach with fixed-time signal, variable sto-

* In constructing this example the work of May et al.[13,14] has been heavily drawn upon.

chastic arrival rate, deterministic variable departure rate.

2. Figures of merit: Individual vehicle delay (maximum and average delay over all vehicles), degree of saturation, load factor, maximum queue length.

3. A relatively simple model is desired.

4. Traffic arrivals will follow the composite exponential model of Eq. 3.10. Figure 9.3 is a flow chart for this generation. Arrival times relative to some reference will be computed by adding the new headway to the previous total. Parameters will be calculated in accordance with the method of Grecco and Sword.[15]

5. The following logic will be used in processing vehicles from their point of arrival through the intersection:

(a) Discharge from the head of the queue will be based on Greenshields' [16] data, as follows:

Position in Queue at Start of Green (M)	Time from Start of Green until Arrival at Entrance to Intersection (sec)
1	3.8
2	6.9
3	9.6
4	12.0
5	14.2
≥ 6	$14.2+2.1(M-5)$

(b) If the signal is green and if there is no queue, at the arrival time of the next vehicle it is processed through the intersection without delay.

(c) If the signal is green and if there is a queue, vehicles at the head of the queue are processed through the intersection in accordance with Greenshields' table. Each arrival joins the queue until it is its turn to be processed.

(d) If the signal is red, arriving vehicles queue until the signal turns green.

(e) On arrival each vehicle's time is recorded until it departs, at which time the difference between departure and arrival times gives the delay.

(f) The clock is advanced in uniform intervals of 0.1 sec.

(g) As each vehicle leaves the intersection its delay is entered into the running average. In addition, its delay is compared with the previous maximum delay and the larger value is retained.

(h) As the end of each signal indication arrives, the current state of the signal is changed within the program.

(i) Before the signal is turned red, a check is made to determine if the full green period was utilized.

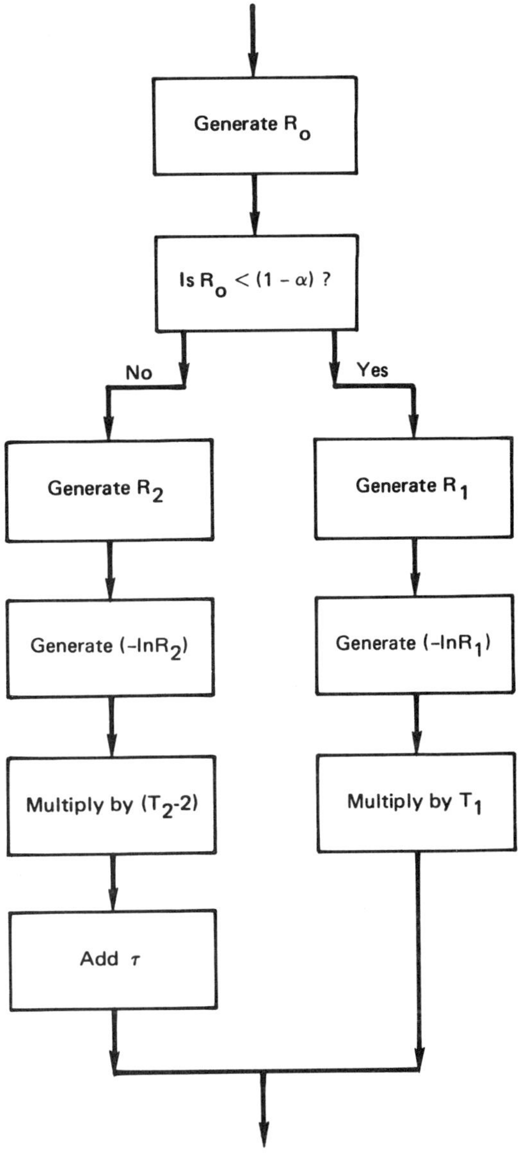

Figure 9.3 Flow chart for generation of composite exponential model of Eq. 3.10.[10]

(j) At the end of an experimental run the various figures of merit are computed and printed out.

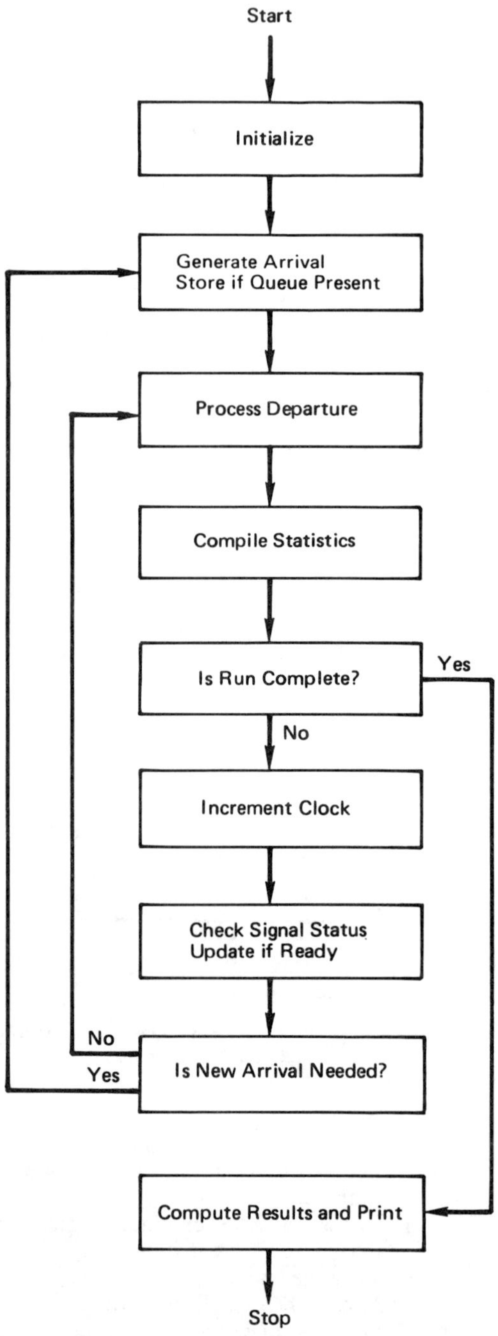

Figure 9.4 Flow chart for simulation example.

6. The program will be written in BASIC.
7. The simulation will be run on a computer having an on-line time-sharing system using a remote terminal.

The flow chart for this program is shown in Figure 9.4. The program listing is given in Appendix D-2. This program is relatively simple in a number of ways; principally, (a) no opposing or cross traffic is considered, (b) time is advanced in uniform intervals (periodic scan), and (c) only one type of random event (arrival time) is generated. The program does, however, illustrate the steps involved in constructing a simulation.

9.5.2 Simulation Example: Simple Four-Way Intersection

It is now possible to progress to the simulation of a more complex situation; namely, the four-way intersection shown in Figure 9.5, as described by Worrall.[17] Several types of event are generated randomly. The intersection consists of two one-way streets, each with a single lane. The east–west street has priority over the north–south street. Low flows are assumed, and the arrival headways are assumed to follow the negative exponential distribution of Eq. 3.8. The value T in the equation is taken as $3{,}600/V$, where V is the arriving volume in vehicles/hr. Each approach is assigned (as input data) probabilities that vehicles will turn or continue straight ahead. A gap acceptance distribution is provided, in tabular form, and the same distribution is used for all gaps. (See Table 9.2.)

Figure 9.5 illustrates the decisions by north–south (N–S) vehicles in accepting or rejecting gaps between east–west (E–W) vehicles. The simulation model may be described with the aid of the flow diagram of Figure 9.6.

The input data consist of volume levels on the two approaches and a gap acceptance table. The input operation is indicated by the first box at the top of Figure 9.6. This includes the initial settings of all counters, etc. The simulation loop starts with the second box, "Generate Next N–S Arrival Gap." This arrival gap is generated by first generating a random fraction and then substituting this fraction in Eq. 9.3. The actual arrival time of the new arrival is obtained by adding the generated time headway to the arrival time of the previous N–S vehicle.

At this point it is appropriate to test whether the undelayed arrival time of the new

N–S vehicle exceeds the maximum duration of the experiment. If the limit has been reached, the program jumps to the calculation and printing of output results; if not, the program continues.

Next the effective arrival time of the new N–S vehicle is computed. This is the time that the vehicle will arrive at the stopline. This calculation starts by determining whether the previous N–S vehicle has crossed the intersection: if it has not, the new vehicle joins the queue; if it has, the new effective arrival time is equal to the new actual arrival time.

Queues may be handled in several ways.

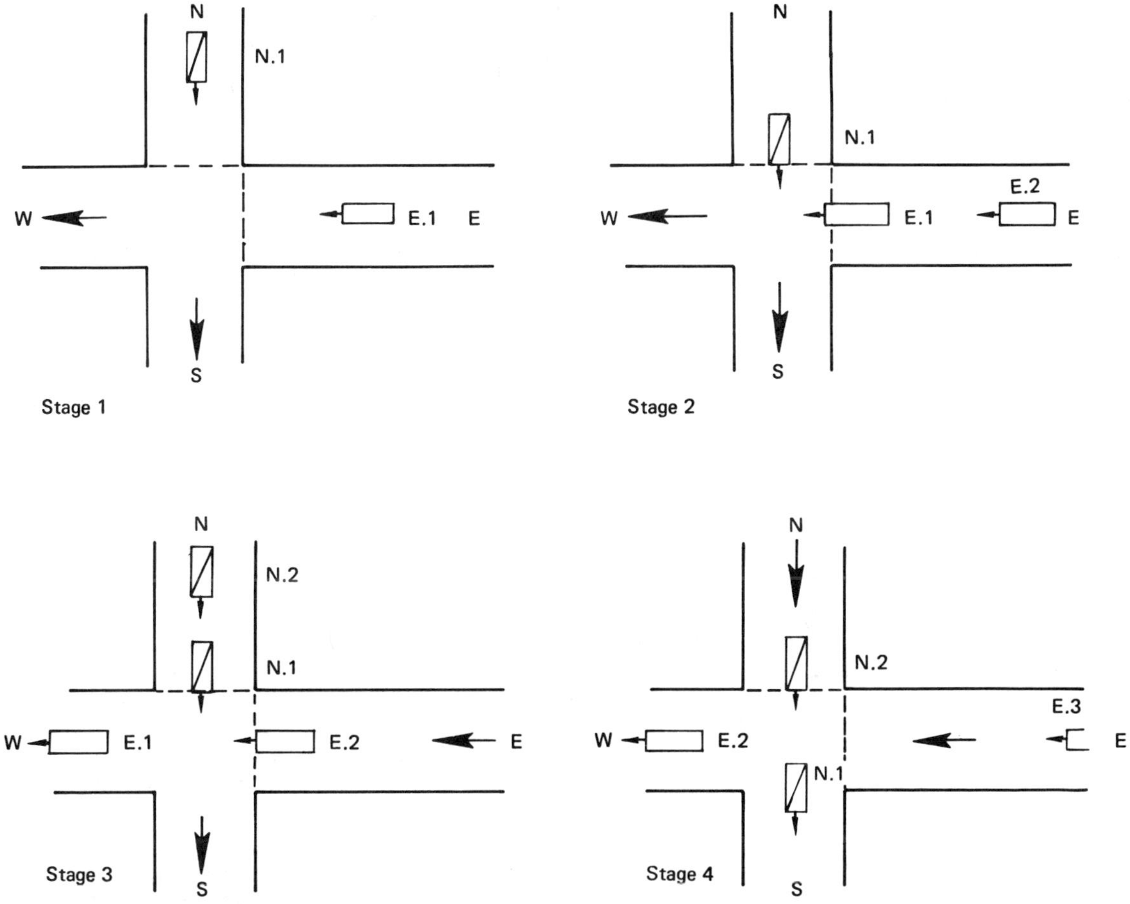

Figure 9.5 Simulated intersection shows schematically the various conditions:[17] Stage 1, vehicles N.1 and E.1 approach intersection; Stage 2, vehicle N.1 reaches intersection and examines E-W stream for acceptable gap, but the available gap between E.1 and E.2 is not acceptable, so vehicle N.1 is delayed; Stage 3, vehicle N.1 continues to examine E-W stream for acceptable gap and vehicle N.2 arrives at intersection and forms a queue; Stage 4, gap between vehicles E.2 and E.3 is acceptable to vehicle N.1, so vehicle N.1 moves off across intersection, and vehicle N.2 commences looking for an acceptable gap. However, the lag between N.1 moving off and E.3 arriving is not acceptable to N.2, so vehicle N.2 is delayed further.

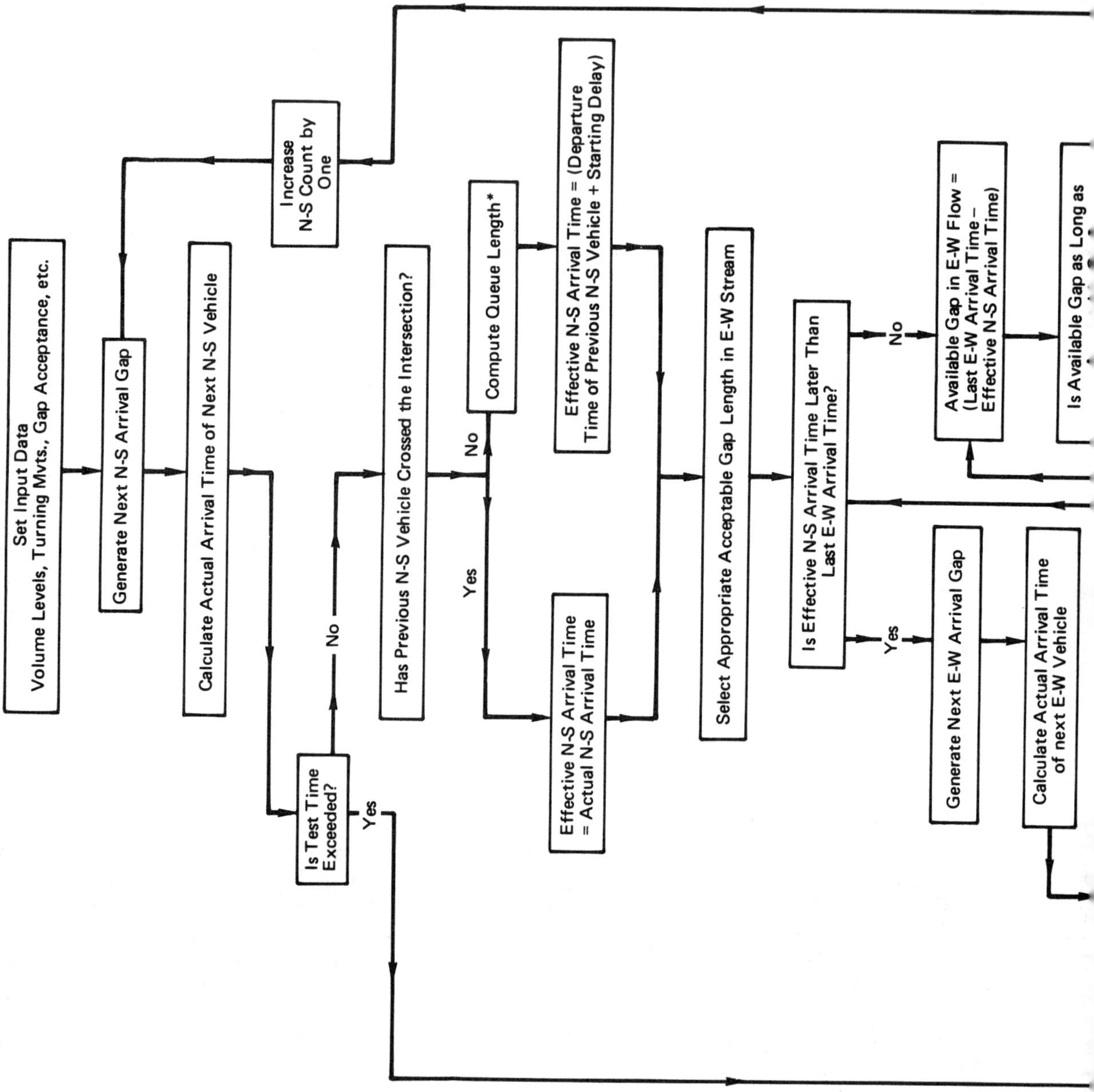
Set Input Data
Volume Levels, Turning Mvts., Gap Acceptance, etc.
Generate Next N-S Arrival Gap
Calculate Actual Arrival Time of Next N-S Vehicle
Increase N-S Count by One
Is Test Time Exceeded?
Yes
No
Has Previous N-S Vehicle Crossed the Intersection?
No
Yes
Compute Queue Length*
Effective N-S Arrival Time = (Departure Time of Previous N-S Vehicle + Starting Delay)
Effective N-S Arrival Time = Actual N-S Arrival Time
Select Appropriate Acceptable Gap Length in E-W Stream
Is Effective N-S Arrival Time Later Than Last E-W Arrival Time?
No
Yes
Available Gap in E-W Flow = (Last E-W Arrival Time – Effective N-S Arrival Time)
Generate Next E-W Arrival Gap
Calculate Actual Arrival Time of next E-W Vehicle
Is Available Gap as Long as

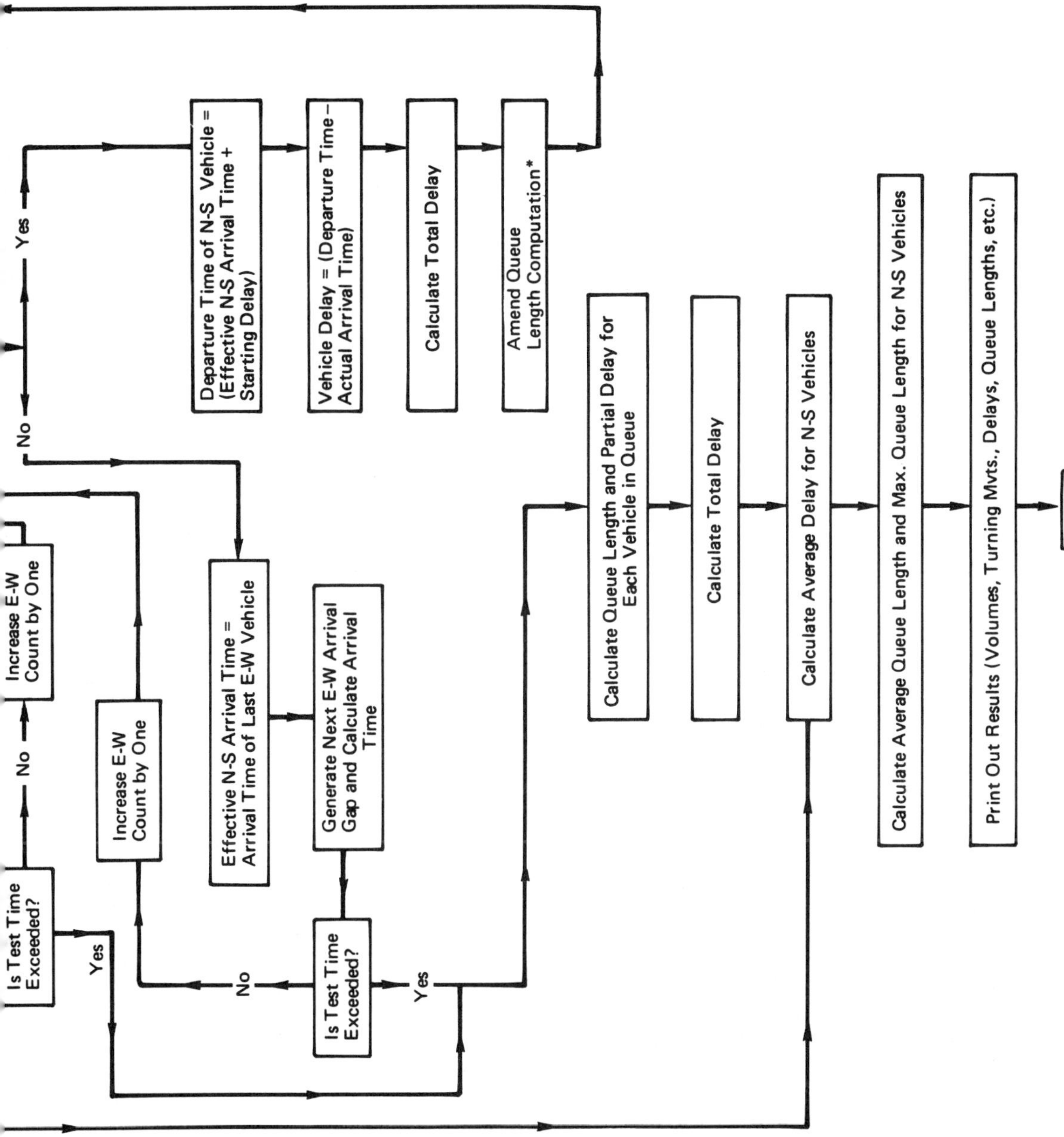

Figure 9.6 Flow chart for intersection simulation.[17]

TABLE 9.2 Gap Acceptance Distribution for Example of Section 9.4.2

Gap (sec)	Percent Acceptance
0–2	0
2–3	9
3–4	18
4–5	22
5–6	40
6–7	59
7–8	61
8–9	80
9–10	90
10–11	92
11–12	93
12–13	94
13–14	95
14–15	96
15–16	97
16–17	98
17–18	99
>18	100

One effective way is to establish within the computer an array three columns wide and of sufficient length to accommodate the longest queue expected. In addition, a queue counter indicates the number of vehicles currently in the queue. As each new vehicle joins the queue, the queue counter is first increased, then the row of the array corresponding to the count is selected. Into the three columns of this row are placed the actual arrival time, the effective arrival time (if it can currently be computed), and the departure time (if it can currently be computed). As the first vehicle in the queue moves out of the queue and across the intersection, the first row in the array is removed, each of the remaining entries is moved up one row, the place vacated by the last vehicle is set to zero, and the queue counter is decreased. At the same time any further computations of departure and actual arrival times are performed.

The gap in the E–W stream acceptable to the first-in-queue N–S vehicle is determined. A random fraction is generated; this fraction is then compared with the percentage values in Table 9.2 (expressed as fractions). The gap corresponding to the random fraction is then designated as the minimum gap that will be accepted by the N–S vehicle.

Now begins an extensive series of tests concerning gaps in the E–W stream. First, test whether the effective arrival time of the (first-in-queue) N–S vehicle is later (greater) than the arrival time of the last E–W vehicle generated. If it is, it is necessary to generate a new E–W vehicle arrival gap and arrival time. The method used is similar to that used for N–S vehicles. If the new E–W arrival time exceeds the time limit for the experiment, the run is terminated (after calculating and printing the output results); otherwise, the E–W traffic count is increased, and the available gap in the E–W traffic is examined for acceptance by the first-in-queue N–S car. The available gap is the arrival time of the last E–W car minus the effective arrival time of the N–S car. After this gap is computed it is compared with the previously computed minimum acceptable gap.

If the gap is not acceptable, the effective arrival time of the first-in-queue N–S vehicle is reset to the arrival time of the last E–W vehicle, and a new E–W vehicle is generated. If the arrival time of this new E–W vehicle does not exceed the time duration of the experiment, the E–W traffic count is increased by one, and the acceptability of the gap in front of this new E–W car is tested. If the available gap (discussed in the previous paragraph) is found acceptable, it is accepted, and the departure time of the N–S vehicle entering the intersection is computed as its effective arrival time plus the appropriate starting delay. Its delay in queue is computed by subtracting its arrival time from its departure time. This delay is added to the cumulative record of delay. After correcting the queue for the departure of one vehicle, the N–S count is increased and the simulation loop starts again by the generation of a new N–S vehicle.

Two important points are illustrated by this example. First, in addition to random generation of arrivals on two approaches there is random generation of gap acceptance. Second, the "clock" is not advanced by a uniform periodic interval; instead the examination moves from one important time to another important time. These differences in the methods of "review" or "scanning" are known, respectively, as "periodic scan" and "event scan."

9.5.3 Simulation Example: Freeway Merging Area

This example, based on a paper by Wohl,[18] consists of a slightly more complex simulation —that of a freeway merging area. Figure 9.7 shows the schematic layout of the merging area.

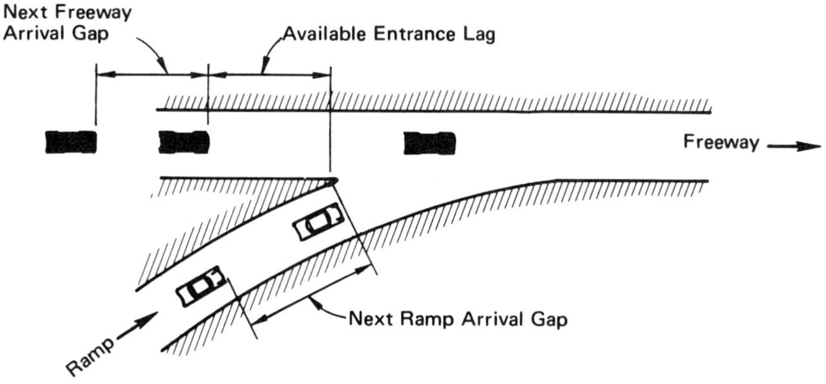

Figure 9.7 Schematic layout of freeway merging area.[18]

The figure of merit to be measured is delay to vehicles entering the freeway from the ramp, as a function of freeway and ramp volumes. The model formulation starts with several assumptions, as follows:

1. The merging area is sufficiently remote from traffic generation sources and traffic controls as to allow the freeway and ramp vehicles to approach the intersection in a random fashion.
2. Freeway vehicles will always be given preference; thus, when conflicts exist (between ramp and freeway vehicles), only ramp vehicles will be delayed on entering the flow of freeway traffic.
3. Freeway vehicles will not shift lanes within the merging area; thus, only the outside lane of traffic (for a multilane freeway) must be considered. This area can be represented by a single-lane freeway and single-lane ramp.
4. When looking for an acceptable gap in the freeway flow, the ramp vehicle will be considered either moving or stopped (delayed), and a different distribution of acceptable gaps will be used for each case.
5. First-in-line ramp vehicles that are stopped while waiting for freeway entrance suffer a 5-sec acceleration loss on entering the freeway flow; ramp vehicles that are delayed but enter directly from a second-in-line position have the same acceleration loss. When a first-in-line vehicle is delayed, there is an additional 3.0-sec starting reaction delay on entering the freeway; for a second-in-line delayed vehicle there is a 2.0-sec starting delay when entering directly from this position.
6. Ramp vehicles that are delayed by the freeway vehicle directly ahead on entering the freeway suffer a delay of 2.0 sec.

These assumptions allow characterization of the freeway merging section in terms of only five elements, as follows:

1. Acceptable low-relative-speed (LRS) entrance gap (i.e., moving vehicle case).
2. Acceptable high-relative-speed (HRS) entrance gap (i.e., stopped vehicle case).
3. Next freeway arrival gap.
4. Next ramp arrival gap.
5. Available entrance gap (or lag).

These various gaps (and lag) are depicted schematically in Figure 9.7. An acceptable low-relative-speed entrance gap is a gap that is accepted by a vehicle that enters without stopping. An acceptable high-relative-speed entrance gap is a gap that is acceptable to a vehicle that has stopped on the ramp before entering. It should be noted, however, that a second-in-line vehicle that may have been stopped may enter as a moving vehicle after the preceding vehicle has entered. The sequence of simulation events is indicated in Figure 9.8, and Figure 9.9 is a flow chart of the computing process.

As described by Wohl [18]:

> The general philosophy and some of the mechanics of the model can best be described by following through the example of Figure 9.8. In this example the experiment starts at some arbitrary time, such as $T = 0000$. The first step is to generate the next ramp arrival gap (8 sec) and the next freeway arrival gap (5 sec) and to determine the times at which the next ramp and freeway vehicles will arrive at the nose of the ramp; these steps are indicated in Figure 9.8(a). The driver of the ramp vehicle bases his decision on whether or not to enter the freeway upon

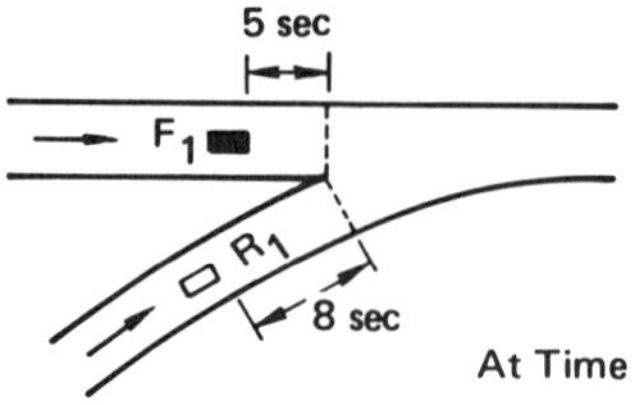

1. Generate next ramp arrival gap (8 sec for R_1)
2. Compute ramp veh. arrival time (0008 sec for R_1)
3. Generate next freeway arrival gap (15 sec for F_1)
4. Compute freeway veh. arrival time (0005 sec for F_1)

At Time T = 0000 sec

(a)

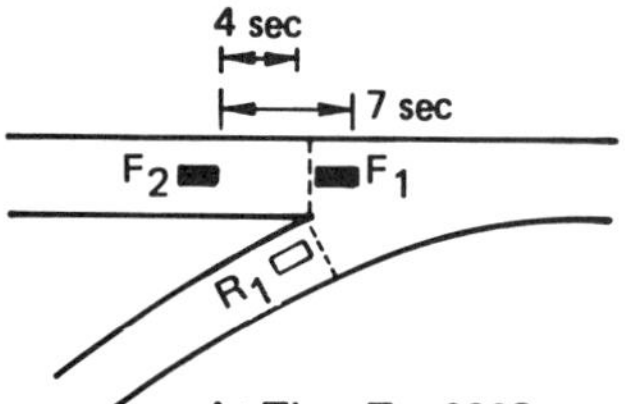

At Time T = 0008 sec

1. Is freeway veh. (F_1) arrival time later than ramp veh. (R_1) arrival time? No.
2. Generate next freeway arrival gap (7 sec for F_2)
3. Compute freeway veh. arrival time (0012 sec for F_2)
4. Is freeway veh. (F_2) arrival time later than ramp veh. (R_1) arrival time? Yes.
5. Compute available entrance gap (4 sec for R_1)
6. Generate acceptable LRS entrance gap (5 sec for R_1)
7. Is the available gap acceptable? No.

(b)

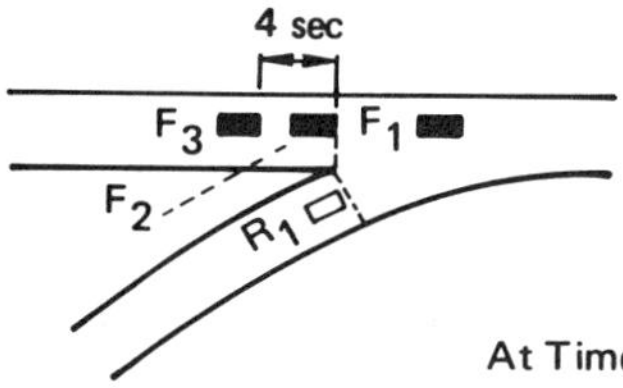

1. Generate next freeway arrival gap (4 sec for F_3)
2. Compute freeway veh. arrival time (0016 sec for F_3)
3. Compute available entrance gap (4 sec for R_1)
4. Generate acceptable HRS entrance gap (6 sec for R_1)
5. Is the available gap acceptable? No.

At Time T = 0012 sec

(c)

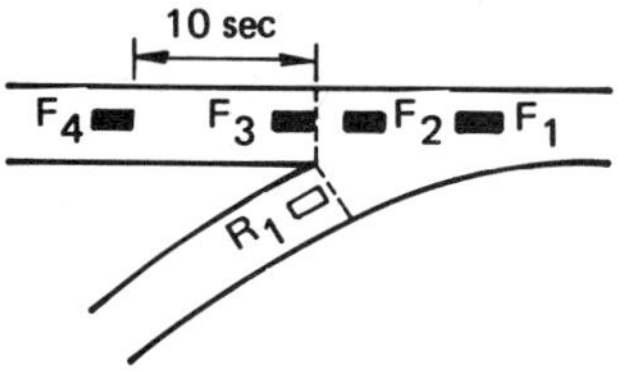

1. Generate next freeway arrival gap (10 sec for F_4)
2. Compute freeway veh. arrival time (0026 sec for F_4)
3. Compute available entrance gap (10 sec for R_1)
4. Is the available gap acceptable? Yes.
5. Compute ramp veh. departure time (0019 sec for R_1)
6. Compute ramp veh. delay (16 sec for R_1)

At Time T = 0016 sec

(d)

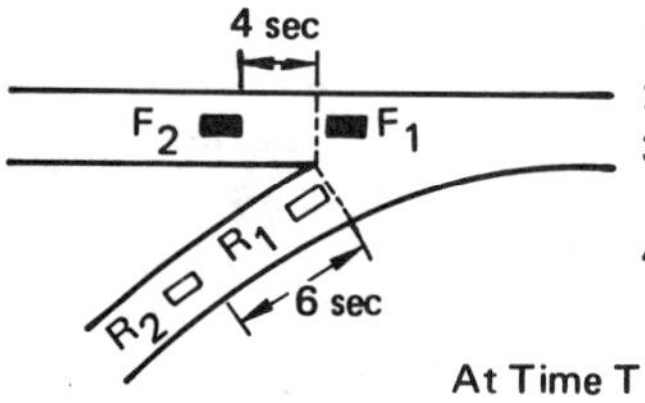

1. Generate next ramp arrival gap (6 sec for R_2)
2. Compute ramp veh. arrival time (0014 sec for R_2)
3. Is the arrival time (of R_2) earlier than the departure time of the previous ramp veh. (R_1)? Yes.
4. Compute looking-for-gap time for R_2 (departure time of previous veh. plus operating delays = 0019 + 3 = 0022 sec)

At Time T = 0008 sec

(e)

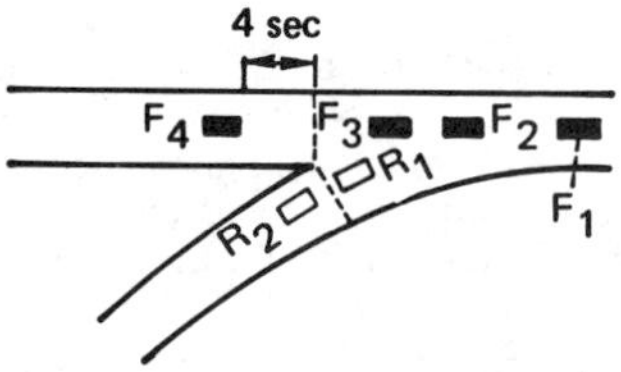

1. Is freeway veh. (F_4) arrival time later than ramp veh. (R_2) looking-for-gap time. Yes.
2. Compute available entrance gap (4 sec for R_2)
3. Generate acceptable LRS entrance gap (3 sec for R_2)
4. Is the available gap acceptable? Yes.
5. Compute ramp veh. departure time (0024 sec for R_2)
6. Compute ramp veh. delay (15 sec for R_2)

At Time T = 0022 sec

(f)

Figure 9.8 Sequence of simulation events.[18]

his expected arrival time at the nose of the ramp and the availability of a proper freeway gap. If at the time the ramp vehicle arrives at the nose there is an acceptable gap in the freeway flow, he will enter. If not, he will be delayed and must wait for a later gap in the freeway flow. Figure 9.8(b) shows the positions of the first ramp and freeway vehicles at the time the ramp vehicle arrives at the nose. Because the first freeway vehicle had passed the nose prior to the arrival of the first ramp vehicle, the next freeway arrival gap (7 sec) and its arrival time at the nose (0012 sec) must be determined; also, a check must be made to ensure that this freeway arrival time (0012 sec) is later than the ramp arrival time (0008 sec). Because it is, the available entrance lag or gap (that is, the gap available for the ramp vehicle to enter the freeway flow) must be computed (it is 4 sec). (If the freeway arrival time is not later than the ramp arrival time, successive freeway arrival times must be determined until the first one with an arrival time later than that of the ramp vehicle is located.) Following this, the acceptable LRS entrance gap (5 sec) for the ramp vehicle must be "generated."

Since the available entrance lag is not acceptable, the ramp vehicle will be delayed at least until the second freeway vehicle passes. Figure 9.8(c) shows the vehicle positions at the arrival time of the second freeway vehicle at the nose (0012 sec); at this time the ramp vehicle examines the next freeway arrival gap (4 sec), which in this case is also the available entrance gap. Since the ramp vehicle has been delayed, we must determine its acceptable HRS gap (6 sec). Again, the available entrance gap is not acceptable to the ramp vehicle, and another freeway gap must be examined.

Figure 9.8(d) shows the vehicles at the time when the third freeway vehicle arrives at the nose. The next freeway arrival gap (10 sec) is determined and the freeway arrival time of the fourth freeway vehicle (0026 sec) is computed. The available entrance gap (10 sec) is larger than the acceptable HRS gap (6 sec); therefore, the ramp vehicle can accept the gap and enter the freeway. The departure time (0019 sec) for the ramp vehicle is the third freeway vehicle arrival (0016 sec) plus the operating delays (3 sec) to account for the starting reaction delay, etc. The final step is to compute the ramp vehicle delay, which is its departure time (0019 sec) minus its original arrival time at the nose (0008 sec) plus 5-sec acceleration delays on entering the freeway flow. Thus the total delay is 16 sec.

The second ramp vehicle cannot enter the freeway until some time following the departure of the first ramp vehicle. However, the second ramp vehicle may have arrived on the ramp earlier than the departure time of the first and been waiting in queue. Since we are primarily concerned with delays to ramp vehicles that enter the freeway flow, the original times of arrival of the ramp vehicles at the nose must be computed as well as the times at which the ramp vehicles actually entered the freeway flow. Thus, the first two steps of Figure 9.8(e) are to "generate" the next ramp arrival gap (6 sec) and compute arrival time of the second ramp vehicle (0014 sec). Following this, it must be determined whether or not the ramp vehicle was delayed by the previous ramp vehicle; that is, Is the arrival time (of the second ramp vehicle) earlier than the departure time of the previous ramp vehicle? If the ramp vehicle was delayed by the previous one, the earliest time that the ramp vehicle can actually begin looking for an acceptable freeway gap must be calculated on the basis of the departure time of the previous ramp vehicle. Consequently, the so-called "looking-for-gap" time of the second ramp vehicle (0022 sec) is equal to the departure time of the previous ramp vehicle (0019 sec) plus operating delays (3 sec) to allow for perception and reaction times, etc.

The next step is to determine the earliest freeway vehicle arrival time that is later than the look-for-gap time of the second ramp vehicle. In this case the last computed freeway vehicle arrival time (0026 sec for the fourth freeway vehicle) is later than the looking-for-gap time of the second ramp vehicle (0022 sec); therefore, the available entrance lag is the difference between these two times, or 4 sec. Based on earlier assumptions, an acceptable LRS entrance gap for the waiting ramp vehicle (3 sec) must be computed. The available entrance gap is acceptable, and the second ramp vehicle enters the freeway; see Figure 9.8(f). Its departure time (0024 sec) is the looking-for-gap time (0022 sec) plus operating delays of 2 sec; its delay (15 sec) is the departure time (0024 sec) minus arrival time (0014 sec) plus 5-sec acceleration delays.

These steps can be repeated for as many simulation trials as are desired; the

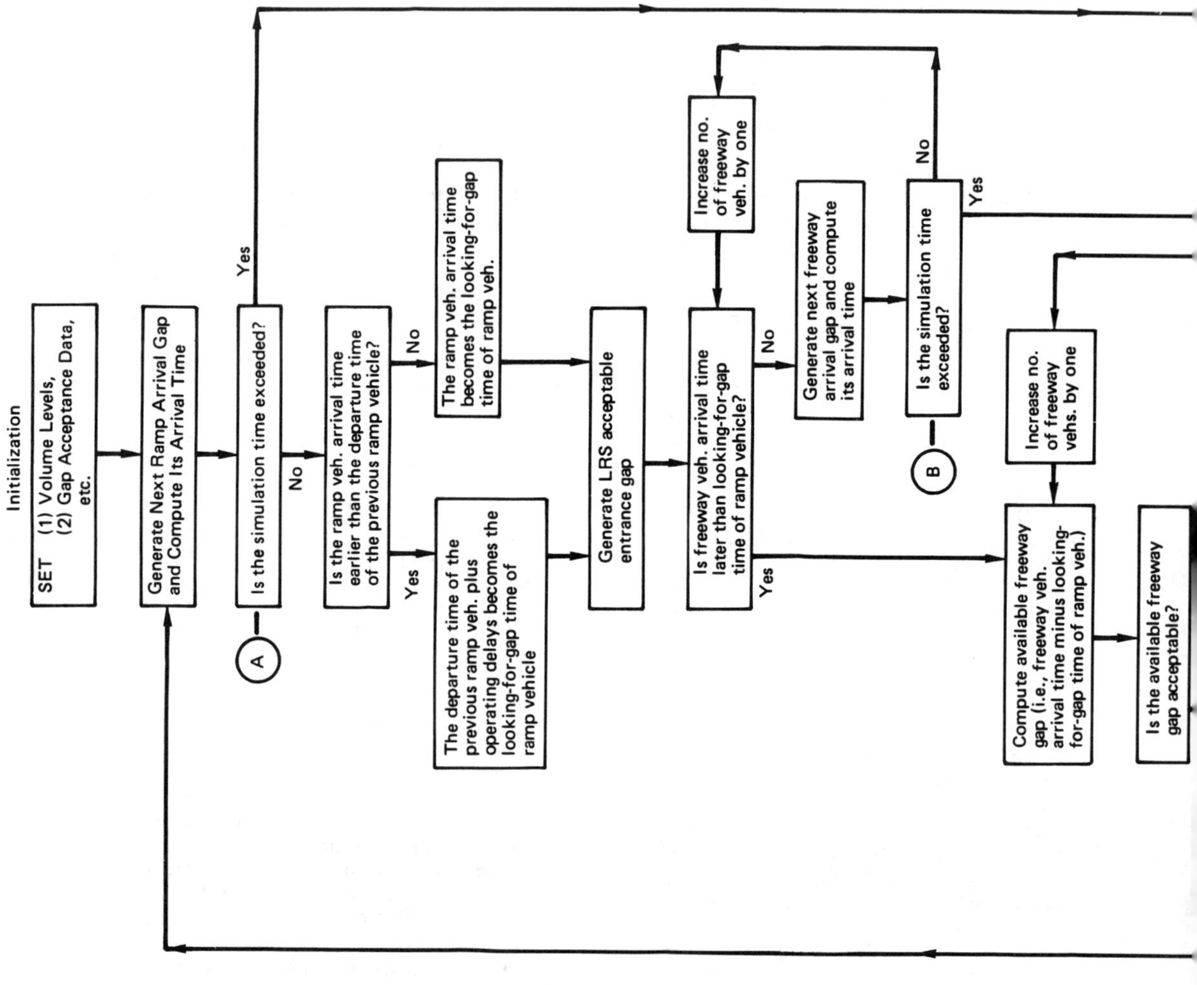
Initialization
SET (1) Volume Levels,
(2) Gap Acceptance Data, etc.
Generate Next Ramp Arrival Gap and Compute Its Arrival Time
Is the simulation time exceeded?
Yes
No
A
Is the ramp veh. arrival time earlier than the departure time of the previous ramp vehicle?
No
Yes
The ramp veh. arrival time becomes the looking-for-gap time of ramp veh.
The departure time of the previous ramp veh. plus operating delays becomes the looking-for-gap time of ramp vehicle
Generate LRS acceptable entrance gap
Increase no. of freeway veh. by one
Is freeway veh. arrival time later than looking-for-gap time of ramp vehicle?
No
Yes
Generate next freeway arrival gap and compute its arrival time
Is the simulation time exceeded?
No
Yes
B
Increase no. of freeway vehs. by one
Compute available freeway gap (i.e., freeway veh. arrival time minus looking-for-gap time of ramp veh.)
Is the available freeway gap acceptable?

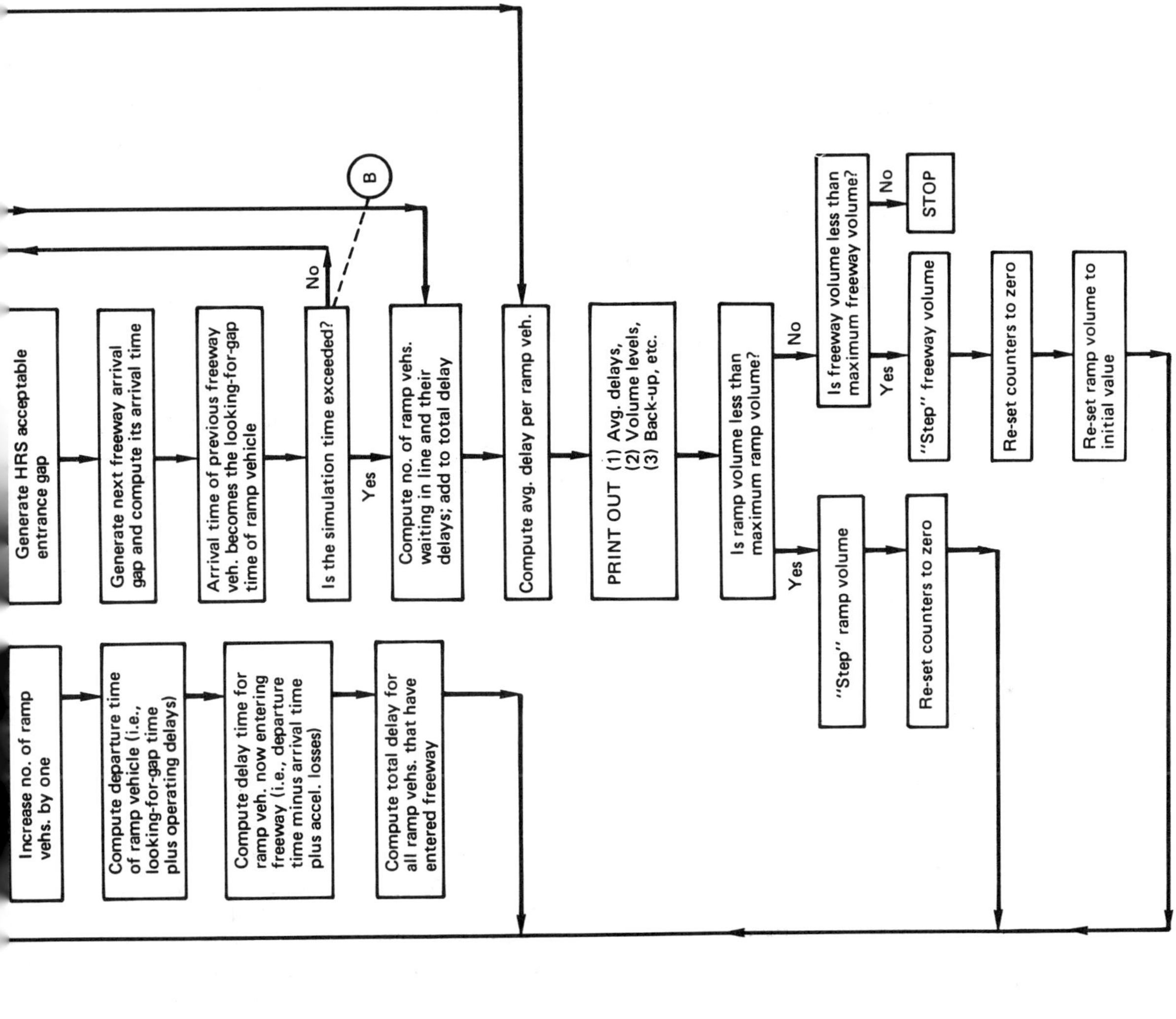

Figure 9.9 Flow chart of computing process.[18]

number of "samples" or trials will, of course, be a function of the accuracy required, and of time and economic consideration.

The flow chart of Figure 9.9 must be broken down into finer details to facilitate actual computer programming. For instance, the generation of vehicle gap acceptance is performed as shown in Figure 9.9 using the techniques of Figure 9.10.

9.6 THE COMPUTER PROGRAM

Having formulated the logic of a simulation problem, the next step is to convert it into a computer program. In the early days of traffic simulation this was done by arduously writing a program in machine language. Now, however, simulations are aided by the availability of a variety of languages to aid the user. There are general languages, such as FORTRAN, that enable the program writer to work with statements that are essentially algebraic in nature. There are also several higher-order languages specifically designed for simulation. The two such languages best known in the United States are SIMSCRIPT and GPSS, both of which are briefly described in Appendix D-1. At present, large simulation programs, especially those that are to be used at different computer centers, can be best written in FORTRAN. Where many small programs are being written to be used on a "one shot" basis, it may be worth considering some special simulation language. (In many cases, of course, the selection of a higher-order language is dependent on the computer available or is made by the management of the computer center. In such situations the user adapts himself to the language available.)

9.7 PROGRAM CHECKOUT

There are two types of checkout of a simulation program: "debugging" and "validation." Debugging determines whether the program is working and, if working, correctly representing the model as defined. Validation is a test to determine whether the model satisfactorily represents the system to be simulated. Validation, a statistical experiment, is discussed further in section 9.8.

Debugging is carried out in a variety of ways. When the simulation program is being written, there will undoubtedly be many subroutines or other special sections that perform particular functions. Each of these should be checked out as it is written. Such testing may require the writing of special test routines to supply data and print the output of the subroutine being tested. Although this entails extra work, the insurance that each subroutine is working properly within itself more than justifies the effort. These tests should, of course, employ test problems or data that lead to known results or results that can be easily checked by hand calculations. Care should be taken to ensure that tests exercise all possible cases in order to avoid some obscure spurious result.

After all of the subroutines have been tested and corrected separately, the complete program is then assembled and tested. If the total program is very long, it is often advisable to put it together a few subroutines at a time, testing at each step. During testing, if other measures fail to find the source of a particular "bug," a trace may be employed, printing out the result after every step in the calculation. (Many computing centers have special programs for tracing.) Once the complete program has been tested and debugged, it is ready for validation experiments.

9.8 EXPERIMENT PLANNING AND DESIGN

As in all experimental undertakings the statistical design of a simulation experiment is important in order to minimize the amount of experimentation and to enable inferences with the desired levels of significance. Simulation experiments have the advantage over field experiments in that simulation enables the experimenter to more easily control the various variables and to perform additional replications at will. This leads some experimenters to disregard that (1) computer time for experiments costs money and (2) any experiment should be designed in such a way that cost of additional measurements is weighed against the value of the additional information to be gained.

With relation to computer simulation experiments, experimental design is considered to include:

1. Consideration of such problems as the

selection of factor levels and factor * combinations, the order of experimentation, the minimization of random error, as well as classical and computer simulation design problems.[27]

2. A plan for starting simulation runs.
3. Validation procedures.

9.8.1 General Simulation Experiment Design

Space does not permit a detailed discourse on experiment design. Simulation experiments make use of all the usual types of experiment design, with emphasis on response surface exploration [28–31] and sequential sampling.[32] Some investigators [33–35] advocate the use of variance reducing techniques for reducing the sample size. A problem that must be given particular attention in the design of simulation experiments is the tendency for stochastic processes to be autocorrelated and hence unanalyzable by traditional statistical methods.[36]

9.8.2 Starting the Simulation

When a simulation is started, the system is usually empty and any measurements made on measures of effectiveness will be essentially worthless. This problem is most simply handled by excluding some initial start-up period from the system evaluation.[37] The difficult question is how to recognize when equilibrium or stability has been achieved. Sometimes stability can be hastened by setting into the system some starting conditions more nearly those anticipated at equilibrium. Tests should be made to assure that the performance is independent of starting conditions. Fishman [44] has recently treated some of these techniques.

9.8.3 Validation

Validation involves a set of experiments on the model in which the results are compared with (historical) measurements on the real system. Although time series analysis is sometimes useful,[45] there are problems in such comparisons and Gafarian and co-workers [38,39,42] have described some of these. The usc of nonparametric statistics is recommended.[40] It should be noted that validation tests are null experiments: A model that fails tests is rejected, but no strong statement can be made about a model that is accepted.[41]

* Here "factor" is used in the statistical experiment design sense.

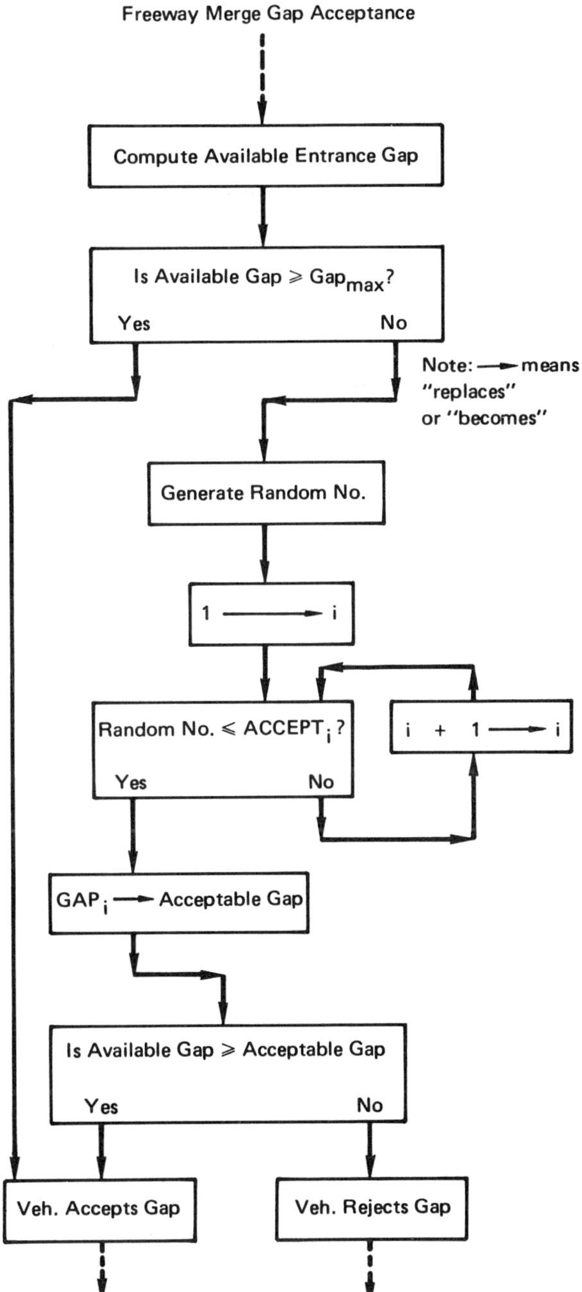

Figure 9.10 Generation of freeway gap acceptance.[18]

9.9 INTERPRETING RESULTS

The interpretation of simulation experiment results is similar to the interpretation of any experimental results. Often the result is the testing of some hypothesis; in other cases the interpretation consists of fitting some curve to the output of the experiment.

9.10 SUMMARY

In formulating a simulation model, one strives for sufficient realism to adequately describe the phenomena of interest. However, one should not go "overboard" in including extra details that will not significantly affect the results.

Programs should be properly checked out and validated.

Traffic simulation programs that are properly modeled and validated constitute important experimental facilities for traffic study.

9.11 REFERENCES

1. Conway, R. W., Some tactical problems in digital simulation. Manage. Sci., 10 (1): 47–61 (1963).
2. Goode, H. H., Simulation—its place in system design. Proc. Inst. Radio Eng., 39 (12):1501–1506 (1951).
3. Davis, H., Simulation with general purpose computers. In D. L. Trautman, H. Davis, J. Heilfron, and E. C. Ho (ed.) *Analysis and Simulation of Vehicular Traffic Flow*. Research Rep. 20. Institute of Transportation and Traffic Engineering, University of California (1954).
4. Gerlough, D. L., *Simulation of Freeway Traffic on a General-Purpose Discrete-Variable Computer*. Ph.D. dissertation. University of California, Los Angeles (1955).
5. Gerlough, D. L., Simulation of freeway traffic by an electronic computer. Proc. Highw. Res. Board, 35:543–547 (1956).
6. Goode, H. H., Pollmar, C. H., and Wright, J. B., The use of a digital computer to model a signalized intersection. Proc. Highw. Res. Board, 35:548–557 (1956).
7. Webster, F. V., *Traffic Signal Settings*. Road Research Technical Paper No. 39. Her Majesty's Stationery Office, London (1958).
8. Wagner, F. A., Barnes, F. C., Stirling, D. P., and Gerlough, D. L., *Urban Arterial and Network Simulation*. Planning Research Corporation Rep. 962 to U.S. Bureau of Public Roads (1966): NTIS No. PB 174629.
9. Chambers, R. P., Random-number generation on digital computers. IEEE Spectrum, 4 (2):48–56 (1967).
10. Gerlough, D. L., Traffic inputs for simulation on a digital computer. Proc. Highw. Res. Board, 38:480–492 (1959).
11. MacLaren, M. D., and Marsaglia, G., Uniform random number generators. J. Assoc. Comput. Mach., 12 (1):83–89 (1965).
12. Hull, T. E., and Dobell, A. R., Random number generators. SIAM Rev., 4 (3): 230–254 (1962).
13. May, A. D., Jr., and Pratt, D., A simulation study of load factor at signalized intersections. Traffic Eng., 38 (5):44–49 (1968).
14. May, A. D., Jr., and Gyamfi, P., Extension and preliminary validation of a simulation of load factor at signalized intersections. Traffic Eng., 40 (1):46–52 (1969).
15. Grecco, W. L., and Sword, E. C., Prediction of parameters for Schuhl's "headway distribution." Traffic Eng., 38 (5): 36–38 (1968).
16. Greenshields, B. D., Shapiro, D., and Ericksen, E. L., *Traffic Performance at Urban Street Intersections*. Tech. Rep. 1. Yale Bureau of Highway Traffic, New Haven, Conn. (1947), p. 27.
17. Worrall, R. D., Simulation of traffic behaviour on a digital computer. Traffic Eng. Control, 5 (2):86–90, 94 (1963).
18. Wohl, M., Simulation—its application to traffic engineering. Traffic Eng., Part I, 30 (11):13–17, 29 (1960); Part II, 31 (1):19–25, 56 (1960).
19. Dimsdale, B., and Markowitz, H. M., A description of the SIMSCRIPT language. IBM Sys. J., 3 (1):57–67 (1964).
20. Tocher, K. D., Review of simulation languages. Oper. Res. Q., 16 (2):189–217 (1965).
21. Freeman, D. E., Programming languages ease digital simulation. Control Eng., 11 (11):103–106 (1964).
22. Krasnowkow, E. S., and Merikallio, R. A., The past, present, and future of general simulation languages. Manage. Sci., 11 (2):236–267 (1964).

23. Teichrow, D., and Lubin, J. F., Computer simulation—discussion of the technique and comparison of languages. Commun. ACM, 9 (10):723–741 (1966).

24. Efron, R., and Gordon, G., A general purpose digital simulator and examples of its application. Part I: Description of the system. IBM Sys. J., 3 (1):22–34 (1964).

25. Blum, A. M., A general purpose digital simulator and examples of its application. Part III: Digital simulation of urban traffic. IBM Sys. J. 3 (1):41–56 (1964).

26. Murphy, J. G., A comparison of the use of the GPSS and SIMSCRIPT simulation languages in designing communication networks. Tech. Memo. 0369/000/00/0/00, Mitre Corp. (26 March 1964).

27. Burdick, D. S., and Naylor, T. H., Design of computer simulation experiments for industrial systems. Commun. ACM, 9 (5):329–339 (1966).

28. Hunter, J. S., and Naylor, T. H., Experimental designs for computer simulation experiments. Manage. Sci., 16 (7): 422–434 (1970).

29. Brooks, S. H., A discussion of random methods for seeking maxima. Oper. Res., 6 (2):244–251 (1958).

30. Hooke, R., and Jeeves, T. A., Comments on Brooks' discussion of random methods. Oper. Res., 6 (6):881–882 (1958).

31. Karson, M. J., Manson, A. R., and Hader, R. J., Minimum bias estimation and experimental design for response surfaces. Technometrics, 11 (8):461–475 (1969).

32. Sasser, W. E., Burdick, D. S., Graham, D. A., and Naylor, T. H., The application of sequential sampling to simulation: an example inventory model. Commun. ACM, 13 (5):287–296 (1970).

33. Maxwell, W. L., Variance reduction techniques. Paper presented at the Simulation Panel of the Department of Defense Logistics Conference held in Warrenton, Virginia, May 26–28, 1965. Rand Corporation (May 6, 1965); NTIS No. AD 615167.

34. Clark, C. E., Importance sampling in Monte Carlo analyses. Oper. Res., 9 (5):603–620 (1961).

35. Kahn, H., and Marshall, A. W., Methods of reducing sample size in Monte Carlo computations. Oper. Res., 1 (4):263–278 (1953).

36. Fishman, G. S., Problems in the statistical analysis of simulation experiments; the comparison of means and the length of sample records. Commun. ACM, 10 (2):94–99 (1967).

37. Conway, R. W., Some tactical problems in digital simulation. Manage. Sci., 10 (1):47–61 (1963).

38. Gafarian, A. V., and Walsh, J. E., Statistical approach for validation simulation models by comparison with operational systems, pp. 702–705. In *Proceedings of Fourth International Conference on Operational Research.* Wiley-Interscience, New York (1966).

39. Gafarian, A. V., and Walsh, J. E., Methods for statistical validation of a simulation model for freeway traffic near an on-ramp. Transp. Res., 4 (4):379–384 (1970).

40. Walsh, J. E., *Handbook of Nonparametric Statistics. II: Results for Two and Several Sample Problems, Symmetry and Extremes.* Van Nostrand, Princeton, N.J. (1965).

41. Conway, R. W., Johnson, B. M., and Maxwell, W. L., Some problems in digital systems simulation. Manage. Sci., 6 (1):92–110 (1959).

42. Gafarian, A. V., Hayes, E., and Mosher, W. W., Jr., The development and validation of a digital simulation model for design of freeway diamond interchanges. Highw. Res. Rec. 208 (1967), pp. 37–78.

43. Green, D. H., Hartley, M. G., and Powner, E. T., Research at the University of Manchester Institute of Science and Technology, pp. 89–96. In *Area Traffic Control.* Institution of Civil Engineers, London (1967).

44. Fishman, G. S., *Concepts and Methods in Discrete Event Computer Simulations.* Wiley, New York (1973).

45. Hsu, D. A., and Hunter, J. S., Validation of computer simulation models using parametric time series analysis. Simuletter (Assoc. Comput. Mach.), 5 (2): 81–82 (1974).

9.12 RELATED LITERATURE

Books

Apter, M. J., *The Computer Simulation of Behaviour.* Harper Colophon Books (1970).

Cashwell, E. D., and Everett, C. J., *A Practical Manual of the Monte Carlo Method for Random Walk Problems.* Pergamon, New York (1959).

Emshoff, J. R., and Sisson, R. L., *Design and Use of Computer Simulation Models.* MacMillan, New York (1970).

Evans, G. W., II, Wallace, G. F., and Sutherland, G. L., *Simulation Using Digital Computers.* Prentice-Hall, Englewood Cliffs, N.J. (1967).

Gordon, G., *System Simulation.* Prentice-Hall, Englewood Cliffs, N.J. (1969).

Hammersley, J. M., and Handscomb, D. C., *Monte Carlo Methods.* Methuen (1964).

Hoggatt, A. C., and Balderston, F. E. (ed.), *Symposium on Simulation Models: Methodology and Applications to the Behavioral Sciences.* Southwestern Publishing Co. (1963).

Martin, F. F., *Computer Modeling and Simulation.* Wiley, New York (1968).

Mise, J. H., and Cox, J. G., *Essentials of Simulation.* Prentice-Hall (1968).

Naylor, T. H., Balintfy, J. L., Burdick, D. S., and Kong, C., *Computer Simulation Techniques.* McGraw-Hill, New York (1966).

Schmidt, J. W., and Taylor, R. E., *Simulation and Analysis of Industrial Systems.* Richard D. Irwin, Inc. (1970).

Tocher, K. D., *The Art of Simulation.* English Universities Press (1963).

Maisel, H., and Gnugnoli, G., *Simulation of Discrete Stochastic Systems.* Science Research Associates (1972).

Bibliographies

International Business Machines Corporation, *Bibliography on Simulation* (1966).

Malcolm, D. G., Bibliography on the use of simulation in management analysis. Oper. Res., 8 (2):169–177 (1960).

Wigan, M. R., *Applications of Simulation to Traffic Problems: An Annotated Bibliography.* (British) Road Research Laboratory, London, Library Bibliography No. 102/MRW (1969).

General Simulation

Meier, R. L., and Dike, R. D., Gaming simulation for urban planning. AIP J., 32 (1):3–22 (1966).

Mihram, G. A., Some practical aspects of the verification and validation of simulation models. Oper. Res. Q., 23 (1):17–29 (1972).

Traffic Simulation

Aitken, J. M., Simulation of traffic at an uncontrolled junction. Traffic Eng. Control, 5 (6): 354–358 (1963).

Ashworth, R., A note on the selection of gap acceptance criteria for traffic simulation studies. Transp. Res., 2 (2):171–175 (1968).

Dickey, J. W., and Montgomery, D. C., A simulation-search technique: an example application for left-turn phasing. Transp. Res., 4 (4):339–347 (1970).

Katz, J. H., Simulation of a traffic network. Commun. ACM, 6 (8):480–486 (1963).

Kell, J. H., Intersection delay obtained by simulating traffic on a computer. Highw. Res. Rec. 15 (1963), pp. 73–97.

Longley, D., A simulation study of a traffic network control scheme. Transp. Res., 5 (1):39–57 (1971).

McClenahan, J. W., and Simkowitz, H. J., The effect of short cars on flow and speed in downtown traffic: a simulation model and some results. Transp. Sci., 3 (2):126–139 (1969).

Morgan, H. L., UTS–I: a macro system for traffic network simulation. AFIPS Conf. Proc., 36:217–222 (1970).

Sakai, T., and Nagao, M., Simulation of traffic flows in a network. Commun. ACM, 12 (6):311–318 (1969).

Saleeb, S., and Hartle, M. G., Simulation of traffic behaviour through a linked-pair of intersections. Transp. Res., 2 (1):51–61 (1968).

Saint John, A. D., Study of automobile accidents through digital simulation, pp. 66–76. In *Proceedings of SAE Analysis and Control of Traffic Symposium* (January 9–10, 1968). (SAE paper 680173).

Schumate, R. P., and Dirksen, J. R., A simulation system for study of traffic flow behavior. Highw. Res. Rec. 72 (1965), pp. 19–39.

Sinha, K. C., and Dawson, R. F., Digital computer simulation of freeway traffic flow. Traffic Q., 24 (2):279–296 (1970).

Thomasson, J. H., Jr., and Wright, P. H., Simulation of traffic at a two-way stop intersection. Traffic Eng., 37 (11):39–45 (1967).

U.S. Bureau of Public Roads, *Proceedings of Highway Conference on the Future of Research and Development in Traffic Surveillance, Simulation, and Control* (Washington, D.C., September 14–15, 1964); NTIS No. PB167200.

Warnshuis, P., Simulation of two-way traffic on an isolated two-lane road. Transp. Res., 1 (1):75–83 (1967).

Wright, P. H., Simulation of traffic at a four-way stop intersection, pp. 44–52. In *Proceedings of SAE Analysis and Control of Traffic Symposium* (January 9–10, 1968). (SAE paper 680170).

Lieberman, E. B., Simulation of corridor traffic: the Scot model. Highw. Res. Rec. 409 (1972), pp. 34–45.

Lieberman, E. B., Worrall, R. D., and Bruggeman, J. M., Logical design and demonstration of UTCS–1 network simulation model. Highw. Res. Rec. 409 (1972), pp. 46–56.

Other Transportation Simulations

Askew, J. R., Beardwood, J. E., and Newby, D., A preliminary investigation of railway signaling methods by computer simulation studies. Transp. Sci., 2 (3):191–212 (1968); Discussion, 3 (2):176–179 (1969).

Statistical Techniques

Ehrenfeld, S., and Ben-Tuvia, S., The efficiency of statistical simulation procedures. Technometrics, 4:257–275 (1962).

Francki, M., The problem of confidence and three methods of variance reduction in the simulation of queueing. Proc. Austr. Road Res. Board, 4:567–581 (1968).

Gafarian, A. V., and Ancker, C. J., Jr., Mean value estimation from digital computer simulation. Oper. Res., 14 (1):25–44 (1966).

Simulation Languages

Dames, R. T., Simulation software. Datamation, 11 (10):29–34 (1965).

Clancy, J. J., and Fineberg, M. S., Digital simulation languages: a critique and a guide, pp. 23–36. In *Proceedings of Fall Joint Computer Conference.* (1965).

Pugh, A. L., III, *Dynamo User's Manual*, 2nd ed. M.I.T. Press, Cambridge, Mass.

Chapter 10

EPILOGUE

Although there is at present no unified theory of traffic flow, there are many theoretical approaches to a variety of traffic phenomena. A substantial portion of these theoretical approaches have made important contributions to practical solutions of traffic problems.

It should be evident from the discussions that traffic flow theory is an evolving science. Each year sees research that results in important theoretical advances. During the course of preparation of this document, for instance, the new literature has been extensive, and there have been several new theoretical developments that could not be adequately covered. Nevertheless, the need for further research is great, and all who find the theory of traffic flow interesting are seriously urged to consider undertaking research.

It is hoped that a unified theory of traffic flow will not be long in coming.

Appendix A
DATA SUPPLEMENTARY TO CHAPTER 2*

A–1 RELATIONSHIPS BETWEEN TIME AND SPACE SPEED STATISTICS

Time Mean Speed from Space Mean Speed

Using the method of Wardrop,[1] segregate total flow into m subflows by speed. Define

$$\bar{u}_t=\frac{\sum_{i=1}^{m} q_i u_i}{\sum_{i=1}^{m} q_i}=\frac{\sum q_i u_i}{q}$$

and

$$\bar{u}_s=\frac{\sum_{i=1}^{m} k_i u_i}{\sum_{i=1}^{m} k_i}=\frac{\sum k_i u_i}{k}.$$

But

$$q_i=k_i u_i.$$

Thus,

$$\bar{u}_t=\frac{\sum (k_i u_i) u_i}{q}=k\frac{\sum k_i u_i^2}{kq}=k\sum\frac{f_i' u_i^2}{q}$$

where

$$f_i'=\frac{k_i}{k}.$$

But

$q=k\,\bar{u}_s$; Thus

$$\bar{u}_t=k\sum\frac{f_i' u_i^2}{k\bar{u}_s}=\frac{1}{\bar{u}_s}\sum f_i'[\bar{u}_s+(u_i-\bar{u}_s)]^2$$

$$=\frac{1}{\bar{u}_s}\Big[\sum f_i'\bar{u}_s^2+2\bar{u}_s\sum f_i'(u_i-\bar{u}_s)+\sum f_i'(u_i-\bar{u}_s)^2\Big]$$

But

$$\sum f_i'(u_i-\bar{u}_s)=0$$

by definition of mean and

$$\sum f_i'(u_i-\bar{u}_s)^2=\sigma_s^2$$

by definition of variance about the space mean speed; therefore,

$$\bar{u}_t=\frac{1}{\bar{u}_s}[\bar{u}_s^2+0+\sigma_s^2]$$

$$=\bar{u}_s+\frac{\sigma_s^2}{\bar{u}_s} \qquad (2.7)$$

Derivation in the Continuous Case

Define

$$\bar{u}_s=\int_0^\infty u f_s(u)\,du$$

$$\bar{u}_t=\int_0^\infty u f_t(u)\,du$$

where

$f_t(u)$ = speed density in time and
$f_s(u)$ = speed density in space.

An important relationship given in Haight and Mosher [11] but proved by Breiman [13] is:

$$\bar{u}_s f_t(u)=u f_s(u)$$

After multiplying both sides by u and integrating over the entire range of u:

$$\bar{u}_s\int_0^\infty u f_t(u)\,du=\int_0^\infty u^2 f_s(u)\,du$$

$$\bar{u}_s\bar{u}_t=\int_0^\infty u^2 f_s(u)\,du \qquad (A)$$

Define

$$\sigma_s^2=\int_0^\infty u^2 f_s(u)\,du-(\bar{u}_s)^2 \qquad (B)$$

Substituting Eq. A in Eq. B gives

$$\sigma_s^2=\bar{u}_s\bar{u}_t-(\bar{u}_s)^2$$

* Reference citations are listed in Chapter 2.

and

$$\bar{u}_t = \bar{u}_s + \frac{\sigma_s^2}{\bar{u}_s}$$

Relationship Between Arithmetic Mean and Harmonic Mean *

To examine this computation, consider the arithmetic and harmonic means without any context of traffic. Define

$$M = \frac{1}{N}\sum_{i=1}^{N} X_i = \text{arithmetic mean}$$

$$V_m = \frac{1}{N}\sum_{i=j}^{N}(X_i - M)^2 = \text{variance about arithmetic mean}$$

$$H = \frac{1}{\frac{1}{N}\sum_{i=1}^{N}\frac{1}{X_i}} = \frac{N}{\sum_{i=1}^{N}\frac{1}{X_i}} = \text{harmonic mean.}$$

Expand in Taylor series:

$$\frac{1}{X_i} = A_0 + A_1(X_i - M) + A_2(X_i - M)^2 + A_3(X_i - M)^3 + \ldots$$

Evaluate constants by differentiating,

$$A_j = \frac{1}{j!}\frac{d^j}{dX^j}\left(\frac{1}{X_i}\right)\Big|_M$$

for

$$j = 0, 1, \ldots$$

$$\frac{1}{X_i} = \frac{1}{M} - \frac{1}{M^2}(X_i - M) + \frac{1}{M^3}(X_i - M)^2 - \frac{1}{M^4}(X_i - M)^3 + \ldots\dagger \quad \text{(C)}$$

Then

$$\frac{1}{N}\sum_{i=1}^{N}\frac{1}{X_i} = \frac{1}{N}\sum_{i=1}^{N}\frac{1}{M} - \frac{1}{NM^2}\sum(X_i - M) + \frac{1}{NM^3}\sum(X_i - M)^2 - \frac{1}{NM^4}\sum(X_i - M)^3 + \frac{1}{NM^5}\sum(X_i - M)^4 + \ldots$$

By the definition of the arithmetic mean

$$\sum^{N}(X_i - M) = 0.$$

Similarly, for any distribution that is approximately symmetrical

$$\sum(X_i - M)^a \approx 0 \text{ for odd values of } a.$$

Thus

$$\frac{1}{N}\sum_{i=1}^{N}\frac{1}{X_i} = \frac{1}{NM}(N) - \frac{1}{NM^2}(0) + \frac{1}{M^3}(V_m) - \left(\frac{1}{M^4 N}\right)(0) + \frac{1}{M^5 N}\sum(X_i - M)^4 \ldots$$

It can be assumed that

$M^5 N \gg \sum(X_i - M)^4$ since Eq. C is converging.

Thus the last term can be neglected, as can all later terms in the expansion.

Then

$$\frac{1}{N}\sum_{i=1}^{N}\frac{1}{X_i} = \frac{1}{M} + \frac{V_m}{M^3} = \frac{M^2 V_m}{M^3}$$

$$H = \frac{1}{(M^2 + V)/M^3} = \frac{M^3}{M^2 + V_m} = \frac{M}{1 + \frac{V_m}{M^2}}$$

$$= \frac{M\left(1 - \frac{V_m}{M^2}\right)}{\left(1 + \frac{V_m}{M^2}\right)\left(1 - \frac{V_m}{M^2}\right)}$$

$$= \frac{M\left(1 - \frac{V_m}{M^2}\right)}{1 - \frac{V_m^2}{M^4}} \approx M\left(1 - \frac{V_m}{M^2}\right) = M - \frac{V_m}{M}.^*$$

Converting to traffic notation gives:

$$\bar{u}_s = \bar{u}_t - \sigma_t^2/\bar{u}_t \quad (2.8)$$

as an approximate method for use in traffic engineering practice. Note that this relationship when combined with Eq. 2.7 implies that

$$u_s/u_t = \sigma_s^2/\sigma_t^2.$$

Thus, in using Eq. 2.8 one must be willing to accept this assumption.

---

* Note the use of approximate methods in this development.

† Eq. C converges for $0 < x < 2M$, which is usually the case for traffic.

* Yule and Kendall [19] suggest this relationship for cases where deviations are small compared to the mean.

A–2 DERIVATION OF FORMULAS FOR MOVING OBSERVER METHOD

Consider the traffic stream to have a constant total flow through the area of observation. Consider further that the stream is composed of m substreams, each having its own uniform speed. Let

q_i = flow of the ith substream;
k_i = concentration of the ith substream;
u_i = speed of the ith substream;
t_i = travel time of the ith substream;
l = length of the roadway section;
u_c = speed of the observer moving with the traffic stream;
u_a = speed of the observer moving against the traffic stream;
t_c = travel time of observer moving with the traffic stream;
t_a = travel time of observer moving against the traffic stream;
t = mean travel time of the traffic stream;
$k = \sum_i^m k_i$ = concentration of the (total) traffic stream;
$q = \sum_i^m q_i$ = flow of the (total) traffic stream;
u = mean speed of the (total) traffic stream;
x = total number of vehicles met by the observer moving against the traffic stream;
y = net number of vehicles that pass the observer while moving with traffic stream (i.e., the number that pass the observer minus the number he passes);
x_i = number of the population x that are in substream i; and
y_i = number of the population y that are in substream i.

Then

$$k_i = q_i / u_i.$$

For Observer Moving Against Stream

$$\text{Flow past observer} = k_i(u_a + u_i) = \frac{q_i}{u_i}(u_a + u_i)$$

$$x_i = (\text{flow})(\text{time}) = \left[\frac{q_i}{u_i}(u_a + u_i)\right]\left[\frac{l}{u_a}\right]$$

$$= q_i l \frac{(u_a + u_i)}{u_a u_i} = q_i\left[\frac{l}{u_i} + \frac{l}{u_a}\right] = q_i[t_i + t_a].$$

For Observer Moving with Stream

$$\text{Flow past observer} = k_i(u_i - u_c) = \frac{(q_i)}{u_i}(u_i - u_c)$$

$$y_i = \left[\frac{q_i}{u_i}(u_i - u_c)\right]\left[\frac{l}{u_c}\right] = q_i l \frac{(u_i - u_c)}{u_i u_c}$$

$$= q_i\left(\frac{l}{u_c} - \frac{l}{u_i}\right) = q_i[t_c - t_i].$$

For Total Stream

$$x = \sum_{i=1} x_i = \sum q_i(t_a + t_i) = t_a \Sigma q_i + \Sigma q_i t_i \qquad \text{(D)}$$

$$y = \sum_{i=1} y_i = \sum q_i(t_c - t_i) = t_c \Sigma q_i - \Sigma q_i t_i \qquad \text{(E)}$$

Adding Eqs. D and E gives

$$x + y = (t_a + t_c)\Sigma q_i = (t_a + t_c) q$$

$$q = \frac{x + y}{t_a + t_c}$$

$$qt = \Sigma q_i t_i$$

$$t = \frac{\Sigma q_i t_i}{q} = \frac{t_c \Sigma q_i - y}{q} = t_c - \frac{y}{q} \qquad \text{(using Eq. E.)}$$

$$u = l / t$$

$$k = q / u$$

Berry and Green [16] have suggested the following number of runs for travel time within a 10 percent range of accuracy: (a) for progressive signal timing (volumes below capacity), 8 runs; (b) for signals not coordinated (volumes at or near capacity), 12 runs; (c) for signals not coordinated (volumes below capacity), 8 runs.

Appendix B

DATA SUPPLEMENTARY TO CHAPTER 3*

B–1 DERIVATION OF THE POISSON DISTRIBUTION †

Consider a line that can represent in a general case either distance or time; for the present purposes consider it to represent time (Figure B.1). Specifically, consider the occurrence of random arrivals where the average rate of arrival (i.e., probability density) is λ. Let $P_i(t)$ = the probability of i arrivals up to the time t, and $P_n(\Delta t) = \lambda \Delta t$ = the probability of one arrival in the incremental period Δt. Because it is assumed that Δt is of such short duration, the probability of more than one arrival in Δt is negligible; therefore, $(1-\lambda\Delta t)$ = the probability of no arrival in Δt. Then,

$P_i(t+\Delta t)$ = the probability that i arrivals have taken place to the time $(t+\Delta t)$
= [Prob ($i-1$ arrivals in t)·Prob (1 arrival in Δt)]+[Prob (i arrivals in t)·Prob (0 arrivals in Δt)];

$$P_i(t+\Delta t) = P_{i-1}(t)\cdot P_1(\Delta t) + P_i(t)\cdot P_0(\Delta t)$$
$$= P_{i-1}(t)\lambda\Delta t + P_i(t)(1-\lambda\Delta t)$$
$$= [P_{i-1}(t) - P_i(t)](\lambda\Delta t) + P_i(t); \text{ and}$$
$$\frac{P_i(t+\Delta t) - P_i(t)}{\Delta t} = \lambda[P_{i-1}(t) - P_i(t)].$$

Letting $\Delta t \rightarrow 0$,

$$\frac{dP_i(t)}{dt} = \lambda[P_{i-1}(t) - P_i(t)] \quad (3.30)$$

Now, $P_{-1}(t) = 0$ (i.e., impossible to have <0),

$P_0(0) = 1$ (i.e., no arrivals up to time $t=0$),

$P_i(0) = 0$ for $i \geq 1$ (zero probability of i arrivals at time $t=0$).

Setting $i=0$ in Eq. 3.30,

$$\frac{dP_0(t)}{dt} = \lambda[0 - P_0(t)]$$

$$\frac{dP_0(t)}{P_0(t)} = -\lambda dt$$
$$\ell n P_0(t) = -\lambda t + c$$
$$P_0(t) = e^{-\lambda t + c}$$

Since $P_0(0) = 1$ and $1 = e^0 = e^c$, $c = 0$ and

$$P_0(t) = e^{-\lambda t}$$

Setting $i=1$ in Eq. 3.30 and inserting the above value for $P_0(t)$,

$$\frac{dP_1(t)}{dt} = \lambda[e^{-\lambda t} - P_1(t)]$$
$$\frac{dP_1(t)}{dt} + \lambda P_1(t) = \lambda e^{-\lambda t}$$

Using method of operators for solving this differential equation *

$$(D+\lambda)P_1(t) = \lambda e^{-\lambda t}$$
$$P_1(t) = \frac{1}{D+\lambda}\lambda e^{-\lambda t}$$
$$= (\lambda t)e^{-\lambda t} + C_2 e^{-\lambda t}$$

But

$$P_1(0) = 0 \qquad \therefore C_2 = 0$$
$$\therefore P_1(t) = (\lambda t)e^{-\lambda t}$$

For $i=2$,

$$\frac{dP_2(t)}{dt} = \lambda[P_1(t) - P_2(t)]$$

* Any standard method may be used for solution of this differential equation. The method of operators is particularly simple. See any standard text, such as Ford.[58] The form $y = \frac{1}{D+A}u(x)$ results in a solution $y = e^{-Ax}\int e^{Ax}u(x)\,dx + c\,e^{-Ax}$.

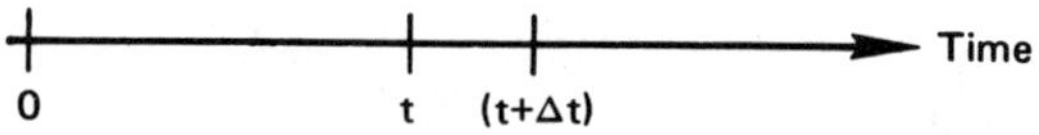

Figure B.1 Schematic representation of uniform probability density.

* Reference citations are listed in Chapter 3.
† This derivation is adapted from derivations by Arley and Buch,[55] Feller,[56] and Fry.[57]

$$\frac{dP_2(t)}{dt}+\lambda P_2(t)=\lambda P_1(t)=\lambda(\lambda t)e^{-\lambda t}$$

$$P_2(t)=\frac{1}{D+\lambda}\lambda(\lambda t)e^{-\lambda t}$$

$$=\frac{\lambda^2t^2}{2}e^{-\lambda t}+C_3e^{-\lambda t}$$

But

$$P_2(0)=0 \qquad \therefore C_3=0$$

$$P_2(t)=\frac{(\lambda t)^2e^{-\lambda t}}{2!}$$

Similarly,

$$P_3(t)=\frac{(\lambda t)^3e^{-\lambda t}}{3!}$$

$$P_4(t)=\frac{(\lambda t)^4e^{-\lambda t}}{4!}$$

$$P_x(t)=\frac{(\lambda t)^xe^{-\lambda t}}{x!} \qquad (3.2)$$

If $\lambda t=m$, the result is the most familiar form of the Poisson distribution:

$$P(x)=\frac{m^x\, e^{-m}}{x!} \qquad (3.1)$$

This relationship states the probability that exactly x arrivals will occur during an interval (of length t) when the mean number of arrivals is m (per interval of t).

Population Mean of Poisson Distribution

In the foregoing m is defined as the mean *arrival rate.* To determine the mean value of the distribution, begin with the definition of the population mean μ for a discrete distribution:

$$\mu=\sum_{x=0}^{\infty}x\,P(x), \qquad \text{for } P(x)=\frac{f(x)}{\sum_{x=0}^{\infty}f(x)} \qquad (3.31)$$

where $f(x)$ is the frequency of occurrence of x. For the Poisson distribution, substitute

$$P(x)=\frac{m^xe^{-m}}{m!}$$

Thus

$$\mu=\sum_{x=0}^{\infty}\frac{xm^xe^{-m}}{m!} \qquad (3.32)$$

$$=0+me^{-m}+\frac{2m^2e^{-m}}{2!}+\frac{3m^3e^{-m}}{3!}\cdots$$

$$=me^{-m}\left[1+m+\frac{m^2}{2!}+\frac{m^3}{3!}\cdots\right]$$

$$=me^{-m}e^{m}$$

$$=m \qquad (3.33)$$

Population Variance of Poisson Distribution

By definition, the *population* variance, σ^2, may be expressed:

$$\sigma^2=\frac{\Sigma f(x)(x-\mu)^2}{\Sigma f(x)} \qquad (3.34)$$

$$=\sum_{x=0}^{\infty}(x-\mu)^2\,P(x) \qquad (3.35)$$

Because the population mean is m, this variance may be stated:

$$\sigma^2=\Sigma(x-m)^2P(x)$$
$$=\Sigma(x^2-2xm+m^2)P(x)$$
$$=\Sigma x^2P(x)-2m\Sigma xP(x)+m^2\Sigma P(x)$$

The last term reduces to m^2 because $\Sigma P(x)=1$. The middle term reduces to $-2m^2$ because $\Sigma xP(x)$ has been shown equal to m in the derivation of the population mean. The first term may be reduced by the following steps:

$$\Sigma x^2P(x)=\Sigma[x(x-1)+x]P(x)$$
$$=\Sigma x(x-1)P(x)+\Sigma xP(x)$$
$$=A+B$$
$$B=\Sigma xP(x)=m$$
$$A=\sum_{x=0}^{\infty}x(x-1)\frac{m^xe^{-m}}{x!}$$
$$=\left[0+0+\frac{2m^2e^{-m}}{2!}+\frac{6m^3e^{-m}}{3!}+\frac{12m^4e^{-m}}{4!}\cdots\right]$$
$$=m^2e^{-m}\left[1+m+\frac{m^2}{2!}\cdots\right]$$
$$=m^2e^{-m}e^{m}=m^2$$
$$\Sigma x^2P(x)=m^2+m$$
$$\sigma^2=[m^2+m]-[2m^2]+[m^2]$$
$$\sigma=m \qquad (3.36)$$

Thus, for the Poisson distribution the population variance equals the population mean.

Derivation of the Distribution of the Sum of Independent Poisson Distributions [4,5]

Consider a population made up of two subpopulations A and B, each distributed according to the Poisson distribution.

For subpopulation A

$$P(x_a) = \frac{m_a^{x_a} e^{-m_a}}{x_a!}$$

For subpopulation B

$$P(x_b) = \frac{m_b^{x_b} e^{-m_b}}{x_b!}$$

If k items occur in a trial from the total population, there may be a mixture of x_a and x_b as follows:

$x_a = k$	$x_b = 0$	$x_a + x_b = k$
$x_a = k-1$	$x_b = 1$	$x_a + x_b = k$
$x_a = k-2$	$x_b = 2$	$x_a + x_b = k$
. . . . . .		
$x_a = 2$	$x_b = k-2$	$x_a + x_b = k$
$x_a = 1$	$x_b = k-1$	$x_a + x_b = k$
$x_a = 0$	$x_b = k$	$x_a + x_b = k$

$$P(k) = P(x_a = k, x_b = 0) + P(x_a = k-1, x_b = 1) \ldots + P(x_a = 1, x_b = k-1) + P(x_a = 0, x_b = k)$$

$$= \frac{m_a^k e^{-m_a} e^{-m_b}}{k!\,0!} + \frac{m_a^{k-1} e^{-m_a} m_b e^{-m_b}}{(k-1)!\,1!} + \frac{m_a^{k-2} e^{-m_a} m_b^2 e^{-m_b}}{(k-2)!\,2!} + \ldots + \frac{m_a e^{-m_a} m_b^{k-1} e^{-m_b}}{1!(k-1)!} + \frac{e^{-m_a} m_b^k e^{-m_b}}{0!\,k!}$$

$$P(k) = e^{-m_a} e^{-m_b} \left\{ \frac{m_a^k}{k!} + \frac{k\, m_a^{k-1} m_b}{k(k-1)!} + \frac{k(k-1) m_a^{k-2} m_b^2}{k(k-1)(k-2)!\,2!} + \ldots + \frac{k(k-1)\ldots 3\cdot 2\cdot 1\, m_a m_b^{k-1}}{k(k-1)\ldots 3\cdot 2\cdot 1(k-1)!} + \frac{m_b^k}{k!} \right\}$$

$$P(k) = \frac{e^{-(m_a+m_b)}}{k!} \left\{ m_a^k + k m_a^{k-1} m_b + \left[\frac{k(k-1) m^{k-2} m_b^2}{2!}\right] + \ldots + k\, m_a m_b^{k-1} + m_b^k \right\}$$

$$P(k) = \frac{e^{-(m_a+m_b)} (m_a + m_b)^k}{k!}$$

When there are subpopulations A, B, . . . , Z, by application of a similar argument the distribution for the whole population is found to be

$$P(k) = \frac{(m_a + m_b + \ldots + m_z)^k e^{-(m_a+m_b+\ldots+m_z)}}{k!}$$

B–2 PEARSON TYPE III DISTRIBUTIONS

Pearson has shown [59,60] that a wide variety of statistical phenomena can be modeled by 12 general distributions. Of these distributions, type III (together with several of its special cases) is particularly useful for certain traffic situations. In its most general form, the probability density function of the type III distribution can be stated (in the notation of this monograph) as follows [14,25]:

$$p(t) = \frac{1}{\beta\Gamma(k)} \left[\frac{t-a}{\beta}\right]^{k-1} e^{-\frac{t-a}{\beta}} \qquad (3.37)$$

where a is a location parameter $a \geq 0$; k is a shape parameter $k > 0$; β is a scale parameter $\beta > 0$. When the location parameter, a, equals zero, the distribution reduces to

$$p(t) = \frac{1}{\beta\Gamma(k)} \left(\frac{t}{\beta}\right)^{k-1} e^{-\frac{t}{\beta}} = \frac{t^{k-1}}{\beta^k \Gamma(k)} e^{-\frac{t}{\beta}} \qquad (3.38)$$

By setting $\alpha = 1/\beta$, this reduces to

$$p(t) = \frac{\alpha^k t^{k-1}}{\Gamma(k)} e^{-\alpha t} \qquad (3.39)$$

If $\beta = T/k$,

$$p(t) = \left(\frac{kt}{T}\right)^{k-1} \frac{k}{T\Gamma(k)} e^{-t/kT} \qquad (3.40)$$

Although some writers still refer to this distribution as type III, it is more commonly known as the gamma distribution. When $k=1$, the gamma distribution reduces to the negative exponential distribution. [Note that $\Gamma(n) = (n-1)!$] When $k \rightarrow \infty$, the distribution is regular (i.e., all headways are equal). When k is restricted to integer values (>1), the resulting distribution is the Erlang. Sometimes k is considered to be a measure of nonrandomness.

B–3 PARAMETERS OF EXPONENTIAL AND SHIFTED EXPONENTIAL DISTRIBUTIONS

Consider the negative exponential distribution as defined by Eq. 3.2

$$P(h < t) = 1 - e^{-t/T}$$

from which the probability density function is

$$p(t)=0-\left(-\frac{1}{T}\right)e^{-t/T}=\frac{1}{T}e^{-t/T}$$

The population mean is

$$\mu=\int_0^\infty tp(t)\,dt$$

where $p(t)\,dt$ is the probability of t. Thus,

$$\mu=\int t\,\frac{1}{T}e^{-t/T}dt$$

and integrating by parts and inserting limits, gives

$$\mu=T \qquad (3.41)$$

Population Variance of Negative Exponential Distribution

$$\sigma^2=\int_0^\infty (t-T)^2\,\frac{e^{-t/T}dt}{T}$$

$$=\int_0^\infty \frac{t^2}{T}e^{-t/T}dt-2\int_0^\infty te^{-t/T}dt+\int_0^\infty Te^{-t/T}dt$$

$$\sigma^2=T^2 \qquad (3.42)$$

Shifted Exponential Distribution

In considering Figure 3.5 in text, assume an origin through the point of contact of the curve and the t (horizontal) axis. About this axis

$$P(h<t)=1-e^{-t'/T'} \qquad (3.43)$$

where t'=abscissa of any point from the origin at the point of contact of the curve and the vertical axis; and T'=abscissa of the mean or center of gravity with respect to the same origin.

Now construct a vertical axis normal to the t axis and a distance τ to the left of the point of contact of the curve and the t axis. Now all abscissas t are measured from this new axis. In Eq. 3.43, each abscissa t' is now replaced by $(t-\tau)$. The value T now represents the abscissa of the center of gravity of the curve from the new axis, and the value of T' in Eq. 3.43 is now replaced by $(T-\tau)$. Thus, for the shifted exponential,

$$P(h<t)=1-e^{-(t-\tau)/(T-\tau)} \qquad \text{for } t\geq\tau \qquad (3.22)$$

Mean of Shifted Exponential Distribution

From Eq. 3.22 the probability density p is obtained as

$$p(t)=\frac{dp}{dt}=\frac{1}{-\tau}e^{-(t-\tau)/(T-\tau)}.$$

By definition, the mean μ is obtained as

$$\mu=\int_{-\infty}^\infty tp(t)\,dt.$$

For the shifted exponential,

$$\mu=\int_{-\infty}^\infty t\,\frac{1}{T-\tau}e^{-(t-\tau)/(T-\tau)}dt$$

$$=0+\int_\tau^\infty t\,\frac{1}{T-\tau}e^{-(t-\tau)/(T-\tau)}dt$$

$$=\frac{e^{\tau/(T-\tau)}}{(T-\tau)}\Big[-\tau(T-\tau)e^{-\tau/(T-\tau)}(T-\tau)^2e^{-\tau/(T-\tau)}\Big]_\tau^\infty$$

$$=T,$$

which could have been derived from the definition of T.

Population Variance Shifted Exponential Distribution (about Origin)

$$\sigma^2=\int_\tau^\infty (t-T)^2\frac{1}{T-\tau}e^{-(t-\tau)/(T-\tau)}dt$$

$$=\frac{e^{\frac{\tau}{T-\tau}}}{T-\tau}\left\{\int_\tau^\infty t^2e^{-\frac{t}{T-\tau}}dt-2T\int_\tau^\infty te^{-\frac{t}{T-\tau}}dt+T^2\int_\tau^\infty e^{-\frac{t}{T-\tau}}dt\right\}.$$

Integrating by parts and simplifying,

$$\sigma^2=(T-\tau)^2 \qquad (3.44)$$

B-4 LOGNORMAL DISTRIBUTION

The *normal* distribution is used to describe systems where the *measured variable* is normally distributed; *lognormal* distribution is used to describe systems where the *logarithm of the measured variable* is normally distributed.[61–63] Because log is not defined for arguments equal to or less than zero, the lognormal distribution is defined only for positive measured variables. (In the present discussion, use of natural logs is assumed; use of common logs (to base 10) would require only a change of scale.) To obtain a straight line for the cumulative lognormal distribution, use graph paper having a log scale on one axis and a normal probability scale on the other. (Figure B.2 shows several curves plotted on normal probability paper; Figure B.3 shows the same curves plotted on lognormal paper.)

The parameters of the lognormal distribution may be computed as follows [61]:

Measured variable$=y_i$; $x_i=\log y_i$, for x_i assumed to be normally distributed.

$$\bar{x}=\frac{1}{N}\sum_{i=1}^{N} x_i=\frac{1}{N}\sum_{i=1}^{N}\log y_i \qquad (3.45)$$

and

$$S_x{}^2=\frac{1}{N-1}\sum_{i=1}^{N}(x_i-\bar{x})^2=\frac{1}{N-1}\sum_{i=1}^{N}(\log y_i-\bar{x})^2 \qquad (3.46)$$

To fit the lognormal distribution, estimation of population parameters by the method of moments [61] uses the equations

$$\hat{\mu}=\text{estimate of population mean}$$
$$=2\,\ln(\bar{x})-\frac{1}{2}\ln(mom_2)$$
$$\hat{\sigma}^2=\text{estimate of population variance}$$
$$=\ln(mom_2)-2\ln(\bar{x}) \qquad (3.48)$$

where $\ln(\bar{x})$ = natural logarithm of $\bar{x}$ and

mom_2 = second moment about origin $=\frac{1}{n}\sum_{i=1}^{n} x_i{}^2$.

Estimation of the population parameters by the maximum liklihood method may be simpler [61]:

$$\hat{\mu}=\frac{1}{n}\sum_{i=1}^{n}\ln(x_i) \qquad \hat{\sigma}^2=S^2 \qquad (3.49)$$

Fitting then proceeds in the same manner as when fitting a normal distribution, making use of normal distribution tables. That is, the probability $\phi(x_i)$ is obtained from normal probability tables using an argument Z_i, where

$$Z_i=\frac{x_i-\mu}{\sigma} \qquad (3.50)$$

B–5 PROBIT ANALYSIS

Probit analysis is a method of treating the percentages of a population making all-or-nothing responses to increasingly severe values of a stimulus. Principally, using a probit removes the need to record a negative deviation from the mean as in a normal probability distribution analysis. If μ is the population mean, σ the standard deviation of the population, and Y the probit of x,

$$Y=5.0+\frac{x-\mu}{\sigma} \qquad (3.51)$$

The median value of the stimulus is the value that produces a response probit of 5.0. Table B.1 shows the transformation of cumulative percentage to probit.[64]

TABLE B.1 Transformation of Cumulative Percentages to Probits [a]

%	0	1	2	3	4	5	6	7	8	9
0	—	2.67	2.95	3.12	3.25	3.36	3.45	3.52	3.59	3.66
10	3.72	3.77	3.82	3.87	3.92	3.96	4.01	4.05	4.08	4.12
20	4.16	4.19	4.23	4.26	4.29	4.33	4.36	4.39	4.42	4.45
30	4.48	4.50	4.53	4.56	4.59	4.61	4.64	4.67	4.69	4.72
40	4.75	4.77	4.80	4.82	4.85	4.87	4.90	4.92	4.95	4.97
50	5.00	5.03	5.05	5.08	5.10	5.13	5.15	5.18	5.20	5.23
60	5.25	5.28	5.31	5.33	5.36	5.39	5.41	5.44	5.47	5.50
70	5.52	5.55	5.58	5.61	5.64	5.67	5.71	5.74	5.77	5.81
80	5.84	5.88	5.92	5.95	5.99	6.04	6.08	6.13	6.18	6.23
90	6.28	6.34	6.41	6.48	6.55	6.64	6.75	6.88	7.05	7.33
	0.0	0.1	0.2	0.3	0.4	0.5	0.6	0.7	0.8	0.9
99	7.33	7.37	7.41	7.46	7.51	7.58	7.65	7.75	7.88	8.09

[a] From Finney.[64]

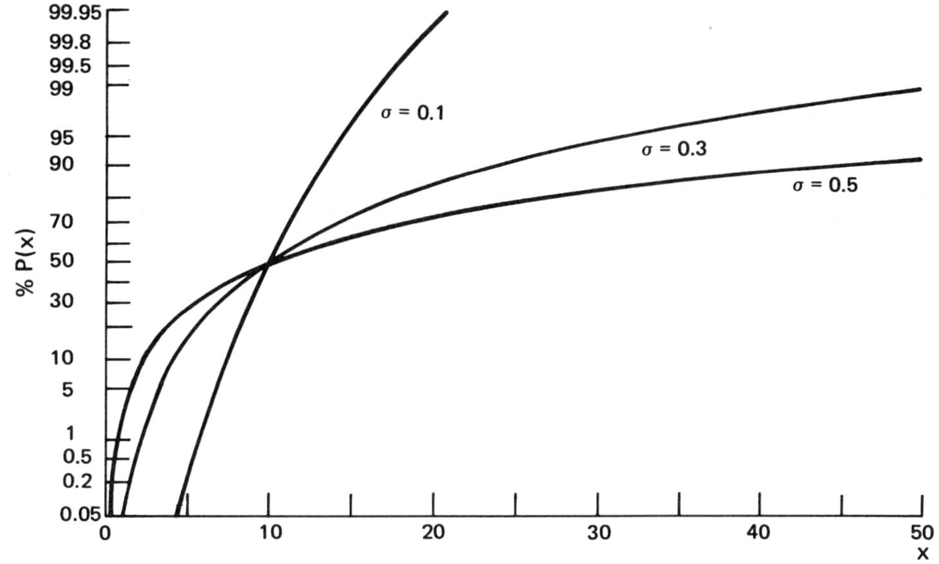

Figure B.2 Three cumulative lognormal distributions, all with log x = 1.0 and standard deviations of log x = 0.1, 0.3, and 0.5 plotted on probability paper.[62]

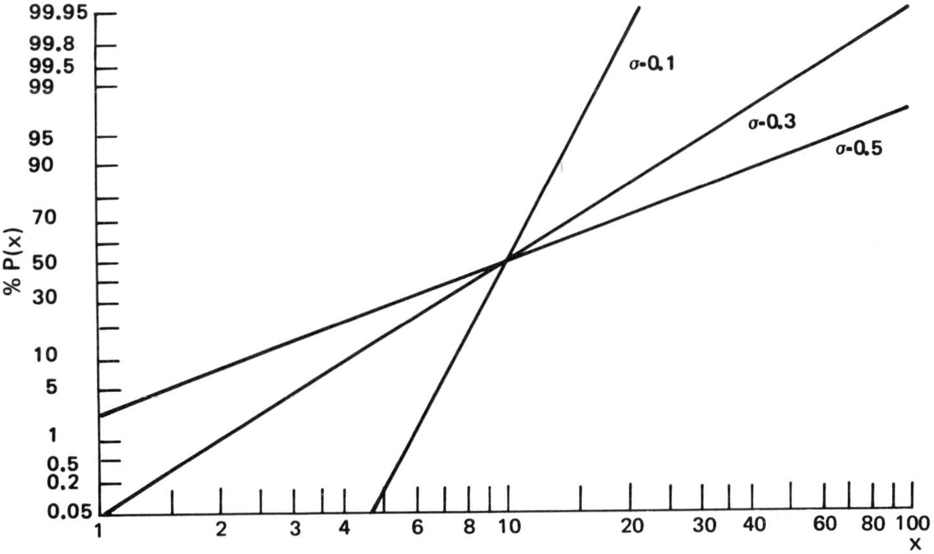

Figure B.3 Same curves as those in Figure B.2 but plotted on logarithmic probability paper.[62]

B–6 FITTING GENERALIZED POISSON DISTRIBUTIONS USING INCOMPLETE GAMMA FUNCTION TABLES

The procedure for fitting generalized Poisson distributions using tables of the incomplete gamma function is as follows:

1. Enter graph (text Figure 3.13) with mean and variance. Select value for k, which need not be an integer but may be any value greater than 1. (If $k=1$, the usual "simple" form of the Poisson should be used.)
2. Compute λ from Eq. 3.29.
3. Form the table:

x	$(p+1)$	u	p	$I(u,p)$	$1-I(u,p)$	$P(x)$
0						
1						
2						
3						
etc.						

The following equations are used:

$$(p+1)=(x+1)k$$
$$u=\lambda/(p+1)^{1/2}$$
$$p=(p+1)-1$$

4. Values of $I(u,p)$ are obtained from tables.[17]

5. The entries in the column $[1-I(u,p)]$ are the cumulative probabilities. The probabilities of the individual values of x—i.e., $P(x)$—are obtained by subtraction:

$$P(0)=1-I(u,p)_0$$
$$P(x)=[1-I(u,p)_x]-[1-I(u,p)_{x-1}]$$
$$=I(u,p)_{x-1}-I(u,p)_x$$

Table 3.11 in text contains observations during 64 15-sec counting intervals on a freeway during the morning peak period. A generalized Poisson distribution has been fitted to the data. Table B.2 illustrates the method of fitting by means of the incomplete gamma function; here the value of k is 2.

B–7 GOODNESS-OF-FIT TESTS

When comparing the fit between a theoretical distribution and a set of experimental data, it is desirable to have some method of quantitative evaluation of the fit. Several tests of statistical significance of the fit are available. These tests permit selection of one of the two decisions: (1) It is not very likely that the true distribution (of which the observed data constitute a sample) is in fact identical with the postulated distribution; (2) the true distribution (of which the observed data constitute a sample) could be identical with the postulated distribution.

It can be seen that either decision can be erroneously made. Decision 1 can be wrong if in fact the postulated distribution is the true distribution. On the other hand, decision 2 can be wrong if the true distribution is in fact different from the postulated distribution. Statistical tests of significance allow for specifying the probability (or risk) of making either of these types of error. Usually, the probability of making the first type of error (incorrectly rejecting the postulated distribution when in fact it is identical with the true distribution) is specified, and no statement is made with regard to the second type of error. The specification of the first type of error is expressed as a "significance level." Common significance levels are 0.01, 0.05, and 0.10. Thus, when a test is made at the 0.05 (5%) level, the engineer takes the chance (risk) that 5% of the rejected postulated distributions are in fact identical with the corresponding true distributions. For a complete discussion of the theory underlying this and other statistical tests of significance, the reader is referred to any standard text on statistics.

Chi-Square (χ^2) Test

The best known test of goodness-of-fit is the chi-square (χ^2) test, which is described as follows:

Let f=observed frequency for any group or interval and F=computed or theoretical frequency for the same group. Then, by definition:

$$\chi^2=\sum_{i=1}^{g}\frac{(f_i-F_i)^2}{F_i} \qquad (3.52)$$

where g is the number of groups.

Expanding,

$$\chi^2=\sum_{i=1}^{g}\left[\frac{f_i^2}{F_i}-\frac{2f_iF_i}{F_i}+\frac{F_i^2}{F_i}\right]$$
$$=\sum_{i=1}^{g}\frac{f_i^2}{F_i}-2\sum_{i=1}^{g}f_i+\sum_{i=1}^{g}F_i$$

In the fitting process, the total number of theoretical observations is set equal to the total number of experimental observations. Thus,

$$\sum_{i=1}^{g} f_i = \sum_{i=1}^{g} F_i = n$$

where n is the total number of observations. Then

$$\chi^2 = \left(\sum_{i=1}^{g} \frac{f_i^2}{F_i} \right) - n \qquad (3.53)$$

Either Eq. 3.52 or Eq. 3.53 may be used for purposes of computation. Usually, Eq. 3.53 will simplify the amount of work involved.

The value of χ^2 obtained by this method is then compared with the value from tables of χ^2, which may be found in any collection of statistical tables. Such tables relate the value of χ^2 and significance level to the degrees of freedom.* The number of degrees of freedom, ν, may be expressed [65–67]:

$$\nu = (g-1) - A$$

where g = number of groups and A = number of parameters estimated in the fitting process. The following table lists information for several distributions:

Distribution	A	ν
Poisson	1	$g-2$
Negative binomial	2	$g-3$
Generalized Poisson	2	$g-3$
Binomial	2	$g-3$

For this value of ν to be valid, however, it is necessary that the theoretical number of occurrences in any group be at least 5. One writer [67] further stipulates that the total number of observations be at least 50. When the number of theoretical occurrences in any group is less than 5, the group interval should be increased. For the lowest and highest groups this may be

* In using a table of χ^2, care should be exercised to note the manner in which the table is entered with the significance level. If the table is so constructed that for a given number of degrees of freedom the value of χ^2 increases with decreasing percentiles (probabilities), the table is entered with the percentile corresponding to the significance level. If the table is such that the value of χ^2 increases with increasing percentiles, the table is entered with the significance level subtracted from one. (In this case, 1.00−0.05=0.95.)

TABLE B.2 Generalized Poisson Distribution Fitted to Data by Incomplete Gamma Function Method
(Vehicle arrivals, Interstate 494 at 24th Avenue, in median lane, at P.M. peak)

Number of Cars per Interval (x_i)	$(x_i+1)k = (p+1)$	$\lambda/(p+1)^{1/2} = u$	p	$I(u,p)^a$	$P(x)$	$\Sigma P(x)$	Theoretical Frequency (F)	Observed Frequency (f)
0	2	10.9	1	1.000	0.000	0.000		
1	4	7.7	3	1.000	0.000	0.000		
2	6	6.3	5	0.998	0.002	0.002	0.1	
3	8	5.5	7	0.987	0.011	0.013	0.7	3
4	10	4.9	9	0.945	0.042	0.055	2.7	0
5	12	4.5	11	0.851	0.094	0.149	6.0	8
6	14	4.1	13	0.669	0.182	0.331	11.7	10
7	16	3.9	15	0.477	0.192	0.523	12.3	11
8	18	3.6	17	0.289	0.188	0.711	12.0	10
9	20	3.5	19	0.150	0.139	0.850	8.9	11
10	22	3.3	21	0.069	0.081	0.931	5.2	9
>10					0.069	1.000	4.4	2
$k=2$							64.0	64

^{a} Values obtained from Pearson's *Tables of the Incomplete Gamma Function*.[17]

TABLE B.3 Binomial Distribution[a] Fitted to Congested Traffic Arrivals
(Interstate 494 at 24th Avenue, median lane, during A.M. peak, for 15-sec intervals)

	Observed Data			Binomial Distribution			Poisson Distribution		
Number of Cars per Interval (x_i)	Observed Frequency (f_i)	Total Cars Observed ($f_i x_i$)	$f_i x_i^2$	Theoretical Frequency (F_i)		f_i^2/F_i	Theoretical Frequency (F_i)		f_i^2/F_i
<3	0	0	0	0.3			1.3		
3	3	9	27	1.0	10.4	11.63	2.5		
4	0	0	0	2.9			4.7	15.5	7.81
5	8	40	200	6.2			7.1		
6	10	60	360	9.8		10.20	8.8		11.37
7	11	77	539	12.3		9.84	9.4		12.87
8	10	80	640	12.1		8.26	8.8		11.37
9	11	99	891	9.4		12.87	7.3		16.58
10	9	90	900	5.8			5.4		15.00
11	1	11	121	2.8	10.0	12.10	3.7		
12	1	12	144	1.0			2.3	8.7	0.46
>12	0	0	0	0.4			2.7		
TOTAL	64	478	3,822	64.0		64.90	64.0		75.46

Observed Data:

$m = 478/64 = 7.469$

$$s^2 = \frac{3822 - \frac{(478)^2}{64}}{63} = 3.999$$

$s^2/m = 0.535$

Binomial Distribution:

$\hat{p} = 0.46$

$\hat{n} = 16.08$

$\Sigma f_i^2/F_i - \Sigma f_i = 64.90 - 64.00 = 0.90$

$\nu = 6 - 3 = 3$

$\chi^2_{0.05} = 7.81$[b]

Poisson Distribution:

$m = 7.469$

$e^{-m} = 0.00057$

$\Sigma f_i^2/F_i - \Sigma f_i = 75.46 - 64.00 = 11.46$

$\nu = 7 - 2 = 5$

$\chi^2_{0.05} = 11.07$[c]

[a] Poisson distribution used for reference.
[b] Accept fit at 0.05 level.
[c] Reject fit at 0.05 level.

accomplished by making these groups "all less than" and "all greater than," respectively.

Table B.3 demonstrates the computations for the χ^2 test.

Kolmogorov–Smirnov Test for Goodness-of-Fit

In addition to the χ^2 test, the Kolmogorov–Smirnov (K–S) test [68,69] is another technique for testing goodness-of-fit. The necessary calculations are relatively simple; and like the χ^2 test, it is nonparametric or distribution-free (i.e., no assumption is made concerning the shape of the population from which the samples are drawn). The test is based on the simple measurement of the maximum vertical difference between the two cumulative probability distributions. This may be done graphically or it may be done in tabular form. It is necessary only to find the maximum absolute difference between the theoretical and observed cumulative distributions. This maximum difference is then compared with the tabulated or computed value of the K–S statistic. Table B.4 gives values of the K–S statistic for various sample sizes and levels of significance.

The K–S test is particularly valuable in cases where the number of observations is small.

This test cannot be used when the population parameters are estimated from the observations, inasmuch as the correction of the critical value (because of such estimation) is unknown. Where adequate data are available, this restriction concerning parameter estimation may be overcome by the following procedure: One half of the data is used to estimate the parameters; the other half of the data is then used for fitting to the population represented by the parameters.* This procedure is illustrated by Tables 3.6 and 3.7 in text.

TABLE B.4 Values of the Kolmogorov–Smirnov Test Statistic "d" [a] for Various Sample Sizes and Levels of Significance [b]

Sample Size (n)	Level of Significance (α)		
	0.10	0.05	0.01
1	0.950	0.975	0.995
2	0.776	0.842	0.929
3	0.642	0.708	0.828
4	0.564	0.624	0.733
5	0.510	0.565	0.669
6	0.470	0.521	0.618
7	0.438	0.486	0.577
8	0.411	0.457	0.543
9	0.388	0.432	0.514
10	0.368	0.410	0.490
11	0.352	0.391	0.468
12	0.338	0.375	0.450
13	0.325	0.361	0.433
14	0.314	0.349	0.418
15	0.304	0.338	0.404
16	0.295	0.328	0.392
17	0.286	0.318	0.381
18	0.278	0.309	0.371
19	0.272	0.301	0.363
20	0.264	0.294	0.356
25	0.24	0.27	0.32
30	0.22	0.24	0.29
35	0.21	0.23	0.27
>35	$\frac{1.22}{\sqrt{n}}$	$\frac{1.36}{\sqrt{n}}$	$\frac{1.63}{\sqrt{n}}$

[a] Maximum vertical difference between two cumulative distributions.

[b] After Massey.[68]

B–8 VARIANCE OF EXPERIMENTAL OBSERVATIONS

Over the years there has been no standardization as to the definition of the variance of a set of experimental observations. Some writers have defined the variance as computed simply with respect to the sample and then have applied a correction to obtain the unbiased estimate of the population variance; others have defined the unbiased estimate of the population variance as the experimental variance. For purposes of the present volume, the latter definition has been adopted; namely,

$$s^2 = \frac{1}{n-1}\sum_{i=1}^{n}(x_i - \bar{x})^2 = \frac{1}{n-1}\sum(x_i - m)^2$$

where s^2 is the unbiased estimate of population variance as computed from sample, here termed the "variance of experimental data."

Estimation of the population variance and correction of the sample variance to give the unbiased population variance (Bessel's correction) have been clearly described by Neville and Kennedy,[70] as follows:

* B. W. Lindgren, personal communication.

Consider a sample of size n drawn from a population with a mean μ and standard deviation σ.

Let x_i be an observation in the sample. Then

$$x_i - \mu = (x_i - \bar{x}) + (\bar{x} - \mu)$$
$$= (x_i - \bar{x}) - \epsilon$$

where $\epsilon = \mu - \bar{x}$ is the "error" or deviation of the sample mean, $\bar{x}$. Squaring, we obtain

$$(x_i - \mu)^2 = (x_i - \bar{x})^2 + \epsilon^2 - 2\epsilon(x_i - \bar{x})$$

For all the observations in the sample, we sum for i from 1 to n and obtain

$$\Sigma(x_i - \mu)^2 = \Sigma(x_i - \bar{x})^2 + n\epsilon^2 - 2\epsilon\Sigma(x_i - \bar{x})$$

But $\Sigma(x_i - \bar{x}) = 0$ by definition of x. Therefore,

$$\Sigma(x_i - \mu)^2 = \Sigma(x_i - \bar{x})^2 + n\epsilon^2$$

If we repeat this calculation for a large number of samples, the mean value of the left-hand side of the above equation will (by definition of σ^2) tend to $n\sigma^2$. Similarly, the mean value of $n\epsilon^2 = n(\mu - \bar{x})^2$ will tend to n times the variance of $\bar{x}$ because ϵ represents the deviation of the sample mean from the population mean. Thus,

$$n\epsilon^2 \rightarrow n\left(\frac{\sigma^2}{n}\right)$$

whence

$$n\sigma^2 \rightarrow \Sigma(x_i - \bar{x})^2 + \sigma^2$$

or

$$\Sigma(x_i - \bar{x})^2 \rightarrow (n-1)\sigma^2$$

Thus

$$\frac{\Sigma(x_i - \bar{x})^2}{n-1} \rightarrow \sigma^2$$

In other words, for a large number of random samples the mean value of $\frac{\Sigma(x_i - \bar{x})^2}{n-1}$ tends to σ^2; that is, it is an unbiased estimate of the variance of the population. The estimate is denoted by s^2. Thus

$$s^2 = \frac{\Sigma(x_i - \bar{x})^2}{n-1}$$

Since the variance of the sample (taken as a finite population with a mean $\bar{x}$) σ^2 is given by

$$\sigma^2 = \frac{\Sigma(x_i - x)^2}{n}$$

Bessel's correction is

$$\frac{s^2}{\sigma^2} = \frac{n}{n-1}.$$

Appendix C
DATA SUPPLEMENTARY TO CHAPTER 7*

C–1 PACEY'S PLATOON DIFFUSION

Pacey[12] has formulated a model for platoon diffusion that may be described as follows: Let

$f(u)\,du$ = the probability distribution of platoon speeds, assumed to be a normal distribution,
τ = travel time between two observation points
$= D/u$ where D = distance between the observation points,
$g(\tau)\,d\tau$ = distribution of travel times.

The probability that a trip time lies between τ and $(\tau+d\tau)$ is the same as the probability that the corresponding speed lies between u and $(u+du)$.

Noting that $u=D/\tau$, $du=\dfrac{-D}{\tau^2}d\tau$.

Thus,

$$g(\tau)\,d\tau=f(u)\,du=f\left(\frac{D}{\tau}\right)\left(\frac{D}{\tau^2}\right)d\tau$$

(The minus sign of the expression for du is dropped because probabilities are always positive.) Now, because speeds are assumed to be normally distributed,

$$f(u)=\frac{1}{\sigma\sqrt{2\pi}}\exp\left[-\frac{(u-\bar{u})^2}{2\sigma^2}\right]$$

where σ is the population standard deviation of speed. Therefore,

$$g(\tau)=\frac{D}{\sigma\tau^2\sqrt{2\pi}}\exp\left[-\frac{\left(\frac{D}{\tau}-\frac{D}{\bar{\tau}}\right)^2}{2\sigma^2}\right]$$

Letting $s=\sigma/D$,

$$g(\tau)=\frac{1}{\tau^2 s\sqrt{2\pi}}\exp\frac{\left[-\left(\frac{1}{\tau}-\frac{\bar{u}}{D}\right)^2\right]}{2s^2}.$$

Considering now the two observation points, the number of cars passing the first point is q_1dt in the interval $[t, (t+dt)]$. Of this flow, the number of cars passing the second observation point will be $q_1(t)g(\tau)dt dT$ in the interval $[(t+\tau), (t+\tau+dT)]$.

The total number of cars passing point two in the interval $[T, (T+dT)]$ is

$$q_2(T)\,dT=\int_t q_1(t)g(T-t)\,dt\,dT,$$

integration being over all values of t for which $q_1(t)>0$.

For real cases histograms result rather than continuous curves. Thus the following expression is employed,

$$q_2(j)=\sum_i q_1(i)g(j-i)$$

where i and j are discrete intervals of the histogram.

It should be noted that the foregoing analysis assumes that passing is possible at will.

---

* Reference citations are listed in Chapter 7.

Appendix D
DATA SUPPLEMENTARY TO CHAPTER 9*

D–1 DISCUSSION OF SIMSCRIPT AND GPSS

Many simulation languages exist, especially when one includes those written in Europe; the two best known in the United States are SIMSCRIPT and GPSS. Before proceeding with brief descriptions and a comparison of these two languages, it will be helpful to consider the following glossary.

Assembler A computer programming language that provides a one-to-one correspondence between mnemonic symbols (source language) and the natural machine language commands (object language), but usually with the automation of such tasks as the assignment of memory addresses. The program written in assembly language is converted to a program in machine language by an assembly program. The assembled program is then used for computer runs.

Interpreter A computer programming language that provides the capability for programming extensive computations by a simple (source) language. The conversion to the machine (object) language usually involves the provision of many machine-language commands for each source command. This conversion is performed *during* the computation run and is repeated each time the source language calls for a particular function or computation. Thus, the interpretation program must occupy part of the computer memory during running, and the interpretation during each iteration can greatly increase the computation time.

Compiler A computer programming language in which programs are written in a source language closely adapted to the type of computation to be performed and in which the conversion to machine (object) language involves the *compiling* of a number of library subroutines. The compilation is performed prior to the computation run.

Endogenous event An event affecting the program from inside the simulation.

Exogenous event An event affecting the program from outside the simulation.

SIMSCRIPT

SIMSCRIPT is a compiler-type language based on FORTRAN [19–23]. In SIMSCRIPT the system is described in terms of *entities, attributes, sets,* and *events.* For instance, cars, intersections, streets, etc., would constitute entities. Attributes of each car would include destination, gap acceptance criteria, normal acceleration, desired cruise speed, etc. Sets would include all cars in queue on the northbound leg of intersection 23, all cars moving as a platoon from intersection 18 to intersection 19, etc. An event represents one or more actions that take place instantaneously at a given time (e.g., main street signal turns green). An activity is an occurrence that takes place between two events (e.g., car has left intersection 14 but has not yet reached intersection 15). The state of the system at any time is given by attribute values and set memberships of all individual entities. The state is changed by the occurrence of an event. Changes in state may be responses to commands such as

> CREATE or DESTROY an individual entity.
>
> ALTER set membership of an individual entity.
>
> CHANGE numerical value of an attribute.

Events are handled by FORTRAN-like subroutines. Endogenous events are triggered dynamically from within the model. Exogenous events may be scheduled prior to the start of the simulation run. Statistics may be accumulated. Output is obtained through the use of a report generator whose format is defined on a format definition form. Input consists of definition cards, initialization cards, and subprograms.

The various inputs are handled as follows: Defined items are converted directly into machine-coded subroutines that will control the

* Reference citations are listed in Chapter 9.

storage and retrieval functions; machine-coded dynamic control (timing) subroutines are also produced directly. SIMSCRIPT subprograms are converted into FORTRAN source language subroutines. These FORTRAN subroutines are then compiled by a standard FORTRAN compiler. Initialization cards are treated as data.

GPSS (General Purpose System Simulator)

GPSS is an interpreter-type simulation language that describes the system in terms of blocks, transactions, and equipment using block diagrams[22–25].

GPSS provides the following elements:

1. Basic elements
   Blocks
   Transactions
2. Equipment elements
   Facilities
   Storages
   Logic switches
3. Statistical elements
   Queues
   Distribution tables
4. Reference elements
   "Savexes"
5. Computational elements
   Arithmetic variables
   Functions

Transactions, temporary elements having eight parameters and eight priority levels, move from one block to another in the model. As a transaction enters a block a subroutine associated with the block type causes a change in the state of the system. Transactions are created by ORIGINATE and GENERATE blocks and are destroyed by TERMINATE and ASSEMBLE blocks. Time is advanced in discrete steps, the value of the step being designated by the user.

Outputs consist of a standard set of statistics written on the output tape at the end of the run.

Inputs consist of definition cards, each of which defines a block, function, table, arithmetic variable, or storage capacity.

Comparison of SIMSCRIPT and GPSS

Table D.1 compares SIMSCRIPT with GPSS, based on the work of Murphy [26] and Teichrow and Lubin.[23] To summarize, SIMSCRIPT is more efficient than GPSS but requires a much higher level of skill on the part of the user; thus, GPSS is easier for use by the beginner.

TABLE D.1 Comparison of SIMSCRIPT and GPSS

Characteristic	SIMSCRIPT	GPSS
General type	Compiler	Interpreter
Basic language	FORTRAN	None
User requirements	Must know FORTRAN	None
Ease of use	Difficult; diagnostics limited; expert help needed	Easy
Can be used for nonsimulation programs	Yes	No
Basic unit of program	Event routine	Block
Time increments	Variable	Fixed
Report generator	Flexible format	Fixed format
Variable names can be assigned to parameters	Yes	No
Memory used for testing problem [a]	Y	2Y
Run time for test problem	3.6 min	26.8 min

[a] In tests by Murphy.[26]

D–2 SIMULATION PROGRAM LISTING

```
00100 REM A=ARRIVAL LIST
00105 REM A1=ALPHA=PROPORTION RESTRAINED
00110 REM A2=(1-ALPHA)=PROPORTION UNRESTRAINED
00115 REM B=GREENSHIELDS' DEPARTURES LIST
00120 REM C1=MAIN CLOCK
00125 REM C2=TIME OF NEXT SIGNAL CHANGE
00130 REM C3=TIME OF ARRIVAL
00135 REM D1=SCAN INTERVAL IN SECONDS
00140 REM D2=DURATION OF RUN IN HOURS
00145 REM D3=TIME OF NEXT DEPARTURE
00150 REM D4=TIME OF LAST DEPARTURE
00155 REM D9= LARGE DEPARTURE TIME
00160 REM G1=SIGNAL STATE:  0=RED; 1=GREEN
00165 REM G2=GREEN+AMBER DURATION IN SECONDS
00170 REM G3=TIME CURRENT GREEN INTERVAL STARTED
00175 REM H1=HEADWAY OF ARRIVAL
00180 REM H2=MINIMUM HEADWAY FOR PASSAGE OF MOVING CAR
00185 REM L1=(LOCATION OF LAST ARRIVAL)-1
00190 REM L2=INDEX OF DEPARTURE
00195 REM N=NUMBER OF DEPARTURES FROM QUEUE
00200 REM M=NUMBER OF CARS PASSING THRU WITHOUT QUEUEING
00205 REM R1=RED INTERVAL DURATION IN SECONDS
00210 REM R2=FIRST RANDOM NUMBER
00215 REM R3=SECOND RANDOM NUMBER
00220 REM S1=SUM OF DELAYS
00225 REM T1=ARRIVAL CONSTANT
00230 REM T2=ARRIVAL CONSTANT
00235 REM U1=UTILIZATION: 1=FULLY UTILIZED; 0=NOT FULLY UTILIZED
00240 REM U2=COUNT OF GREEN INTERVALS NOT FULLY UTILIZED
00245 REM U3=COUNT OF ALL GREEN INTERVALS
00250 REM Q1=MAX QUEUE LENGTH
00255 REM Q2=FLOW LEAVING INTERSECTION
00260 REM X1:   1=MOVING DEPARTURE;  0=DEPARTURE FROM QUEUE
00265 REM INITIALIZE
00270 DIM A(30),B(30)
00275 MAT A=ZER
00280 U2=U3=0
00285 L1=C1=C3=G1=S1=M=N=X1=0
00290 Q1=0
00295 PRINT "TYPE VOLUME IN VEHICLES PER HOUR";
00300 INPUT V
00305 IF V>869 THEN 00930
00310 PRINT "TYPE DURATION OF RED AND DURATION OF (GREEN+AMBER)"
00315 PRINT "IN SECONDS";
00320 INPUT R1,G2
00325 PRINT "TYPE DURATION OF RUN IN HOURS";
00330 INPUT D2
00335 PRINT "TYPE SCAN INTERVAL IN SECONDS";
00340 INPUT D1
00345 PRINT "TYPE MIN. HDWY. FOR MOV. PASSAGE IN SEC";
00350 INPUT H2
00355 T1=2.5
00360 T2=24-0.0122*V
00365 D9=3600*D2+1
00370 PRINT
00375 PRINT
00380 D3=D9
```

```
00385 E1=1.0
00390 A1=0.00115*V
00395 A2=1-A1
00400 B(1)=3.8
00405 B(2)=6.9
00410 B(3)=9.6
00415 B(4)=12.0
00420 B(5)=14.2
00425 FOR I=6 TO 30
00430 B(I)=14.2+2.1*(I-5)
00435 NEXT I
00440 C2= C1+ R1
00445 D4=C2
00450 REM
00455 REM
00460 REM GENERATE ARRIVAL
00465 R2=RND(X)
00470 R3=RND(X)
00475 IF R2<=A1 THEN 00490
00480 H=T2*(-LOG(R3))
00485 GO TO 00495
00490 H=T1*(-LOG(R3))+E1
00495 C3=C3+H
00500 GO TO 00685
00505 REM ENTER ARRIVAL INTO QUEUE LIST
00510 A(L1+1)=C3
00515 L1=L1+1
00520 IF L1=<Q1 THEN 00530
00525 Q1=L1
00530 GO TO 00535
00535 REM PROCESS DEPARTURES
00540 IF G1=0 THEN 00670
00545 IF D3>C1 THEN 00670
00550 IF D3>C2 THEN 00670
00555 D4=D3
00560 REM CHECK IF MOVING
00565 IF X1=1 THEN 00655
00570 S1=S1+D3-A(1)
00575 N=N+1
00580 FOR I=1 TO Q1
00585 A(I)=A(I+1)
00590 NEXT I
00595 L2=L2+1
00600 IF L1=0 THEN 00615
00605 IF L1<0 THEN 00920
00610 L1=L1-1
00615 GO TO 00620
00620 IF L1>0 THEN 00640
00625 X1=1
00630 D3=D9
00635 GO TO 00665
00640 REM
00645 D3=G3+B(L2)
00650 GO TO 00665
00655 M=M+1
00660 D3=D9
00665 REM
00670 REM CHECK IF TIME TO GENERATE NEW ARRIVAL
00675 IF C1>=C3 THEN 00460
```

```
00680 REM
00685 REM CHECK COMPLETION OF RUN AND OUTPUT RESULTS
00690 IF C1<D2*3600 THEN 00770
00695 Q2=(N+M)/D2
00700 PRINT "    SIGNALIZED INTERSECTION SIMULATION"
00705 PRINT "    INPUT VOLUME="V;"VEHICLES PER HOUR"
00710 PRINT "SIG.CYCLE";R1;"SEC. RED",G2;"SEC. GREEN+AMBER"
00715 PRINT "MIN. HDWY FOR MOVING PASSAGE=";H2;"SECONDS"
00720 PRINT "                RESULTS"
00725 PRINT
00730 PRINT "FLOW LEAVING INTERSECTION=";Q2;"VEHICLES PER HOUR"
00735 PRINT "MAXIMUM QUEUE LENGTH=";Q1
00740 PRINT "TOTAL DELAY";S1;"VEHICLE SECONDS FOR RUN OF";D2;"HOURS"
00745 PRINT "TOTAL GREEN INTERVALS=";U3
00750 PRINT "NUMBER OF GREEN INTERVALS NOT FULLY UTILIZED=";U2
00755 GO TO 00925
00760 REM
00765 REM INCREMENT CLOCK. CHECK SIGNAL STATUS; CHANGE IF READY.
00770 C1=C1+D1
00775 IF C2>C1 THEN 00855
00780 IF G1=0 THEN 00815
00785 REM CHANGE TO RED
00790 G1=0
00795 X1=0
00800 C2=C2+R1
00805 GO TO 00860
00810 REM CHANGE TO GREEN
00815 G1=1
00820 U1=1
00825 U3=U3+1
00830 L2=1
00835 G3=C2
00840 C2=C2+G2
00845 D4=G3
00850 D3=G3+B(1)
00855 REM CHECK IF TIME TO PROCESS NEW ARRIVAL
00860 IF C1<C3 THEN 00535
00865 IF X1=0 THEN 00505
00870 IF L1>0 THEN 00905
00875 IF C3<=(D4+H2) THEN 00905
00880 IF U1=0 THEN 00895
00885 U1=0
00890 U2=U2+1
00895 D3=C3+T3
00900 GO TO 00535
00905 X1=0
00910 GO TO 00505
00915 REM
00920 PRINT "QUEUE INDEX ERROR"
00925 STOP
00930 PRINT "VOLUME TOO GREAT"
00935 STOP
00940 END
```

Appendix E

ILLUSTRATION CREDITS AND COPYRIGHT PERMISSIONS

Grateful acknowledgment is made to the respective publishers for permission to use the following items as the basis of certain of the illustrations and other material prepared for use in this publication. The permissions are generally for copyrighted material, are for this use only, and in no way are to be construed as extending to further use or reproduction of the material covered without express written permission of the copyright holder.

Item	*Page*	*Data Source*
Figure 3.3	24	Daniel L. Gerlough and Frank C. Barnes, "The Poisson and Other Probability Distributions in Highway Traffic." *Poisson and Other Distributions in Traffic,* Eno Foundation for Transportation, Saugatuck, Conn. (1971), Figure 2, p. 37.
Figure 3.6	26	M. Wohl and B. V. Martin, *Traffic System Analysis for Engineers and Planners.* McGraw-Hill, New York (1967), Figure 15.4, p. 506.
Table 3.8	29	Daniel L. Gerlough and Frank C. Barnes, "The Poisson and Other Probability Distributions in Highway Traffic." *Poisson and Other Distributions in Traffic,* Eno Foundation for Transportation, Saugatuck, Conn. (1971), Example 18, p. 48 (as corrected by the authors Mar. 8, 1974).
Figure 3.10	32	W. L. Grecco and E. D. Sword, "Prediction of Parameters for Schuhl's Headway Distribution." *Traffic Engineering,* 38(5):36-38, Figure 1.
Figure 3.13	35	F. A. Haight, *Mathematical Theories of Traffic Flow.* Academic Press, New York (1963), Figure 10, p. 97.
Figure 4.4	53	H. Greenberg, "An Analysis of Traffic Flow." *Operations Research,* 7(1):79-85 (Jan. 1959), Figure 3.
Figure 4.5	53	R. T. Underwood, "Speed, Volume and Density Relationships." *Quality and Theory of Traffic Flow,* Bureau of Highway Traffic, The Pennsylvania State University (1961), Figure V-2, p. 148.
Figure 4.6	53	P. K. Munjal and L. A. Pipes, "Propagation of On-Ramp Density Perturbations on Unidirectional and Two- and Three-Lane Freeways." *Transportation Research,* 5(4):241-255 (Dec. 1971), Figure 1.
Figure 4.10	55	L. C. Edie, "Car-Following and Steady-State Theory for Noncongested Traffic." *Operations Research,* 9(1):66-76 (Jan. 1961), Figure 3.

Item	*Page*	*Data Source*
Figure 4.11	55	R. T. Underwood, "Speed, Volume and Density Relationships." *Quality and Theory of Traffic Flow,* Bureau of Highway Traffic, The Pennsylvania State University (1961), Figure V-3, p. 150.
Figure 4.12	56	A. C. Dick, "Speed/Flow Relationships Within an Urban Area." *Traffic Engineering and Control* (29 Newman St., London W1P 3PE, England), 8(6):393-396 (Oct. 1966), Figure 4.
Figure 4.13	57	J. Drake, J. Schofer, and A. D. May, Jr., "A Statistical Analysis of Speed-Density Hypotheses." *Vehicular Traffic Science,* American Elsevier, New York (1967), Figure 1, p. 113.
Figure 4.16	59	L. C. Edie, "Car-Following and Steady-State Theory for Noncongested Traffic." *Operations Research,* 9(1):66-76 (Jan. 1961), Figure 4.
Figure 4.17	59	H. Greenberg, "An Analysis of Traffic Flow." *Operations Research,* 7(1):79-85 (Jan. 1959), Figure 1.
Figure 4.18	60	L. C. Edie, "Car-Following and Steady-State Theory for Noncongested Traffic." *Operations Research,* 9(1):66-76 (Jan. 1961), Figure 6.
Figure 4.22	62	J. G. Wardrop, "Some Theoretical Aspects of Road Traffic Research." *Proceedings of The Institution of Civil Engineers* (London), Part II, 1(2): 325-362 (1952), Figure 3.
Figure 4.28	66	N. W. Guerin, "Travel Time Relationships." *Quality and Theory of Traffic Flow,* Bureau of Highway Traffic, The Pennsylvania State University (1961), Figure III-12, p. 101.
Figure 4.29	67	R. H. Haase, "Decreasing Travel Time for Freeway Users." *Memo RM-3099-FF,* The Rand Corporation, Santa Monica, Calif. (Oct. 1962), Figure 14, p. 5.
Figure 4.30	68	R. J. Smeed, "Some Circumstances in Which Vehicles Will Research Their Destinations Earlier by Starting Later." *Transportation Science,* 1(4): 308-317 (Nov. 1967), Figure 2a.
Figure 5.1	72	T. Rockwell, "Skills, Judgment and Information Acquisition in Driving." In (ed. T. W. Forbes) *Human Factors in Highway Traffic Research,* Wiley, New York (1971), Figure 1, p. 135.
Figure 5.3	76	J. Szafran, "Limitations and Reliability of the Human Operator of Control Systems to Process Information." *Aerospace Medicine,* 37(3):239-242 (Mar. 1966), Figure 2.
Figure 5.5	79	G. Johansson and H. Rumar, "Drivers' Brake Reaction Times." *Human Factors* (The Johns Hopkins University Press), 13(1):23-27 (Feb. 1971), Figure 1.

Item	*Page*	*Data Source*
Figure 5.6	80	M. L. Braunstein and K. R. Laughery, "Detection of Vehicle Velocity Changes During Expressway Driving." *Human Factors* (The Johns Hopkins University Press), 6(4): 327-331 (Aug. 1964), Figure 2.
Table 5.5	82	T. Rockwell, "Skills, Judgment and Information Acquisition in Driving." In (ed. T. W. Forbes) *Human Factors in Highway Traffic Research,* Wiley, New York (1971), Table 3, p. 148.
Figures 6.2, 6.3, 6.4, 6.10	92, 93, 106	R. Herman, E. W. Montroll, R. B. Potts, and R. W. Rothery, "Traffic Dynamics: Analysis of Stability in Car-Following." *Operations Research,* 7(1): 86-106 (Jan. 1959), Figures 4, 5, 6, 7.
Table 6.6	102	J. W. McClenahan and H. J. Simkowitz, "The Effect of Short Cars on Flow and Speed in Downtown Traffic: A Simulation Model and Some Results." *Transportation Science,* 3(2):126-139 (1969), Table III.
Figure 6.9	105	R. Herman and R. W. Rothery, "Microscopic and Macroscopic Aspects of Single-Lane Traffic Flow." *Journal of the Operations Research Society of Japan,* 5(2):74-93 (Dec. 1962), Figure 10.
Figures 7.7-7.13	117, 118, 119, 120	M. H. Lighthill and G. B. Whitham, "On Kinematic Waves. II. A Theory of Traffic Flow on Long Crowded Roads." *Proceedings of The Royal Society* (London), Series A, 229 (1178): 317-345 (May 10, 1955), Figures 5, 6, 7, 9, 10, 11, 12.
Table 7.2	124	E. F. Graham and D. C. Chenu, "A Study of Unrestricted Platoon Movement of Traffic." *Traffic Engineering,* 32(7):11-13 (Apr. 1962), Table I.
Figures 7.20, 7.21, Table 7.3	126, 127	M. J. Grace and R. B. Potts, "A Theory of the Diffusion of Traffic Platoons." *Operations Research,* 12(2):255-275 (Mar. 1964), Figures 6 and 4, and Table II.
Figures 7.22, 7.23, 7.24	130, 131	I. Prigogine and R. Herman, *Kinetic Theory of Vehicular Traffic.* American Elsevier, New York (1971), Figure 4.1 (p. 44), Figure 4.4 (p. 47), Figure 4.5 (p. 48).
Tables 7.5, 7.6	132, 133	R. Herman, L. Lam, and I. Prigogine, "Kinetic Theory of Vehicular Traffic: Comparison with Data." *Transportation Science,* 6(4):440-452 (1972).
Figures 8.14, 8.15	156	D. E. Blumenfeld and G. H. Weiss, "Merging from an Acceleration Lane." *Transportation Science,* 5(2):161-168 (1971), Figures 2 and 3.
Figure 8.16	157	R. E. Allsop, "Delay at a Fixed Time Traffic Signal. I. Theoretical Analysis." *Transportation Science,* 6(3):260-285 (1972), Figure 1.
Figure 8.18	161	G. F. Newell, "Approximation Methods for Queues with Applications to the Fixed-Cycle Traffic Light." *SIAM Review,* 7(2):223-240 (Apr. 1965), Figure 4.

Item	*Page*	*Data Source*
Figures 8.19-8.23	162, 163, 164, 165	T. P. Hutchinson, "Delay at a Fixed Time Signal. II. Numerical Comparisons of Some Theoretical Expressions." *Transportation Science,* 6:286-305 (1972), Figures 1, 2, 3, 4, 8.
Figures 9.5, 9.6	181, 182	R. D. Worrall, "Simulation of Traffic Behavior on a Digital Computer." *Traffic Engineering and Control* (London), 5(2):86-90, 94 (June 1963), Figures 4 and 3.
Figures 9.7-9.10 and Text excerpt pp. 185-190	185-190, 192	M. Wohl, "Simulation—Its Application to Traffic Engineering. II." *Traffic Engineering,* 31(1):19-25, 26 (Oct. 1960), Figures 1, 2, 3, 4; and text excerpt, pp. 20-21.
Table B.1	206	D. J. Finney, *Probit Analysis.* Cambridge University Press, 3rd edition (1971), Table 1, p. 22.
Figures B.2, B.3	207	A. Hald, *Statistical Theory with Engineering Applications.* Wiley, New York (1952), Figures 7.3 and 7.4, p. 163.

SPONSORSHIP OF THIS SPECIAL REPORT

GROUP 3—OPERATION AND MAINTENANCE OF TRANSPORTATION FACILITIES

LLOYD G. BYRD, *Byrd, Tallamy, MacDonald and Lewis* (Chairman)

Committee on Traffic Flow Theory and Characteristics

Kenneth W. Crowley, *Polytechnic Institute of New York* (Chairman)
Robert F. Dawson, *University of Vermont* (Vice-Chairman)
Edmund A. Hodgkins, *Federal Highway Administration* (Secretary)

Patrick J. Athol, John L. Barker, Martin J. Beckmann, Martin J. Bouman, Kenneth A. Brewer, Donald E. Cleveland, Lucien Duckstein, Leslie C. Edie, H. M. Edwards, A. V. Gafarian, Denos C. Gazis, Daniel L. Gerlough, John J. Haynes, James H. Kell, John B. Kreer, Leonard Newman, O. J. Reichelderfer, Richard W. Rothery, August J. Saccoccio, A. D. St. John, William C. Taylor, Joseph Treiterer, William P. Walker, Sidney Weiner, W. W. Wolman

K. B. Johns, *Transportation Research Board staff*

The organizational units and the chairmen and members are as of December 31, 1974.